MOBILE TELEPHONES: NETWORKS, APPLICATIONS, AND PERFORMANCE

MOBILE TELEPHONES: NETWORKS, APPLICATIONS, AND PERFORMANCE

ALVIN C. HARPER
AND
RAYMOND V. BURESS
EDITORS

Nova Science Publishers, Inc.
New York

For permission to use material from this book please contact us:
Telephone 631-231-7269; Fax 631-231-8175
Web Site: http://www.novapublishers.com

LIBRARY OF CONGRESS CATALOGING-IN-PUBLICATION DATA
Mobile telephones : networks, applications, and performance / edited by Alvin C. Harper and Raymond V. Buress.
p. cm.
Includes bibliographical references and index.
ISBN 978-1-60456-436-5 (hardcover)
1. Cellular telephone systems. 2. Network performance (Telecommunication) I. Harper, Alvin C. II. Buress, Raymond V.
TK5103.2.M6325 2008
621.3845'6--dc22 2008005362

Published by Nova Science Publishers, Inc. ✣ New York

CONTENTS

PREFACE

The mobile phone or mobile, also called a cellular phone, or cell phone is a long-range, portable electronic device used for mobile communication that uses a network of specialized base stations known as cell sites. In addition to the standard voice function of a telephone, current mobile phones can support many additional services such as SMS for text messaging, email, packet switching for access to the Internet, and MMS for sending and receiving photos and video. Most current mobile phones connect to a cellular network of base stations (cell sites), which is in turn interconnected to the public switched telephone network (PSTN) (the exception is satellite phones. This new book presents the latest research in this revolutionary field.

Chapter 1 concentrates upon the technological aspects of the development of 21st century mobile communications. Most of the chapter is devoted to mobile communications via handsets as this remains for the foreseeable future much the most popular method. However, since the purpose of the chapter is to discuss the most recent variants of mobile communications, which for now generally involve laptop connectivity via data cards and modems, the latter are also covered. In the latter regard, the chapter also covers Wi-Fi and WiMAX as well as other technologies that are as yet poorly established. The development of these technologies – assuming they do not fall by the wayside – covers the period of the next decade, beyond which point, with technological developments proceeding apace, it is difficult to make predictions.

The initial sections of the chapter cover in considerable detail the sequential development of mobile communications, defining terms and distinguishing between developments in different parts of the world. Hence, the chapter should be of interest to readers irrespective of their location. The underlying structure is based upon a sequential explanation in terms of both so-called 'generations' – from 1G to 4G – and data transfer speeds, with each technology allocated a separate section..

There follows a brief introduction to the core services and platforms such as SMS, e-mail and PTT, and the chapter concludes with a discussion of Wi-Fi, WiMAX and related technologies that have yet to become fully established such as mesh and ad hoc networks.

On 1 May 2004, the European Union (EU) witnessed its single largest expansion when ten countries joined. The accession of these ten countries – Cyprus (South), Czech Republic, Estonia, Hungary, Latvia, Lithuania, Malta, Poland, Slovakia and Slovenia was followed by that of two further countries, Bulgaria and Romania, on 1 January 2007. These countries may well not be the last to join the EU, but their accession represents a useful point at which to assess just how big an effect EU expansion has had upon its industrial structure and the strategies of its companies, in this case with respect to mobile telecommunications.

The purpose of Chapter 2 is to explore two particular issues in the context of the mobile telecommunications market: Firstly, whether as a consequence of EU expansion, the accession countries will indeed be the recipients of inward investment from those companies that have, until now, largely ignored them; and secondly, whether those companies already active in the accession countries will seek to reinforce their existing positions.

To this end, the paper is structured as follows. The initial section charts the evolution of telecommunications regulation within the EU and addresses the accession process. In the main section, the ownership of mobile communication licences across the enlarged EU is described. In addition, the geographical footprint of operators is established, with a distinction being made between those operators that have invested in the accession countries and those that have not. In the second main section the focus shifts onto the accession countries, with the analysis being driven by the issues outlined above. Conclusions are drawn in the final section.

A number of serious non thermal biological effects, ranging from changes in cellular function like proliferation rate changes or gene expression changes to cell death induction, decrease in the rate of melatonin production and changes in electroencephalogram patterns in humans, population declinations of birds and insects, and small but statistically significant increases of certain types of cancer, are attributed in our days to the radiations emitted by mobile telephony antennas of both handsets and base stations. Chapter 3 reviews briefly the most important experimental, clinical and statistical findings and presents more extensively a series of experiments, concerning cell death induction on a model biological system. Mobile telephony radiation is found to decrease significantly and non thermally insect reproduction by up to 60%, after a few minutes daily exposure for only few days. Both sexes were found to be affected. The effect is due to DNA fragmentation in the gonads caused by both types of digital mobile telephony radiation used in Europe, GSM 900MHz, (Global System for Mobile telecommunications), and DCS 1800MHz, (Digital Cellular System). GSM was found to be even more bioactive than DCS, due to its higher intensity under equal conditions. The decrease in reproductive capacity seems to be non-linearly depended on radiation intensity, exhibiting a peak for intensities higher than 200 $\mu W/cm^2$ and an intensity "window" around $10\mu W/cm^2$ were it becomes maximum. In terms of the distance from a mobile phone antenna, the intensity of this "window"corresponds under usual conditions to a distance of 20-30 cm. The importance of different parameters of the radiation like intensity, carrier frequency and pulse repetition frequency, in relation to the recorded effects are discussed. Finally, this chapter describes a plausible biophysical and biochemical mechanism which can explain the recorded effects of mobile telephony radiations on living organisms.

Evolution in the world of mobile telephones is moving toward a full functional and rich in features device. Examples of novel functionalities are the possibility to play and handle different multimedia contents, to have a wide number of communication capabilities through both remote (i.e. the Internet) and local network access. These examples testify to an evolutionary trend that will likely lead to mobile phones more and more comparable to portable computers. One of the most appealing features for a mobile phone appearing on the market is the number of built-in network interfaces (the most recent devices include interfaces to different networks, such as HSDPA, UMTS, EDGE, GPRS, WLAN, Bluetooth, and Infrared). This both pushes the providers to offer a wider portfolio of novel applications, which best exploit the augmented possibilities, and paves the way to Research and Development (R&D) activities in the field of wireless communications towards the deployment of real "Personal Communications" scenarios. *Cooperative Wireless Applications*, which the authors will focus on in the present chapter, is a significant example of a R&D area fuelled by the availability of heterogeneous wireless network interfaces, based on both cellular and short-range technologies. *Cooperation* covers many different aspects, which are being studied at the different protocol layers. The authors focus on a scenario where more devices can cooperate with each other on a short-range communication link basis. Usually a costly, high power demanding cellular network is the way users access a service (e.g. web content browsing, file download). A short range link has, instead, the following appealing features: high data rate, low power consumption, and low costs. By exploiting the synergy between a *long-range* (cellular) and a *short-range* network, different entities may share service costs through cooperation with peer devices. Currently, smartphones offer the possibility to implement the users own application to be run on the device. This offers us the opportunity of performing an analysis of telephone network interfaces behaving in mutual co-operation and the possibility of evaluating the advantages actually attainable from the aforementioned innovative communication paradigm. In Chapter 4 the authors first introduce the reader to the concept of cooperation in wireless scenarios and, subsequently, try to summarise the main implementation efforts made until now on mobile telephones. A performance evaluation campaign and a report on the main research results is given as a proof of theoretical advantages in cooperating under different coverage and traffic load conditions. Also a description of the ad-hoc implemented test-beds are given. Performance studies and relevant implemented evaluation tools are specifically oriented to Symbian Operating System mobile phones, although results are general and valid for any mobile terminal.

The method of analysis applied in Chapter 5, Fast Fourier transform (FFT), allows us to discern between the brain integrated biological signal and noise, invalid impure signal.

Alterations induced by mobile telephones in the EEGs of healthy subjects are the outcome of electromagnetic oscillation.

The non-linear analysis, Higuchi algorithm, of electromagnetic alterations, due to the use of mobile telephones, yields a statistically relevant decrease of the fractal dimension (D)

There is a correlation between the biological brain signal and fractal dimension decrease.

Auditory cortex activation does not affect integration of oscillation from a mobile telephone.

Oscillations from active mobile telephones alter EEG recordings, at least in the fronto-parieto-temporal areas, at the frequency band of beta brain waves.

Artefacts introduced by the system (subject and recording devices) alter EEG recordings in all areas of the cranial calotte, at the frequency band of beta brain waves.

The extensive community adoption of mobile phones coupled with the development of its own specific slang is seen by some as an emerging new phenomena. Nevertheless existing psychological approaches to the impulse control disorders and the addictions can definitely serve as useful starting points for understanding phone use. In Chapter 6 the authors discuss whether heavier mobile phone use could be viewed as an addiction. Building upon previous research on internet use, studies demonstrate that specific psychological traits can explain mobile phone use and misuse. For instance, both mobile phone use and internet use are linked to lower levels of the agreeableness trait. Nevertheless, in contrast to the introverted stereotype of the heavier internet user, mobile phone use is more strongly linked to extraversion. As the mobile phone is a communication device, it is not surprising that its use can be predicted by psychological traits such as extraversion, and as extraversion is also associated with sensation seeking, this trait can also explain the self stimulatory behaviours and risk taking behaviour associated with mobile phone (e.g. use while driving). In particular, the mobile phone elicits problems for heavier users, with a section of the community reporting difficulty managing their time, financial problems, and aches and pains. The development of problems can be understood in terms of distraction and traits like procrastination. Tendencies towards procrastination can explain messaging behaviour, distraction from important tasks, and hence loss of resources (i.e. time or money) that influences quality of work. It is likely that some elements of problematic use will be resolved by innovations such as Bluetooth (hands free while driving), gesture or handwriting recognition (aches and pains), and video (antisocial behaviour). But other innovations such as location aware advertising and gambling are likely to create new problems such as distraction and financial loss for a proportion of the community.

A Wireless Sensor Network (WSN) is a power constrained system, because nodes run on limited power batteries. When a node switches between modes (i.e., transmit, receive, and sleep), there are also changes related to the power consumed by the node not only after the node is in a different mode, but also during the transition process. Although the switching energy is considered negligible most of the time, it is strongly dependent on the technology (e.g., the radio) deployed on the nodes. In particular, for low-duty cycle networks such as WSNs, the switching energy may impact on the overall energy consumption. Network simulators usually employ the linear discharge model which is simple, but it does not capture the battery capacity recovery effect. Getting a more accurate measure for the energy consumption of a sensor network may lead to better estimates for the network lifetime. Chapter 7 explores the application of a more realistic battery model (i.e., the Rakhmatov-Vrudhula battery model) for the simulation of WSNs.

The "chain of survival" is a series of steps linking a person whose life is in danger to medical service providers. It currently functions appropriately in relation to the resources available. The authors suggest a simple improvement to the survival chain: Chapter 8 presents a method for managing emergency calls received from a user/citizen, which can be used in a cellular telecommunication system.

A database should be instituted, on a voluntary basis, for trained personnel. Such database should comprise resuscitation specialists, physicians, but also anyone who has undergone recognized training (e.g.: Basic Life Support, Advanced Life Support, Advanced Trauma Life Support). Upon receiving a medical emergency call, an off-duty operator can be selected from the database (in a cooperative, non-competitive manner with the dedicated emergency services), based on current position of her/his cellular phone. If the call includes information as to the type of emergency (e.g.: trauma, cardiac arrest) the selection of the operator can take into account a predefined compatibility with the specific emergency. Consequently, the operator who is both closest to and best prepared for the emergency is contacted via cellular phone. If the operator is available she/he can (again on a voluntary basis) precede or join the mobile unit on site, managing the emergency according to her/his training standard, possibly in cooperation with the ambulance personnel and even up to hospital admission.

A simple software system able to identify the nearest volunteers in the existing database is needed, and the spread of Basic Life Support training among the population, along with the distribution of semi-automatic defibrillators in public places, could improve first aid in cardiorespiratory circulatory arrest, trauma or catastrophes.

The spectacular advances in wireless communications technology and the increased use of mobile phones in the past decades have led to much of the research into the effects of electromagnetic fields on humans and animals being centred on the study of the biological effects of mobile phones. Among the numerous studies carried out in this field, many involve the design of experimental radiofrequency systems that are manageable in the laboratory and that can be applied to biological models of interest for the study and measurement of the effects of this type of non-ionizing radiation on humans and animals, particularly on their nervous systems.

Chapter 9 assessed the validity of an experimental model of standing wave radiation of GSM radio frequency (similar to that emitted by mobile phones) applied to the biological study of the effects on cerebral activity of rats exposed to acute doses of this radiation after having been converted into a model of seizure-proneness by injection of low doses of picrotoxin. Various studies have taken into account physical parameters (frequency, polarization, modulation, pattern and uniformity of the field, type of transmission and radiation from the equipment, as well as the cavity dimensions and material) and biological parameters (dielectric properties of the tissue under study, orientation relative to polarization, and also age, sex, weight, application of anaesthetics, drugs and electrodes), as well as possible artefacts in the calculation of the rate of energy absorption. Analysis of the behaviour of the rats during exposure to the radiation and of postmortem morphological effects, indicated by the c-Fos protein, with specific absorption rates (SAR) always below the thermic levels at 900 MHz (as outlined in the prevailing legislation) and application of different potentials, revealed an important degree of modification of neuronal activity in the brain. The validity of this and other experimental models currently used in the study of the effects of mobile phones on the nervous system is discussed in light of the results. Finally, a personal view is offered of the future of experimental research into mobile phone signals and of the methodology of biological tests used to detect possible adverse effects.

One of the most popular and fastest growing technologies of our times is that of mobile phones.

As explained in Chapter 10, accurate forecast of the size of the mobile phone market is essential for both regulators, and corporate players in order to make the right decisions, and embrace effective strategies in a continuously changing market place. S-shaped logistic curves have been used successfully in measuring technological diffusion for the Internet, computers and other areas of technological innovation. S-curves work better when applied to populations with natural barriers that will eventually decelerate their growth as they move closer to their upper limit. In the case of mobile phones there are indeed barriers such as the number of people in a certain region, the number of people with sufficient income to finance mobile use, and finally the number of people that are neither too young nor too old to use mobile phones.

Recently, in 2005, this methodology was applied to actual mobile accounts for years up to 2004, in order to produce forecasts concerning the future of the mobile phone business. These projections concerned the whole World and selected regions such as China.

In order to better illustrate the suitability of S-curves in describing the dynamics of the mobile phone market, the authors test the forecasts made back in 2005 with actual data for recent years. In the case of World mobile accounts, the emergence of a second growth wave after the completion of a first one is confirmed, thus leading to a new estimate of approximately 3.5 billion accounts in the near future. On the other hand, China's projections with limited data up to year 2001 lead to relatively accurate estimates of mobile accounts with less than 8% error for 2006 which is considerably smaller than the expected statistical error.

In Chapter 11 the authors propose a new system of an optical wireless link for telephone networks, especially, for mobile telephone applications. The message is kept in secret under a signal scramble and coding by using nonlinear effects of light in a micro ring resonator. The message is multiplexed with the chaotic signals and transmitted to the end users via an optical wireless link. The original message can be performed the secret by using the synchronous encryption before transmission via the optical wireless link. By these techniques, the messages can be kept in secret and retrieved by the specific user by the synchronous decryption, filtering and cancellation the scramble codes. The potential of using such a proposed device and system to perform the mobile telephone networks security is discussed in details.

EXPERT COMMENTARY

In: Mobile Telephones…
Editors: A.C. Harper and R.V. Buress, pp. 3-7
ISBN: 978-1-60456-436-5

THE CONVERGENT MOBILE TELEPHONE: AN EMERGING BRIDGING MEDIUM

Ran Wei [*]
School of Journalism and Mass Communications, University of South Carolina, Columbia, SC 29208, USA

The Mobile Phone as an Emerging Media Platform

In 2004, Susan Herring, Editor of *Journal of Computer-Mediated Communication*, wrote in an essay appearing in *New Media & Society* about Internet technologies that were "slouching toward the ordinary." To some extent, the same thing can be said about the mobile phone. With millions of people worldwide use it for work and play everyday (Ito, Okabe, & Matsuda, 2005; Wei, 2001, 2008; Wei & Lo, 2006), the mobile phone has become ordinary, no longer a novelty. Children at the age of ten carry it. Teen users found it indispensable; they feel lost without it (Wei, 2007). Among adult users, mobile lifestyle is a distinguished feature in industrialized as well as developing courtiers (Katz & Sugiyama, 2006; Wei, 2006).

On the other hand, the global popularity of the mobile phone continues to grow. According to the International Telecommunication Union (ITU), the number of mobile phone users worldwide reached 2.7 billion in 2006, twice as large as the number of Internet users of 1.1 billion. In China alone, the number of mobile phone users totaled 488 million in mid-2007. It is not surprising that the mobile phone was considered as the most successful new communication technology in the history of technology diffusion (Levinson, 2004). When debuting on June 29, 2007, over 200,000 i-Phones, *Apple*'s cool smart phone, were sold. One million Americans now own an i-Phone. The best in wireless telecommunications technologies, such as mobile video conferencing and wireless computing to be available on G4 super smart phones, is yet to come.

The changing regulatory environment favors rapid advances in mobile telephony technology and launching of new applications and services. One of the consequences of the *1996 Communication Act* is coming of the so-called "3-in-1 service" era (e.g.,

[*] E-mail address: ran.wei@sc.edu. Tel: (803) 777-5762; Fax: (803) 777-4103.

telecommunications, broadcasting, and the Internet), in which the mobile phone has emerged as a media platform (Nilsson, Nuldén, & Olsson, 2001; Wei, 2008). For example, the platform of DVB (Digital Video Broadcasting)-MHP (Multiple Home Platform). Through DVB-MHP, a multicast can send programming to thousands of mobile phone users, who can play back personalized content on their phone anywhere anytime.

New Applications and Services via Mobile Platform

Mobile network operators (MNO), otherwise known as wireless carriers, sought to leverage on the global popularity of the mobile phone and the converged functionality it offers to expand services and generate new revenues. In the U.S., *MobiTV* started offering streaming video on *Sprint* networks in 2003. In 2006, terrestrial-based DMB (Digital Multimedia Broadcasting) went into service in South Korea. The European Union endorsed DVB-H (Digital Video Broadcasting – Handheld) mobile video broadcasting standard in July 2007; DVB-H based mobile television was widely available in Italy and Finland. Most shows on mobile television are short, ranging from three to six minutes. Industry research shows that there are seven million mobile phone users who view videos on their phone; the total is expected to grow to 24 million by 2010 (The Economist, 2007). No wonder, the mobile phone is touted as the third screen after television and PC. Mobile television has a great potential to be a viable broadcasting platform.

However, the pressing question is: Will the small screen really make a big difference? That is, will mobile phone delivered television be the next killer application after SMS (short message services)? In this commentary, I will address this question from a user perspective to seek a conceptual understanding of the mobile phone as a brand-new media platform. I will discuss what sort of medium the mobile phone is, what its characteristics are, and how it differs from broadcasting and computer-based media. My thesis is that traditional mobile telephony concepts developed with 2G and 3G are dated; a concept of the convergent mobile phone as something new is badly needed.

The Mobile Phone as a Bridging Medium

The key attribute of mobile communications via the mobile phone is mobility, which allows users to maintain perpetual contact on the move (Katz & Aakhus, 2002). Short (2006) described mobility as a distinctive element of the contextual user environment (CUE) of mobile communications. Some scholars defined the mobile phone as a mobile medium (Levison, 2004). True, past studies identified mobility as a strong motive for using the mobile phone (Leung & Wei, 2000; Wei, 2008). Moreover, the gratification of mobility was found a significant predictor of mobile phone use at on-the-go locations such as on the street, in buses, in subways, and at shopping malls. Others (Ozcan & Kocak, 2003) also found that being away from home (e.g., at the workplace) and on the move (e.g., in the car) was an important factor affecting the level of mobile phone usage.

Nevertheless, defining the mobile phone as a mobile medium suffers a narrow conception because it disregards the dimension of time and the contextual user environment (CUE). The metaphor of the mobile phone as the third screen is thus misleading. It implies the small mobile screen will rival television and PCs. A recent analysis (Wei & Huang, 2008) of using the mobile phone to view television indicates that mobile television is viewed the most at certain time of the day and under circumstances in which the user is physically in a mobile environment such as buses or taxi. The mobile phone as a broadcasting platform fills the gap between fixed localities that people traverse routinely. Users tend to use the first two screens (e.g., TV and PC) at a fixed location; no matter it is workplace or home. A report (Red Bee Media, 2006) confirms that mobile phone users categorically prefer to watch full-length shows on television at home.

Thus, the mobile phone is not simply a mobile medium; it is characteristically a bridging medium in nature. Instead of rivaling television and pc, it supplements the first two established screens.

Critical Factors for Success and Major Constraints of the Third Screen

Thus, new applications and services, such as mobile television delivered via the mobile phone as a bridging medium, suit the lifestyle of millions who commute and spend a great deal of time on the move. This is good news to mobile network operators and their content providers. Recent research (Wei & Huang, 2008) found a linkage between mobility and interest in viewing television delivered via the mobile phone. Mobile phone users favor mobile television for its portability and mobility. Moreover, younger and better-educated male mobile phone users tend to be more interested in the third screen than the general mobile phone population. This trend is favorable to wireless carriers because younger users grew up as the Net generation. They tend to use the mobile media for Internet connection as well. M-commerce will likely be attractive to this segment of the nearly three billion of mobile phone users.

On the other hand, interest in using the mobile phone as a television viewing platform is low (Wei & Huang, 2008). The mobile phone has yet to be established as the third screen. Bandwidth is one problem; pricing is another. Younger users who are interested in mobile television have not reached their earning potential to make sufficient money needed to use the mobile phone as an all-in-one medium. In addition, small screen size is a major constraint to making the mobile phone the third screen a success. Before these constraints are removed, text messaging in the form of SMS, which far outnumbers e-mail sent everyday, remains the No. 1 application.

Future Trends

Several future trends in wireless telephony technologies will affect the evolution of the mobile phone from a personal communication device connected to a wireless telecommunications network into a bridging media platform.

First, network technologies in accessing the mobile phone-based platform improves with upgrade from 3G to 3.5G—HSPA (high-speed packet access), which promises speeds up to 14.4 Mbit/s (vs. 3G at 384 kps). Over 130 mobile network operators in 61 countries offer HSPA services, accounting for one-fifth of all GSM operators worldwide (Lunden, 2007).

Even better, 4G technologies are on the horizon after *AT&T* and *Verizon* chose LTE (long Term Evolution) as their common access platform in late 2007. Meanwhile, Wi-Max, a rival ad hoc networking technology, is being deployed in Korea, Japan, and Taiwan. Regardless of which standard will prevail, 4G technologies will provide seamless connection with the wireless broadband networks.

Second, the wireless telecommunications market will become more competitive. In the auction of wireless spectrum to be made available in 2009, the FCC will reserve up to a third of the newly allocated spectrum to bidders who will be "receptive to all kinds of devices and applications" from phone makers and third-party software providers (Stross, 2007).

Third, the power in the wireless telecommunications business will shift from wireless carriers to hardware and software providers. For example, *Google* will bid with $4.6 billion to launch "Google Mobile." It differentiates itself from existing carriers on an "open access" proposal, which allows users to choose phones, functions, and software. *Verizon* quickly followed suit. It promises to open its networks so that American mobile phone users can freely choose phones and applications in 2008.

References

The *Economist*, (July 19, 2007). Mobile TV: The third screen. Retrieved on October 15, 2007 from http://www.ebusinessforum.com/index.asp?layout=rich_story&channelid=5&categoryid=18&title=Mobile+TV%3A+The+third+screen&doc_id=11097.

Herring, S. (2004). Slouching toward the ordinary: Current trends in computer mediated communication. *New Media & Society*, **6**(1), 26-36.

Ito, M., Okabe, D., & Matusda, M. (2005). *Personal, portable, pedestrian: Mobile phones in Japanese life*. Cambridge, MA: MIT Press.

Katz, J., & Aakhus, M. (2002). *Perpetual contact: Mobile communication, private talk, public performance*. NY: Cambridge University Press.

Katz, J. E., & Sugiyama, S. (2006). Mobile phones as fashion statements: Evidence from student surveys in the US and Japan. *New Media Society*, **8**(2), 321-337.

Levison, P. (2004). *Cellphone: The story of the world's most mobile medium and how it transformed everything*! NY: Palgrave.

Leung, L. & Wei, R. (2000). More than just talk on the move: A use-and-gratification study of the cellular phone. *Journalism & Mass Communication Quarterly*, **77**(2), 308-320.

Lunden, I. (November 21, 2007). Mobility diary: Fulfilling the promise of broadband on the move. *Financial Times*.

Nilsson, A., Nuldén, U., & Olsson, D. (2001). Mobile media: The convergence of media and mobile communications. *Convergence*, **7**(1), 34-38.

Ozcan, Y., & Kocak, A. (2003). A need or a status symbol? Use of cellular telephones in Turkey. *European Journal of Communication*, **18**(2), 241-254.

Red Bee Media. (2006). *Small screens need big ideas: Study highlights need for more innovative TV on mobile phones.* Retrieved on November 24, 2007 from http://www.redbeemedia.com/press/2006/smallscreens.shtml.

Short, M. (2006). CUE: The contextual user environment. Retrieved on November 26, 2007 from http://www.mobileuserexperience.com/?p=275.

Stross, R. (July 22, 2007). When mobile phones aren't truly mobile. *New York Times*.

Wei, R. (2001). From luxury to utility: A longitudinal analysis of cell phone laggards. *Journalism & Mass Communication Quarterly,* **78**(4), 702-719.

Wei, R. (2006). Lifestyles and new media: Adoption and use of wireless communication technology in China. *New Media & Society*, **8**(6), 991-1008.

Wei, R. (2007). The high-tech cell phone as self phone: Social identity and goal attainment in individual cell phone dependency. *Asian Communication Research*, **3**(1), 4-25.

Wei, R. (2008). Motivations for use of the cell phone for mass communications and entertainment, *Telematics & Informatics,* **25**(1), 36-46.

Wei, R., & Huang, J. (2008). Profiling user responses to mobile TV: Effects of individual differences, mobility and technology cluster on critical mass. *Paper submitted to the 53rd Convention of Broadcasting Education Association,* Los Vegas, NV, April 16-19.

Wei, R. & Lo, V. (2006). Staying connected while on the move: Cell phone use and social connectedness. *New Media & Society*, **8**(1), 53-77.

SHORT COMMUNICATION

In: Mobile Telephones...
Editors: A.C. Harper and R.V. Buress, pp. 11-26

ISBN: 978-1-60456-436-5

ALTERATION OF ENZYMIC ELECTRON TRANSFER REACTIONS INDUCED BY MICROWAVES EMITTED BY GSM MOBILE PHONE

Roberta De Carolis[1], Fiorenzo Marinelli[2] and Mario Barteri[1*]
[1]Department of Chemistry, University of Rome "La Sapienza", Italy
[2]Inst. of Molecular Genetics, Bologna, Italy

Abstract

Enzymatic activity and conformational features of bovine lactoperoxidase (LPO), ascorbate oxidase (AO) and laccase (LAC) are modified when aqueous solutions of these enzymes are exposed to the microwave field emitted by a mobile phone. Several kinetic and spectrophotometric measurements were carried out comparing native and irradiated samples of the same enzyme solution. Experimental data obtained demonstrate that mobile phone radiations alter the active site conformational state. It has been also proved that electromagnetic fields (EMFs) interfere with free radical intermediate species of the electron transfer enzymatic reaction, providing a considerable modification of the kinetic parameter K_M, V_M and k_{CAT}.

Introduction

The global diffusion of mobile communications is enormously increased into our everyday life, such that it is estimated that, nowadays, there are more than 1.5 billions mobile phones in use. Due to the microwave character of mobile telephone emissions, it was suggested that phone emissions might affect biological systems. A number of studies have attempted to determine if (and how) exposure to microwaves (MWs), under conditions similar to GSM emissions, can induce alterations in the kinetic of the folding process of globular proteins [1-2]. In our laboratory it has been shown as MWs significantly modify kinetic and structure of acetylcholinesterase (AChE), an important enzyme which regulates the activity

* Author to whom correspondence should be addressed.

of the neurotransmitter acetylcholine in the brain [3]. The conceptual starting point of these our previous studies foresaw the use of highly purified bio-macromolecules (proteins, enzymes and nucleic acids) as molecular probes, to check possible kinetic and/or conformational effects induced by MWs exposure. We supposed that any possible biological damages could start at *"molecular level,"* involving biological macromolecules. We believe conceivably true that the *in vivo* effect of conformational and/or kinetic alteration in proteins and enzymes may be progressively propagated to higher and to more complex levels of the biological systems, modifying biochemical reactive pathways or membrane activities. Many prestigious laboratories have been interested in studying biological effects induced by exposure to MWs in the working range of mobile phone. These studies have often obtained conflicting results leading to contrasting conclusions. Furthermore, the use of commercial mobile phones as a MWs source has been criticized, since phone emissions vary widely [4]. This variability is generally considered as a huge drawback originating irreproducible experimental data and providing incongruent results. However, several studies carried out using certified emission sources (TEM chambers and/or antennas) on biological systems, such as specific cell strain, selected tissues and genetically modified mice, also yield irreproducible effects. The main reason of this incongruence should be probably searched in the intrinsic complexity of biological systems. Even a single cell is a very complicated system made up of a precise and synergic assembly of phospholipids, nucleic acids, proteins and enzymes. The specific and peculiar distribution of these compounds rules all the cellular biological activity, including the ability of repairing or masking possible damages. Thus, the topological distribution of cellular biomolecules, with different chemical composition and different conformational states, are important in the specific interaction with electromagnetic fields (EMFs). For instance, the electrostatic interactions of EMFs with phospholipid-bilayers of cellular membrane are different if compared to those with a protein. On these macromolecules the charge distribution can cause conformational and functional modification of the biological activity.

Our research strategy consisted in using aqueous solutions of specific and highly purified metallo enzymes and proteins to study the EMFs effects at a molecular level. We selected proteins that fitted at least one of the following criterions: proteins with a high dipole momentum ($\geq$ 500 Debye), metallo-enzymes containing at least one paramagnetic metal ion, or enzymes which catalyzes electron-transfer reactions with a mechanism involving paramagnetic intermediate species, such as free radicals.

In this chapter, we report some results concerning the effects of mobile phone microwaves on structure and activity of metallo-enzymes. Our studies were devoted to demonstrate that the mobile phone emission interferes with electron transfer processes, which take place during enzymic reactions catalyzed by Oxidases and Peroxidases. Biochemical reactions catalyzed by these enzymes proceed generating free radicals intermediate compounds [5], which are paramagnetic species sensitive to electromagnetic fields. Here we show experimental evidences proving that the microwaves emitted by a dual band mobile phone (915-1822 MHz) alter both conformational and configurational features of steady-state transition complex of the enzymatic reaction mechanism. Figure 1 shows the electric field emissions of the mobile phone recorded operating in both receiving and calling mode. In receiving mode, mobile phone needs a lower amount of energy to be connected, while operating in calling mode oscillations of the emitted electric field increases. All our experiments were carried out operating in receiving mode by dialling from a remote phone to the phone positioned at 5 cm far from the enzyme solution. As soon as the phone we utilized received a call, the emitted intensity of electric field rapidly increased in the first 10 s, reaching a maximum of 9 V m^{-1}, before decreasing to a lower standby asymptotic value of about 3 V m^{-1}.

In this context, we carried out experimental kinetic measurements by comparing the enzymatic activity of native protein to aliquots of the same solution exposed 20 minutes and during the enzymatic reaction. Experimental measurements were made on native and irradiated before samples of the same enzyme solution to verify whether irradiation induce alterations on protein structure. Collecting kinetic data in absence and in presence of MWs we revealed as irradiation perturbs the reaction mechanisms. Our studies were made using bovine lactoperoxidase (LPO), ascorbate oxidase (AO) and laccase (LAC). Lactoperoxidase (LPO, EC 1.11.1.7) is a mammalian enzyme which is the product of exocrine gland secretion, *e.g.* into milk, saliva, and tears [6-8]. Bovine LPO consists of a single polypeptide chain containing 612 amino acid residues. The amino acid sequence is known and the molecular weight is approximately 78 kDa. LPO is a basic protein having an isoelectric point of 9.6 and it has about 10% of carbohydrate, localized on five potential N-glycosylation sites. Its secondary structure is composed of 23% α-helices, 65% β-sheet and 12% unordered structure. Figure 2 shows the Lineaweaver-Burk plots obtained following experimental procedure described above to perform kinetic measurements of guaiacol oxidation by hydrogen peroxide (H2O2) catalyzed by LPO [9]. The kinetic parameters KM, VM and kCAT, which are summarized in Table 1 were calculated from the intercept and the slope of the linear fit. For native LPO these parameters are significantly lower than those of the irradiated samples. This difference is more marked in the case of LPO solution irradiated during the measurements, proving that exposure has considerably modified the enzyme activity.

The reaction proceeds with a very complex mechanism [10], but it can be simplified as follows:

$$
\begin{aligned}
& H_2O_2 \underset{k_2}{\overset{k_1}{\rightleftarrows}} LPO\,(Fe^{IV+\bullet}) + H_2O \\
& LPO\,(Fe^{IV+\bullet}) + GUA \xrightarrow{k_3} LPO\,(Fe^{IV}) + GUA^{\bullet} \\
& LPO\,(Fe^{IV}) + GUA \xrightarrow{k_4} LPO\,(Fe^{III}) + GUA^{\bullet}
\end{aligned}
\tag{1}
$$

Table 1. Kinetic activity parameters of lactoperoxidase samples calculated from the slope and intercept of the straight lines depicted in Figure 1

Sample	K_M (μM)	V_m (μM *s^{-1})	k_{CAT} (s^{-1})
Native	603 ± 13	5.5 ± 0.3	1809 ± 80
Irradiated before	2033 ± 41	12.1 ± 0.6	3980 ± 199
Irradiated during	14250 ± 285	69 ± 3	22697 ± 1135

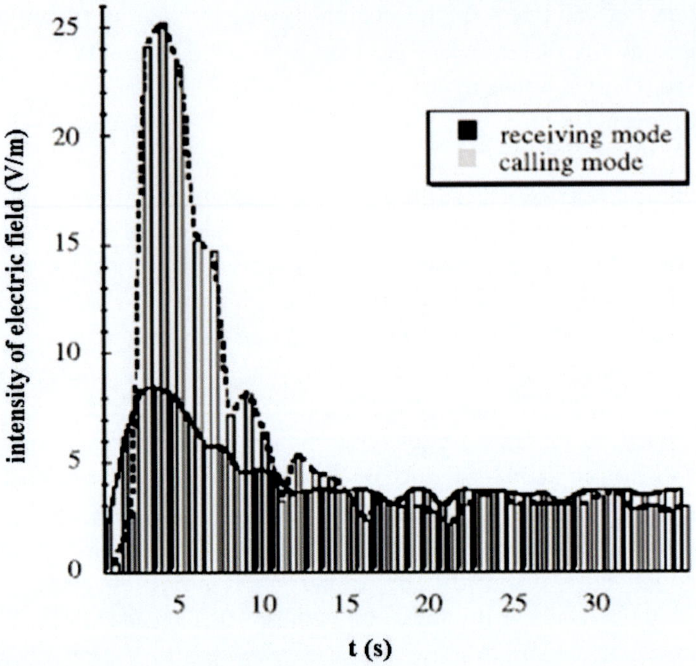

Figure 1. EMF emission of the mobile phone recorded with PMM 8053 analyzer in both receiving (black histograms) and calling (grey histograms)mode [3].

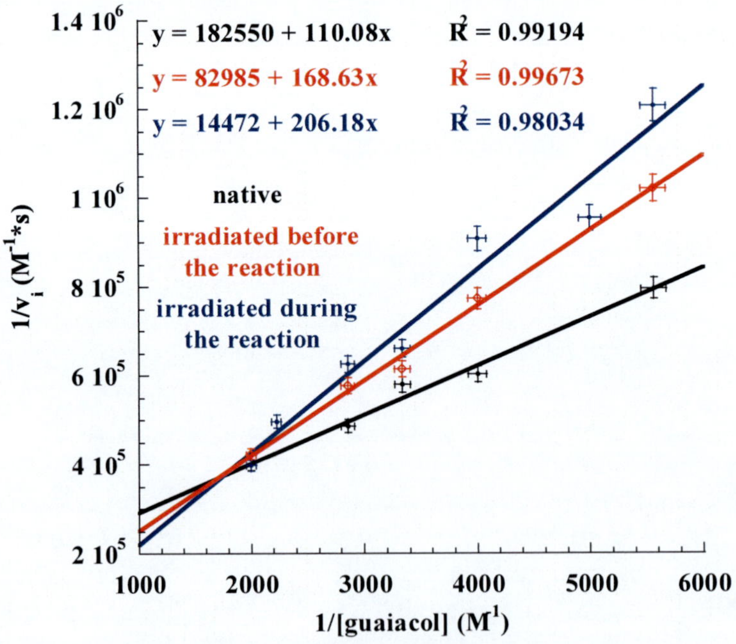

Figure 2. Lineweaver-Burk plot of lactoperoxidase samples. Figure shows the plot of the inverse of initial reaction rate (1/v) versus the inverse of substrate concentration (1/[S]). [LPO]= 1.49×10^{-8}M, PBS 0.1 M (pH = 7.2), T = (25.0 ± 0.1) °C. Data have been obtained spectrophotomerically by recording the absorbance at 470 nm of reaction product (tetraguaiacol).

where $Fe^{IV+\bullet}$ and $GUA^{\bullet}$ represent the free radical species of iron and guaiacol respectively. The final product of the reaction is tetraguaiacol, obtained from guaiacol radicals ($GUA^{\bullet}$) combination. The kinetic equations which describe this process are:

$$v = -\left(\frac{d\,[Gua]}{dt}\right) = \frac{V_M\,[Gua]}{[Gua] + K_M} \tag{2}$$

where:

$$V_M = k_{CAT}[Enz]_0\,; \tag{3}$$

$$K_M = \frac{k_4(k_2 + k_3)}{k_1(k_3 + k_4)} = \frac{k_{cat}}{k_1} + \frac{k_4 k_2}{k_1(k_3 + k_4)} \tag{4}$$

$$k_{CAT} = \frac{k_3 k_4}{(k_3 + k_4)} \tag{5}$$

Kinetic measurements were collected spectrophotomerically at the tetraguaiacol absorption wavelength (470 nm). Data obtained demonstrated that exposure alters the kinetic parameters, probably modifying either conformational features of the enzyme active site or interfering with free radical intermediates species. To obtain further evidences, a set of electrochemical experiments (cyclic voltammetry) were carried out (Figure 3).

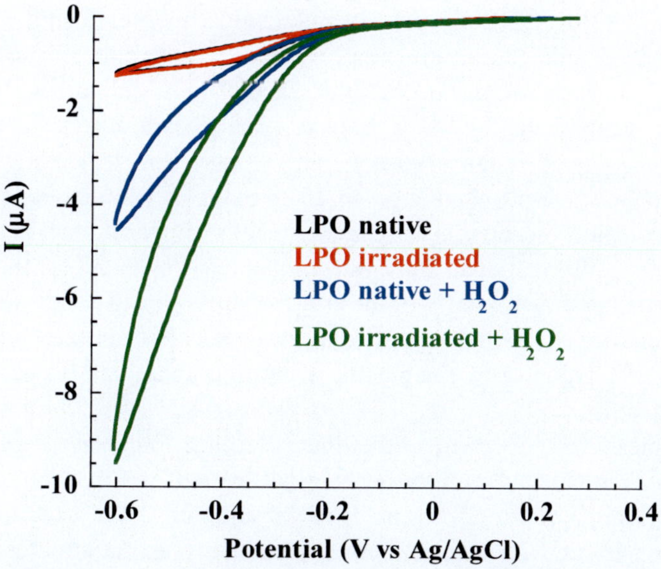

Figure 3. Voltammograms of the LPO enzymatic activity experiments [LPO] = 2.93×10^{-6}M, [H_2O_2] = 3.20×10^{-3}M, PBS 0.1 M, (pH = 7.2), T = (25.0 ± 0.1) °C.

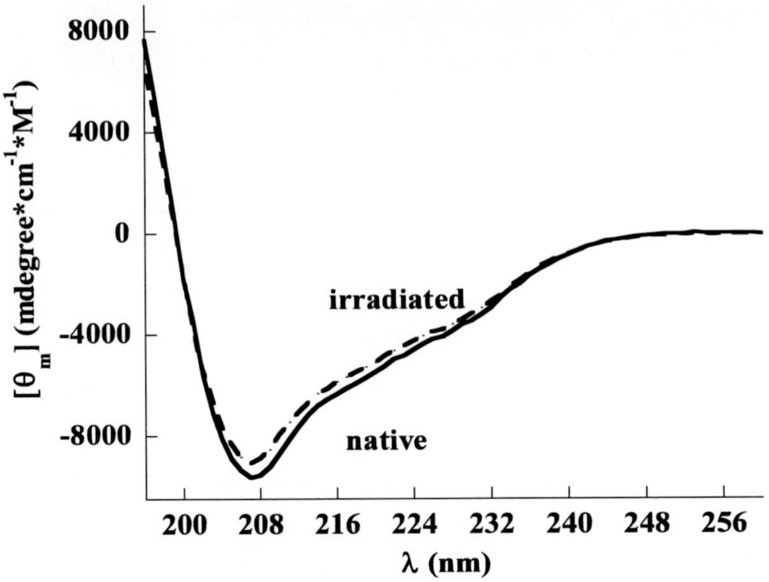

Figure 4. Far-UV CD spectra of lactoperoxidase samples; the spectra were normalized with peptide concentration. [LPO]= 6.11x10-7M, PBS 0.1 M (pH = 7.2).

The voltammograms were recorded in the range of potential where electron transfer between LPO (in solution) and gold electrode does not occur. In this case, reproducible and superimposable measurements of both native and irradiated samples (red and black curves) have been obtained. This result point out that irradiation does not enable altered protein to give rise electron transfer with electrode. Conversely, in the presence of H_2O_2, the increase of the cathodic current (in the range between -0.25 V and -0.60 V) arises from the reduction of H_2O_2 catalyzed by LPO. In this case, the electrochemical measurements with native (curve blue) and irradiated enzyme (curve green) provide different trend in the voltammograms, proving that the increase of LPO catalytic activity is in direct relationship with conformational changes induced by MWs. Indeed, circular dichroism spectra (CD), recorded in peptidic bond absorption range (190-260 nm) (Figure 4), demonstrate that secondary structure of the enzyme doesn't change up to 20 minutes of continuous irradiation. On the contrary, steady-state fluorescence measurements (Figure 5) show that irradiation provokes modification of the conformational states of tryptophan residues. In the LPO structure, tryptophans are placed in the proximity of the enzyme active site as schematically represented in Figure 6. This picture shows heme, proximal and distal His and the Trp residues placed close to the prosthetic group drawn using the crystallographic data extracted from Protein Data Bank (PDB) facility.

Fluorescence measurements confirm that the difference found in the cyclic voltammetry measurements can be attributed to significant modification of LPO active site induced by mobile phone emission. After 20 minute of irradiation, LPO structural changes are irreversible and consistent with a new thermodynamically stable structure of the enzyme, providing reproducible spectroscopic and kinetic data. However these results do not explain the dramatic variation observed in kinetic measurements carried out exposing the enzyme-substrate mixture during the reaction. In this case, K_M, V_M and k_{CAT} values (see Table 1) change of 1-2 orders of magnitude merely because the system was exposed to the mobile phone during the reaction development. This effect gives a further demonstration that pulsed and variable EMFs alter the mechanism of electron transfer processes [10-11].

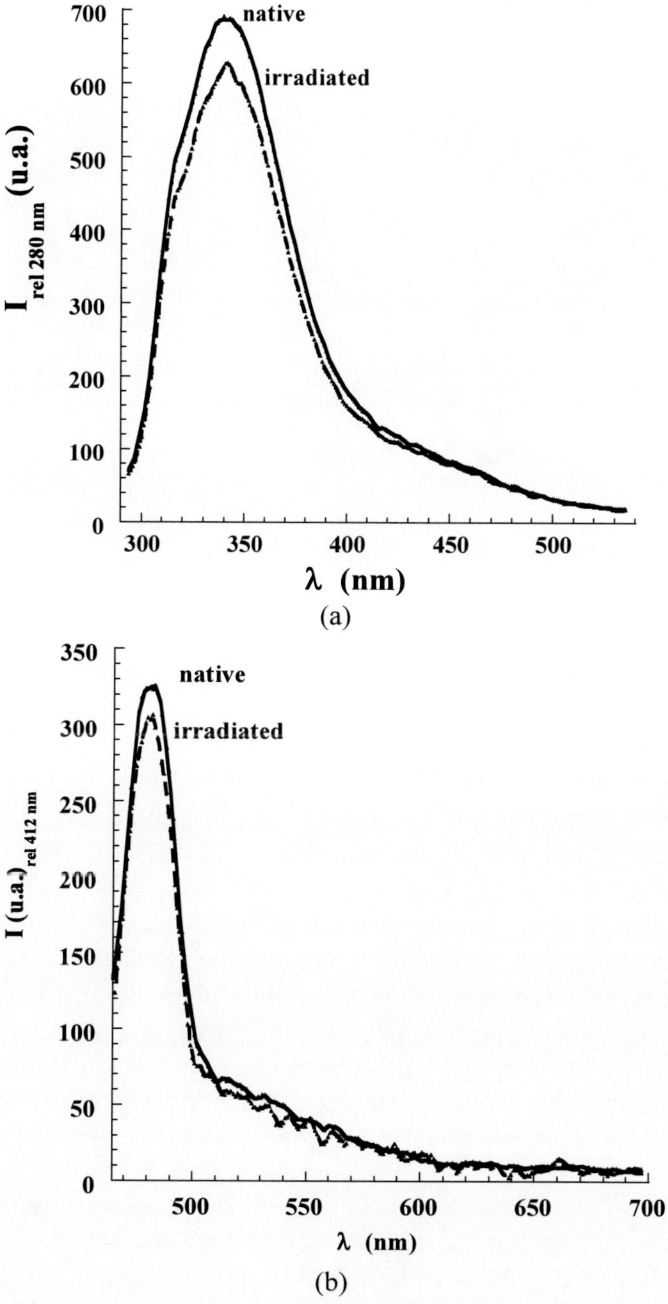

Figure 5. Steady-state fluorescence spectra of lactoperoxidase samples recorded exciting sample at 280 nm (aromatic residues absorption) (a) and exciting at 412 nm (active site absorption) (b). Both in the cases irradiation determines fluorescence quenching (dotted line) showing that EMFs provoke a shift of Trp residues from the inside towards the solvent and an alteration of active site conformation. [LPO] = $6.11x10^{-7}$M, PBS 0.1 M (pH = 7.2), T = (25.0 ± 0.1)°C.

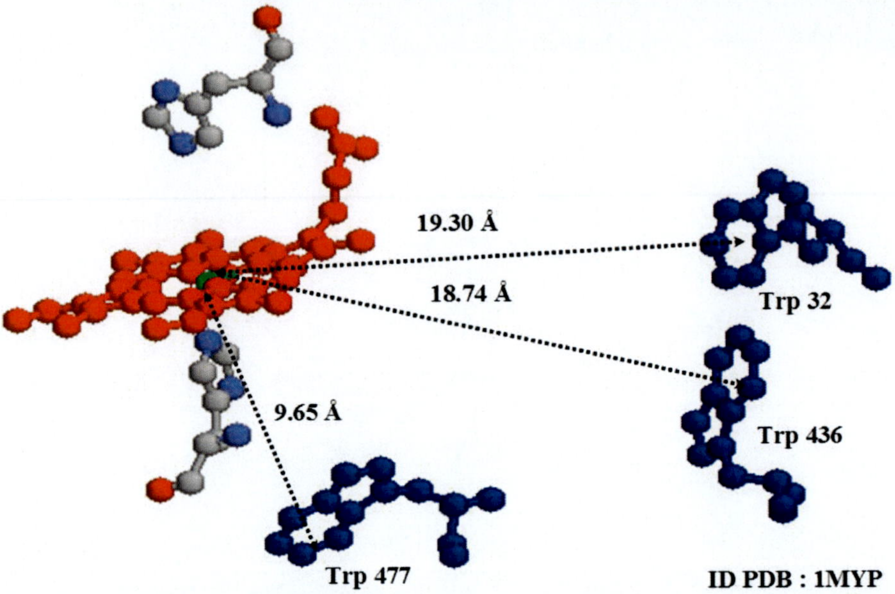

Figure 6. “Ball and sticks” representation of LPO active site. In the picture the heme group is represented in red, His in CPK colour, while Trp 32, 436, 477 are blue.

At this stage, we wondered whether data obtained for LPO may be extended to other metallo-proteins. The question was if enzymes with different amino-acid sequence and active site can be modified by MWs irradiation. For this purpose, we studied the influence of mobile phone emission on two copper active site enzymes: ascorbate oxidase (AO) and laccase (LAC). These enzymes are member of the multicopper oxidases family, which catalyze the oxidation reaction of suitable substrates with oxygen through electron transfer mechanism. They have got the same active site, containing four tightly bound copper ions distributed in three different redox sites, namely type-1, type-2, type-3 and classified on the basis of their spectroscopic features. In particular, they are composed of one type-1 (T1) copper, which is the site of substrate oxidation (E^0 = +790 mV), of one type-2 (T2) and two type-3 (T3) copper ions arranged in a trinuclear cluster. The electrons yielded by substrate oxidation are transferred to trinuclear cluster and used to reduce molecular oxygen to water. This reaction occurs by outer-sphere electron transfer mechanism, through electronic tunnelling mediated by aromatic amino acids residues [12]. The effects of irradiation were carried out using the same experimental procedure we adopted for LPO. Figure 6 shows the Lineaweaver-Burk plot for the enzymatic reaction of ascorbic acid oxidation by oxygen. Kinetic data were collected by means of electrochemical measurements (chronoamperometry) using the same AO solution, either irradiating the enzyme before for 20 minutes (red line) or under continuous irradiation during the reaction (blue line). The kinetic parameters (K_M, V_m and K_{CAT}) obtained are summarized in Table 2, where a significant alteration of the oxidation process, induced by exposure of the reaction bulk to the MWs, appears evident. These parameters, calculated from the slopes and the intercepts of the linear plots (Figure 7) reveal that a general inhibition mechanism was induced, since V_m and k_{CAT} values are lower than those of native sample. Somewhat surprisingly, the enzyme sample irradiated before to be employed in the catalytic reaction, exhibits a peculiar behaviour, showing an evident deviation from Michaelis-Menten mechanism. Data do not give a linear Lineaweaver-Burk plot, indicating that the reaction mechanism is deeply modified. Our opinion is that MWs probably induce conformational modification of the AO catalytic site giving rise to a number of different conformers. Each of them is characterized by peculiar catalytic activity due to changes in the metal ion environment, which may concern non reproducible enzymic activity. This last observation is corroborated by steady-state fluorescence measurement of the intrinsic tryptophan emission spectra showed in Figure 8 (panel a), as irradiation determines a quenching of tryptophan emission intensity of about 9%. Commonly, the decrease of the intensity is associated to a conformational change of Trp residues which are shifted from the protein inside towards the solvent [12]. Furthermore, the fluorescence spectrum of type-3 copper ions (panel b Figure 8), recorded exciting at 330 nm, shows an increase of the intensity of about 15%, confirming that MWs induce structural changes in the AO active site.

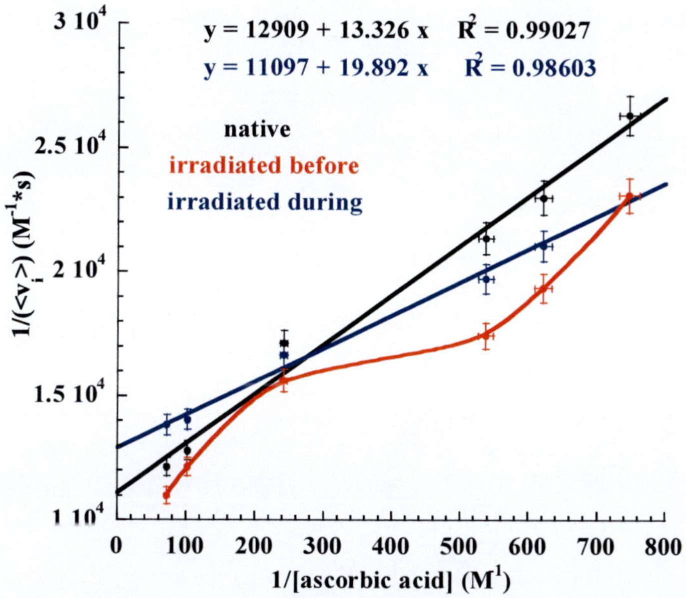

Figure 7. Lineweaver-Burk plot of ascorbate oxidase samples. Figure shows the plot of the inverse of initial reaction rate (1/v) versus the inverse of substrate concentration (1/[S]). [LPO]= 3.60×10^{-8}M, PBS 0.1 M (pH = 7.2), T = (25.0 ± 0.1) °C. Data have been obtained using an electrochemical techniques (chronoamperometry) by recording the oxidation of ascorbic acid (substrate) at 700 Mv.

Table 2. Kinetic activity parameters of ascorbate oxidase samples calculated from the slope and intercept of the straight lines depicted in Figure 6

Sample	K_M (μM)	V_m (μM *s^{-1})	k_{CAT} (s^{-1})
Native	1792 ± 70	90 ± 3	1782 ± 69
Irradiated before	non-detectable	non-detectable	non-detectable
Irradiated during	1032 ± 50	77 ± 2	1525 ± 42

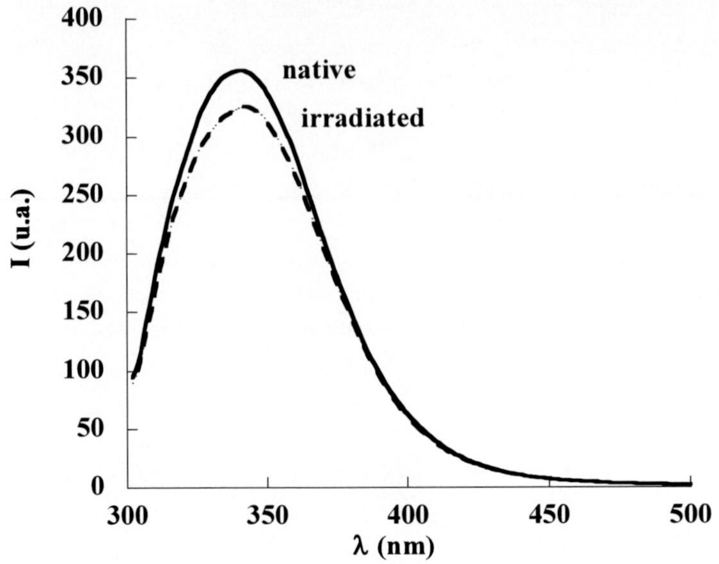

(a)

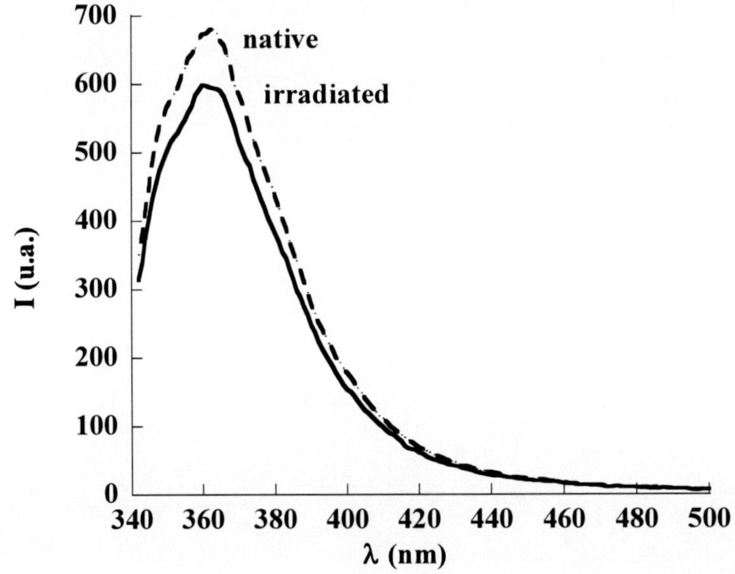

(b)

Figure 8. Steady-state fluorescence spectra of ascorbate oxidase recorded exciting sample at 295 nm (a) (tryptophan residues absorption) and exciting at 330 nm (type-III copper (b). Irradiation determines fluorescence quenching (dotted line) exciting at 295 nm showing that EMFs provoke a shift of Trp residues from the inside towards the solvent and a fluorescence increase (dotted line) exciting at 330 nm showing an alteration of active site conformation due to MWs. [AO] $=1.65\times10^{-7}$M, PBS 0.1 M (pH = 7.2), T = (25.0 ± 0.1)°C.

As it has already found for LPO, also in this case, we observed a modification of enzyme active site without any modification of secondary structure (see far UV CD spectra, Figure 9).

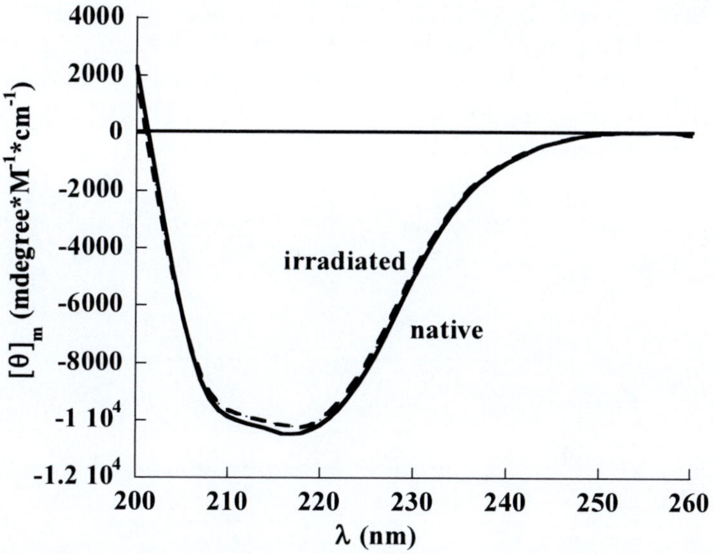

Figure 9. Far-UV CD spectra of ascorbate oxidase samples; the spectra were normalized with peptide concentration. [AO]=1.65x10^{-7}M, PBS 0.1 M (pH = 7.2).

Figure 10 shows the AO active site and Trp residues surrounding copper metal ions which are subject of structural modifications.

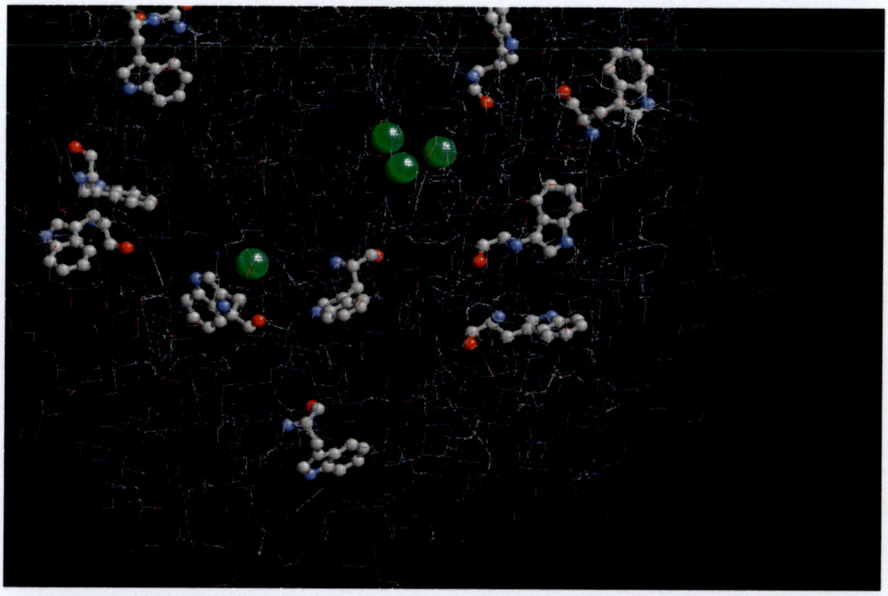

Figure 10. Schematic representation of AO active site got by crystallographic data (ID PDB:1AOZ). Copper ions are pictured as green balls and Trp residues surrounding the active site are reported as "ball and stick" model.

Laccases and AO have got very similar active sites, but different protein structures originated from different amino acid sequences. Generally, laccases have a single polypeptidic chain, while AO is a dimer. To carry out comparative evaluation of the MWs effects, we chose the laccase extracted from Agaricus Bisporus (LAC) because in water solution it is a dimer, as well as AO. The first results we obtained was that irradiation did not change secondary and the tertiary structure of LAC. Far-UV CD and steady-state fluorescence spectra of native and irradiated before (20 minutes) LAC revealed that any structural modification occurred to the protein samples. These experimental evidences are shown in Figures 11a and 11b respectively, where both CD and fluorescence spectra are neatly superimposable.

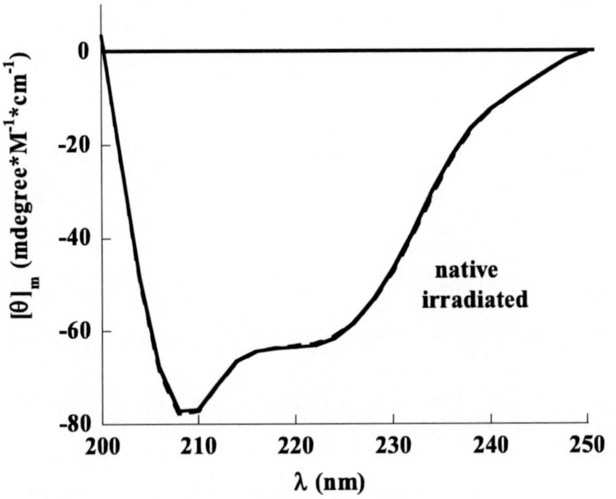

(a)

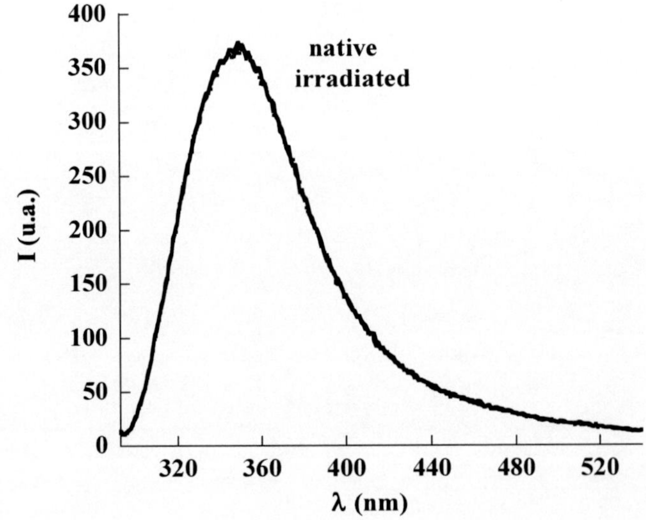

(b)

Figure 11. Far-UV circular dichroism (a) and Steady-state fluorescence spectra (b) of laccase samples recorded exciting sample at 280 nm (a) (aromatic residues absorption). [LAC] =9.70x10^{-6}M, PBS 0.1 M (pH = 7.2), T = (25.0 ± 0.1)°C. CD spectra were normalized with peptide concentration.

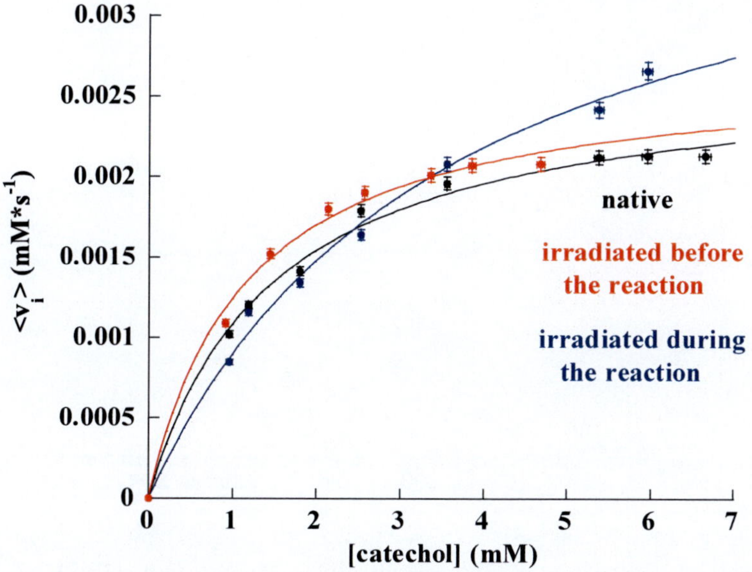

Figure 12. Michaelis-Menten plot of laccase samples. [LPO]= 2.37x10^{-7}M, PBS 0.1 M (pH = 7.2), T = (25.0 ± 0.1) °C. Data have been obtained spectrophotomerically by recording the absorbance at 450 nm of reaction product.

Table 3. Kinetic activity parameters of laccase samples calculated from the parameters of Michaelis-Menten fit depicted in Figure 12

Sample	Km (μM)	Vm (μM *s-1)	kCAT (s-1)
Native	1500 ± 100	2.7 ± 0.1	11.2 ± 0.1
Irradiated before	1300 ± 100	2.8 ± 0.1	11.6 ± 0.1
Irradiated during	3600 ± 400	4.1 ± 0.3	16.9 ± 0.3

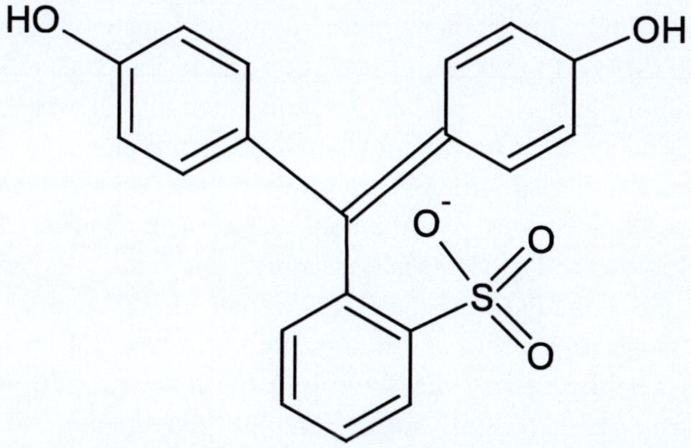

Figure 13. Molecular structure of phenol red.

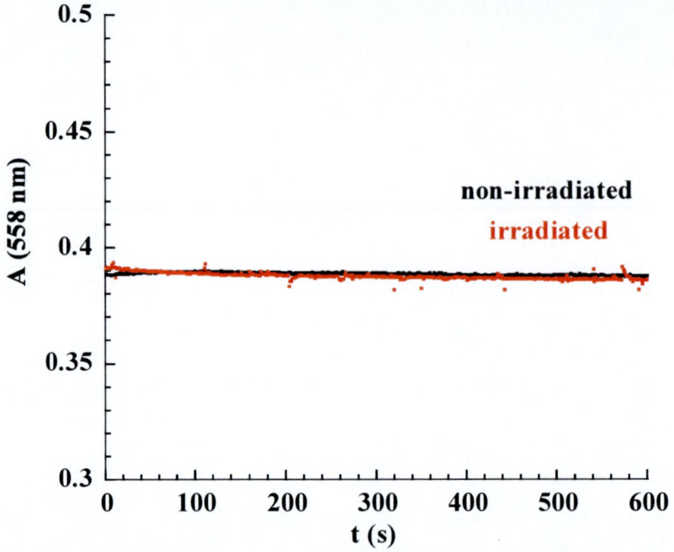

Figure 14. Phenol red in TRIS buffer with and without irradiation with mobile phone ([phenol red]) = 1.05×10^{-6}M, TRIS buffer 0.04 M (pH = 8.3), T = (25.0 ± 0.1) °C.

Kinetic measurements of the LAC enzymic activity provided unchanged parameters which confirmed the lack of effects due to MWs exposure. On the contrary, the same reaction carried out under MWs field is significantly modified. These results are shown in Figure 12 and the kinetic parameters K_M, V_m and K_{CAT}, calculated by means of the best fit of the Michaelis – Menten plot, are summarized in Table 3.

The findings that a kinetic effect has been obtained irradiating the reaction bulk put in evidence that the electromagnetic field interferes with the reaction mechanism mediated by free radical intermediate species. To understand the difference between AO and LAC catalytic activity, we considered that despite the proteins are both dimers with the same active site, they have got a different overall dipole momentum (μ_{AO} = 1156 Debye, μ_{LAC} = 629 Debye). It seems reasonable to suppose that the molecular overall dipole momentum plays an important role in determining conformational changes in the protein structure. On the other hand, also LPO has a high dipole momentum (μ = 1055 Debye) and it shows structural modifications as well as AO. In the case of LAC there is instead a close relationship between protein structural stability and low value of dipole momentum. Nevertheless, the hypothesis that the EMFs could induce a "instantaneous" thermal effect should be considered, since an increase of temperature due to irradiation can explain the observed changes of the enzyme activities. Temperature fluctuations in the kinetic experiments can be determined using the TrisHCl - Phenol Red method [13]. This procedure represents a "chemical thermometer" because it exploits the sensitivity of the TRIS buffer to the temperature. A change of temperature varies the buffer pK and, in turn, the solution pH which can be relieved measuring the optical density of a suitable concentration of the indicator (see Figure 13). Phenol red has a strong absorbance band at 558 nm (red) and it shows a pH dependent dissociation equilibrium:

$$D_Y \underset{}{\overset{K}{\rightleftarrows}} D_R + H^+$$

where D_Y and D_R are the phenol red indissociated and dissociated form respectively. Figure 14 shows the $A_{558\ nm}$ vs time plots of non-irradiated (black) and irradiated samples (red) phenol red in TRIS buffer solution. It appears evident that, since any absorbance fluctuation has been recorded, no thermal effects have been reviled up to 600 seconds of mobile phone irradiation.

Conclusion

In this chapter the experimental data prove that the mobile phone emission alters the features and the catalytic activities of metallo-enzymes. These phenomena have been attributed to the electrostatic interactions between the oscillations of the electric field associated to MWs and charge distribution in the protein structure. It has been also observed that electron-transfer mechanisms, as well as free radical intermediates species, are sensitive to the "external" fluctuation of electromagnetic field. Similar structural modifications have been already observed in protein solutions containing low concentration of denaturating agents, such as guanidine or urea [12]. Among different possible protein denaturated states, molten globule has been defined as such a thermodynamically stable feature occurring when tertiary protein structure changes without secondary structure modifications [12]. The spectroscopic data we recorded demonstrate that mobile phone emission can induce structural changes ascribable to a molten globule conformation. Far-UV CD spectra gathered in the peptide absorption range show that irradiation does not affect secondary structure of the enzyme, while evidence of tertiary structure modification is revealed by the decrease of tryptophan fluorescence.

Despite these results cannot be used to conclude that the mobile phone emission can lead to hazardous effect for the human health, data here showed may represent a useful starting point to improve our knowledges on the effects of the MWs exposure. The use of highly purified protein samples represents a useful tool to obtain reliable and reproducible experimental data even using a widely variable radiation source.

References

[1] Mancinelli F., Caraglia M., Abbruzzese A., d'Ambrosio G., Massa R., Bismuto E., *J. Cell. Biochem.* 2004, 93, 188.

[2] Copty A. B., Neve-Oz Y., Barak I., Golosovsky M., Davidov D., *Biophys. J.* 2006, 91, 1413.

[3] Barteri M., Pala A., Rotella S., *Biophys. Chem.* 2005, 113, 245.

[4] Malaric K., Bartolic J., Malaric R., IMTC 2004, *Instrumentation and Measurement Technology Conference*, Corno, Italy, 18-20 May 2004.

[5] van Holde K.E., Johnson W.C., Shing Ho P., *Principles of Physical Biochemistry*, ed. by Prentice-Hall Inc., Upper Saddle River, New Jersey, USA, 1998, 28-51.

[6] K. M. Pruitt, J. O. Tenovuo, , *The Lactoperoxidase System: Chemistry and Biological Significance*, ed. by Marcel Dekker, Inc., New York, USA, 1985.

[7] J. Everse, K. E. Everse, M. B. Grisham, *Peroxidases in Chemistry and Biology*, ed. by CRC Press, Inc., Boca Raton, FL, 1991, Vol. I, pp 143-180.

[8] E. L. Thomas, P. M. Bozeman, D. B. Learn, *Lactoperoxidase. New Recognition of an "Old" Enzyme in Airway Defenses*, ed. by CRC Press, Inc., Boca Raton, FL, 1991 Vol. I, pp 123–142.
[9] Doerge D.R., Divi R.L. , Churchwell M.I., *Anal. Biochem.* 1997, 250, 10.
[10] M.Blank, L. Soo, *Bioelectrochemistry* 2003, 61, 93.
[11] M. Blank, L. Soo, *J. Cell. Biochem.* 2001, 81, 78.
[12] K.E. van Holde, *Principles of Physical Biochemistry*, ed. by Prentice-Hall, Inc., Upper Saddle River, New Jersey, USA, 1998, pp 28-42.
[13] B. Chance, A. C. Maehly, *Methods in Enzymology*, ed. by S.P. Colowick and N.O.. Kaplan, Academic Press, New York, USA, 1964, Vol. II, pp 764.

RESEARCH AND REVIEW STUDIES

In: Mobile Telephones...
Editors: A.C. Harper and R.V. Buress, pp. 29-75
ISBN: 978-1-60456-436-5

Chapter 1

MOBILE COMMUNICATIONS IN THE 21ST CENTURY

Peter Curwen and Jason Whalley
University of Strathclyde, Glasgow, Scotland

Abstract

This chapter concentrates upon the technological aspects of the development of 21st century mobile communications. Most of the chapter is devoted to mobile communications via handsets as this remains for the foreseeable future much the most popular method. However, since the purpose of the chapter is to discuss the most recent variants of mobile communications, which for now generally involve laptop connectivity via data cards and modems, the latter are also covered. In the latter regard, the chapter also covers Wi-Fi and WiMAX as well as other technologies that are as yet poorly established. The development of these technologies – assuming they do not fall by the wayside – covers the period of the next decade, beyond which point, with technological developments proceeding apace, it is difficult to make predictions.

The initial sections of the chapter cover in considerable detail the sequential development of mobile communications, defining terms and distinguishing between developments in different parts of the world. Hence, the chapter should be of interest to readers irrespective of their location. The underlying structure is based upon a sequential explanation in terms of both so-called 'generations' – from 1G to 4G – and data transfer speeds, with each technology allocated a separate section..

There follows a brief introduction to the core services and platforms such as SMS, e-mail and PTT, and the chapter concludes with a discussion of Wi-Fi, WiMAX and related technologies that have yet to become fully established such as mesh and ad hoc networks.

Introduction

The 21st century is yet to complete its first decade and a decade does not superficially appear to be a long time in terms of most human endeavours. But there are exceptions and mobile communications is one of them. Ask most members of the public how long they have had a mobile handset and it rapidly becomes clear that affordable handsets capable of doing something other than convey voice – that is, using digital as against analogue technology – have only been commonplace for roughly a decade. In 1997, there were roughly 100 million digital subscribers throughout the world; ten years on there are roughly three billion. This chapter is largely concerned with what is generally known as third-generation mobile communications, a development of digital technology that was launched almost without exception only after the turn of the century. However, it also of necessity discusses the evolution of mobile communications prior to 2000 and how that evolution is likely to progress beyond 2008. It encompasses developments throughout the entire world although, for obvious reasons, it tends to concentrate upon those countries/regions where developments are 'ahead of the curve'. For reasons of space, this chapter concentrates upon technological issues.

Background

The mobile – alternatively wireless or cellular radio – communications market encompasses radio paging and a variety of radio telephone services that are provided by connecting together two or more mobile devices or terminals via a base station. For the purposes of the discussion below these devices will for the most part be taken to be handsets since they dominate current usage, although they are constantly evolving both in form and function – Motorola, for example, has taken to referring to 'the device formerly known as the cell phone'. Suffice it to say for now that whatever form a device may take, including data cards and modems in computers – it is currently already capable of incorporating some or most of the following: blogs, calculator, calendar, camera with flash, clock, diary, digital music player, digital purse, e-mail facility, games, instant messaging, modem, satellite tracking, spreadsheet, text messaging, torch, television, video camera and word processor. This list is incomplete, and becomes ever more incomplete with each passing day. The device is also useful for conveying voice, which remains its primary function in its guise as a phone handset. Crucially, all such devices are intended to be more or less permanently to hand – the urban dweller's equivalent of a Swiss Army knife enabling its possessor to do without, for example, a watch, alarm clock and increasingly a camera or music player – and there is growing evidence that the loss of a handset is regarded as more distressing than that of a wallet or credit card.

Mobile communications have a long and complex history [Houseman, 2002], and have been commercially available in some form since 1946, although the initial format in the USA involved a single transmitter covering an entire city. Such a network had very little capacity and delivered calls via a fixed-wire link to the home. Mobile networks only truly developed once a cellular structure was adopted. Modern cellular networks commenced services in the USA in 1983. Interestingly, although the Nordic Mobile Telephone analogue system – NMT-450 operating in the 450 MHz band where MHz is short for megahertz or million hertz – was first introduced in Scandinavia as early as 1981, it was not adopted throughout Europe and there were multiple incompatible standards in use by 1985 – another was the Total Access Communication System (TACS) – at which point the Federal Communications Commission (FCC), the main regulatory body for telecommunications in the USA, was imposing a common analogue Advanced (sometimes Analog) Mobile Phone Service (sometimes System) (AMPS) technology in the 800 MHz band. However, this situation completely reversed itself when digital networks became feasible. For its part, the U.S. government took the view that competition would determine which standard was best, whereas the European Union (EU) set in hand, commencing as early as 1982, a process that would produce a single EU-wide digital standard.

Control over technology is one of the issues addressed by a system of licensing. Because spectrum is finite, and certain spectrum bands are subject to excess demand, it has long been the custom to allocate spectrum via licences although defining property rights for spectrum is a complicated matter, especially when considered in an international/supranational context [Gruber, 2004; Faulhaber, 2006]. Among the issues to be addressed are:

- the size of the spectrum band to be allocated for each licence;
- the area to be covered by each licence;
- the life span of each licence;
- the uses to which each licence may be put;
- whether each licence is to be transferable or tradeable;
- whether transmission power is to be restricted.

As technology has evolved, so licensing conditions have had to evolve in tandem. Throughout the world, the standard digital technology is currently known as second-generation or 2G. The original, analogue, technology (1G), which uses Frequency Division Multiple Access (FDMA) technology such that each user has exclusive use of a specific frequency, is still to be found but is being phased out wherever possible and the spectrum transferred to 2G networks which, because they are digital, use the spectrum much more efficiently.

Categorisation of Technologies by Region

There is some disagreement about how to represent the evolution of mobile technologies over time, partly because, for example, the changeover from a circuit-switched to a packet-based digital network can be a long-winded process and hence aspects of the two can co-exist for long periods. For example, distinctions can be made in terms of 'generations', commencing with 1G, and they can be made in terms of data transfer speeds. What follows is an attempt to impose some kind of sensible structure, based essentially upon the comparative position of the more common technologies as seen from the modern-day perspective. In terms of data transfer speeds, 2G broadly provides – greater precision is not possible since even the various trade associations fail to agree on this – a maximum downlink of between 10 kbps (kilobits per second) and 30 kbps, 2.5G a maximum lying between 100 kbps and 200 kbps, 2.75G a maximum lying between 300 kbps and 500 kbps and true 3G a maximum of at least 2 Mbps (mega or million bits per second). Compare these with the then world record speed for data transfer which was achieved in 2004 with both Japan's NTT DoCoMo (in August) and Siemens (in December) claiming to have delivered 1,000 Mbps over the 5 GHz spectrum band in laboratory conditions. Such speed is potentially attractive even if it is going to be unavailable commercially for some time, but the more contentious issue at the present time is, not surprisingly, what *minimum* speed can reasonably be delivered while claiming that a service falls into a particular category. The reality is that the average speeds attained fall well short of the claimed maxima.

Technology has not developed in a consistent way throughout the world. What follows is a brief overview of the main differences which are defined and explained in detail in what follows.

Europe

As noted, *1G* mainly comprised NMT and TACS and is essentially defunct, while *2G* is the Global System for Mobile (GSM) first launched by Radiolinja on 1 July 1991. GSM can be separately overlaid – that is, enhanced using the same spectrum – to produce *2.5G*, known as the General Packet Radio Service (GPRS) and *2.75G* known as Enhanced Data [Rates] for GSM Evolution (EDGE). GSM can also be extended to deal with higher data transfer speeds, but almost invariably so far using *different* spectrum in order to create *3G*, known as Wideband Code Division Multiple Access (W-CDMA). In its turn, W-CDMA can be enhanced to produce *3.5G*, known as High-Speed Packet Access (HSPA).

The Americas

In many parts of the Americas (comprising the USA, Canada, Central America and South America), GSM and its variants are also (increasingly) to be found. However, the most common route, commencing with the AMPS version of *1G* launched in 1983, proceeds successively through *2G* Code Division Multiple Access (CDMA) technology, known as IS-95 or cdmaOne; *2.5G* known as cdma2000 1xRTT (Radio Transmission Technology) or IS-2000; and *3G* known as cdma2000 1xEV-DO (Evolution-Data Optimized) or cdma2000 1xEV-DV (Evolution-Data and Voice).

There is no strict equivalent to 2.75G since cdma2000 1xRTT is typically faster that GPRS but slower than EDGE, although it is sometimes (reasonably) referred to as 2.75G or (incorrectly) as 3G. The CDMA route for a given operator is usually a continuous overlay commencing with 2G since the same spectrum is used consistently although an operator may own spectrum in more than one band and allocate different services to individual bands. It is also worth noting that cdma2000 is a registered trademark of the Telecommunications Industry Association of the USA rather than a generic term, as is CDMA. The spectrum and hardware allocated to cdma2000 1xEV-DO provides a data-only service whereas other technologies provide both data and voice. Furthermore, 1xEV-DO is optimised for downlink connectivity (at up to 2.4 Mbps) and its uplink connectivity speed in its Release 0 format is typically only 154 kbps although Revision A is considerably faster – 3.1 Mbps and 1.8 Mbps respectively. This is why it was deemed to be desirable eventually to move on to 1xEV-DV, although it now looks increasingly as though this will be ignored in favour of EV-DO Revision A (Rev. A) which is the cdma2000 version of *3.5G*.

It is also important to note that an operator such as AT&T started out using IS-54/IS-136 Digital AMPS (D-AMPS), which is commonly referred to as Time Division Multiple Access (TDMA) although TDMA is strictly a technique used in other protocols as well as D-AMPS. D-AMPS/TDMA is a technology using the 824-849 and 869-894 MHz bands that has no direct route through 2.5G to 3G. Hence, what needs to happen is that TDMA is overlaid with GSM (a similar technology) and the GSM route is followed thereafter. It is also possible to switch in the other direction – that is, to convert from GSM to cdma2000 using GSM 1x technology, although only China appears to have an interest in this possibility.

Japan

The situation in Japan is confused by the fact that there is a technology called the Personal Handyphone System (PHS), now largely provided by Willcom, which is used in confined geographic areas such as cities. This has no direct equivalent in any of the above – the nearest equivalent in Europe is Digital Enhanced Cordless Telecommunications (DECT). PHS initally operated alongside the *1G* technology known as Japanese Mobile Cellular System in respect of the two operators, NTT DoCoMo and Softbank (the former Vodafone K.K.) not using the CDMA route favoured by KDDI. *2G* for DoCoMo and Softbank is known as Personal Digital Cellular (PDC). *2.5G* tends to be known by the individual operator brand such as i-mode (DoCoMo) or Ezweb (KDDI). While KDDI uses cdma2000 1xEV-DO for *3G*, DoCoMo and Softbank (somewhat confusingly) use W-CDMA.

China

There is a roughly equivalent service to PHS known, *inter alia*, as a Personal Access System (PAS). In other respects, Chinese operators use either the GSM route or the CDMA route. However, a third alternative *3G* technology is under development known as Time Division Synchronous Code Division Multiple Access (TD-SCDMA).

Rest of World

Countries comprising the rest of the world essentially follow either the GSM route which is much the most popular, the CDMA route or both. However, it is worth observing that there is a *2G* technology used by Sprint Nextel in the USA and South America called Integrated Digital (sometimes Dispatch) Enhanced Network (iDEN) that is also occasionally to be found elsewhere in the world.

Table 1. Subscribers by technology (million). End-year[1]

	1995	1996	1997	1998	1999	2000	2001	2002	2003	2004	2005	2006
GSM	13.0	32.8	71.1	138.4	258.4	456.1	626.2	809.3	1012.0	1296.0	1719.7	2,143.8
CDMA	-	1.0	7.4	22.4	52.6	80.3	110.9	104.2	98.9	87.4	41.2	20.8
cdma 1x								36.1	80.1	131.9	233.4	279.6
cdma 1xEV-DO								0.2	4.6	12.3	26.8	50.5
US TDMA	0.7	2.6	6.3	15.9	38.0	67.6	94.1	101.1	100.1	90.0	47.8	20.4
PDC	3.3	13.9	26.8	38.1	44.8	50.8	56.8	56.1	58.1	54.2	46.3	31.7
iDEN	-	0.3	1.4	3.1	5.1	8.2	11.1	11.0	13.4	16.8	22.6	25.6
W-CDMA	-	-	-	-	-	-	-	0.2	2.8	16.3	48.2	98.0
Total digital	17.0	50.6	113.0	217.9	398.9	663.0	899.1	1118.2	1375.7	1704.9	2186.0	2,670.4
Total analogue								19.7	12.9	9.2	6.6	3.1
Total wireless								1137.9	1388.6	1714.1	2,192.6	2,673.5

[1] There are two series available which are not entirely compatible. The second series breaks down CDMA data as from 2002 and has accordingly been used from that point. It should be noted that a considerably different set of figures is available (patchily) at www.3gamericas.com

Source: derived from data available at www.gsmworld.com/news/statistics/substats

Growth of Mobile Subscriptions

The development of telecommunications in both wealthy and poor countries has relied increasingly upon mobile telephony either as a complement to, or substitute for, fixed-wire connections. Some idea of the growth of subscriptions by technology can be obtained from Table 1 which concentrates upon the digital technologies that concern us in this chapter. As can be seen, analogue had effectively ceased to play any real role by 2002, during which year the billionth subscription was taken out. 2002 was also the year when 3G technologies first put in an appearance of any consequence. It may also be noted that, because it was much easier to introduce cdma1xEV-DO (an overlay) compared to W-CDMA (which needed new infrastructure), W-CDMA subscriber numbers did not overtake those for 1xEV-DO until 2005. W-CDMA is now growing much faster than 1xEV-DO, which is hardly surprising given the evident dominance of GSM as the world's most popular 2G technology.

Evolution of Cellular Networks

Since some readers will be largely unfamiliar with mobile telephony, it is necessary briefly to review how a 2G cellular network is configured. At the heart of a modern mobile network are to be found a number of base stations which link call-originating handsets to such mobile and fixed-wire networks as are capable of receiving the transmitted signals. The signals are picked up and passed on by antennas attached to the base stations, each of which sits at the heart of a hexagonal-shaped cell which is adjacent to, and overlaps, other cells, and which covers a specific geographic area. Cells are generally described as 'macro' when the antenna is installed on a mast or building above the average roof level, as 'micro' when the antenna is under the average roof level and as 'pico' when it is installed indoors. Switching centres connect up the base stations, keep track of calls and ensure, for example, that if the sender is moving around that the call is picked up by the nearest cell. Base stations connect up to mobile switches either via fibre-optic cable or, if there is line of sight, by a radio link. In turn, a mobile switch may connect up to a fixed-wire switch in order to pick up or deliver a call to a fixed handset. As most readers will know from experience, this is not an infallible system and connections can be broken when, for example, a subscriber wanders to the extreme limit of the connected base station and cannot for whatever reason be picked up by any other.

The size of a cell is dependent upon a number of factors, and in particular the radio frequency in use since the higher the frequency, the smaller the cell. When a subscriber dials in, he or she is assigned a tiny portion of the available spectrum and the task of the system is to keep the assigned spectrum connected up irrespective of the movement of the caller. A further issue of significance is consequently the volume of traffic to be handled since a base station has a finite capacity, and when it is reached an additional base station will need to be built and the size of cells reduced to compensate. Also, physical features such as buildings or natural aspects of the terrain may need to be circumvented by creating additional cells. In general, urban cells will be located no less than 0.2 km and no more than 0.5 km apart, while rural cells will be ten times as far apart.

Since the spectrum licensed to a specific network is strictly limited, it is important that a particular frequency is utilised efficiently. This is done by taking account of the fact that a signal weakens as it travels, so although it will be strong in the nearest cell to an antenna, it will be very weak two cells further away. Hence, more distant cells can use the same frequency without fear of interference. Increasing the number of cells accordingly increases the number of times a frequency can be used within a geographic area.

The original analogue technology was subject to a number of limitations, most particularly its inability to be compressed without loss of clarity and the fact that, whereas it was satisfactory for voice telephony, it was unsatisfactory for data transmission which had to be converted from a digital to an analogue format prior to transmission and then back again at the receiving end. Analogue transmission tends to suffer from crackle, lack of security and vulnerability to fraud, although these shortcomings are partly compensated by its virtue of not fading out. Digital signals were initially approved during the 1980s. Where these continue to coexist with analogue signals most handsets need to be dual-mode to cope with both types of signal.

GSM, which is the common digital system in the European Union, was licensed widely from the beginning of 1992. It uses the TDMA standard which allows each call to run along a channel within the available bandwidth and assigns it a fraction of the time available within that channel.. In contrast, CDMA assigns a special electronic code to each signal and hence allows the entire frequency band to be occupied simultaneously. This is claimed to provide greater capacity, better sound quality, lower power consumption and a decreased potential for fraudulent use.

Mobile networks have largely changed over from a circuit-switched to a packet-switched basis. Packet-switching divides data up into individual packets of a specified size and in a specified format before sending them along the network. The correct sequence for delivery and the address for delivery are determined via control information attached to the packets before being sent. As a packet-switched network is permanently available to all data from any source, these inevitably get jumbled up together and may take a variety of routes from any one source to any one destination. However, the control instructions should ensure that the data is reassembled into the correct order before delivery. Since this may be subject to delays, it is advisable to keep voice data – which are time-critical – separate from non-voice data.

It is evident that the system must recognise the identity of individual subscribers, and this is normally achieved via the Subscriber Identity Module (SIM) which comes as a small card to be inserted in a mobile device. It contains a unique International Mobile Subscriber Identity used to identify the subscriber when logging into the system.

Most of the 2G licences in Western Europe were issued between 1992 and 1996, and utilised spectrum in the 900 MHz GSM band. In the 900 MHz band, the uplink is 890-915 MHz and the downlink is 935-960 MHz. This 25 MHz band is divided into 124 carrier frequency channels, each spaced 200 kHz apart, with eight speech channels per frequency channel. The term GSM is also used generically in relation to spectrum in the 1800 MHz band (1710-1785 and 1805-1880 MHz) otherwise known as a Personal Communications Network (PCN). PCN spectrum was typically licensed in Western Europe after 1997, either to existing GSM licensees to ameliorate spectrum shortages as mobile telephony surged in popularity or to permit the entry of new operators. A crucial factor was that the administrative arm of the European Union, the European Commission, enforced the exclusive usage of these two spectrum bands via the auspices of the European Telecommunications Standards Institute (ETSI).

Elsewhere in the world a variety of spectrum bands are in use for 2G. Overall, however, by far the most popular technology is GSM although outside Europe it is generally used in alternative spectrum bands – typically the 800 MHz band (824-849 and 869-894 MHz) and/or the 1900 MHz band (1850-1910 and 1930-1990 MHz) known in the USA as Personal Communications Services (PCS). The use of GSM is fostered by the factor of international 'roaming' – that is, the ability of a GSM handset owner to take it to another country where it can still be used both internally and to make international calls to other countries where GSM is in use (although a multi-band handset will usually be required).

Given the use of GSM throughout Europe, as well as its partial use in the USA, there is accordingly an obvious advantage for countries switching from analogue to digital to adopt it. However, some countries adopted one of the alternative standards to GSM before it became so well-established. In particular, the U.S. government made no attempt to enforce a national standard, and hence three different technologies were adopted. In addition to GSM, there was widespread use of both CDMA transmitting in the 450 MHz, 800/850 MHz, 1700 MHz and 1900 MHz bands (with 2100 MHz under development), and Time Division Multiple Access (TDMA).

In Japan, PHS operating in the 1880-1930 MHz band was introduced in 1995 to utilise a dense network of low-powered, and hence cheap, base stations when in mobile mode. PHS handsets can also be used via a connection to a fixed-wire network when in the home or office, and PHS is hence roughly equivalent, as noted, to the DECT standard used in Western Europe. It is also of note that in 2002, China Telecom, despite the absence of a cellular licence, set up what is variously known as a personal access system or public access service (PAS), 'Go-to-phone' service or Xiaolingtong (smart little connector or 'Little Smart') which allows mobile subscribers to switch calls to their handsets through to fixed lines, thereby avoiding the charges made to mobile call recipients. The service – there is also an equivalent known as Haoyitong in Tibet – which is used primarily for voice telephony and simple text messages although Internet browsing and downloads are now available, only works within urban environments – antennas fixed to rooftops have a range of up to two kilometers and hand-off is less than perfect. It is technically a fixed-wire technology because it uses fixed-wire infrastructure for its backbone, hence allowing China Telecom to provide competition for China Mobile and China Unicom without acquiring a mobile licence. A significant point is that PAS handsets did not contain SIM cards, hence preventing any kind of roaming. However, in November 2004, China Netcom announced that it would be introducing handsets with a personal identity module (PIM) card which would at least allow a subscriber with a PAS handset to use it in another city by changing the PIM card.

Third-generation mobile telephony (3G) is essentially an extension of 2G capabilities, although, as discussed below, the developmental process has in practice required an intermediate stage, known commonly as 2.5G, to be introduced. The W-CDMA version of 3G, incorporating both terrestrial and satellite components, is known in Western Europe as the Universal Mobile Telecommunications System (UMTS), and that term has incrasingly become used in preference to W-CDMA. UMTS has been described by ETSI as 'personalised globally accessible high quality mobile communications services', but it is more commonly described as 'mobile interactive multimedia services'. The official European Commission version is that UMTS is an advance on GSM capable of combining the use of terrestrial and satellite components [Council of the European Communities, 1998 and 1999]. UMTS did not strictly represent a major technological breakthrough because GPRS already provided many services that were not originally expected to evolve until a later stage of development. Furthermore, upgraded or 2.5G variants of GSM will co-exist with UMTS for an extended period. However, UMTS and HSPA enable the provision of more sophisticated services that require greater bandwidth – effectively multimedia-on-the-move. 3G always utilises a packet-switched network which enables subscribers to have an 'always on' connection to the Internet, and hence one that permits charges to be levied per packet sent rather than according to the time the circuit is open. The relationship between the various cellular systems so far described is set out in Table 2.

2.5G

General Packet Radio Service (GPRS)

GPRS was slow to establish itself, partly because of the somewhat adverse reaction to the Wireless Application Protocol (WAP) in Europe. It was always intended to be an intermediate technology, and it was initially unclear how long it would have the marketplace to itself before 3G took over. In practice, the reality was that GPRS came into its own as the prospects for 3G receded somewhat and some network operators around the world are no longer convinced that they wish to proceed beyond 2.5G. GPRS-enabled networks are now the norm in Western Europe, and have become commonplace elsewhere in the world where GSM is the 2G standard.

The GPRS infrastructure and handsets support 'always-on' data transmission at up to 21 kbps per channel over a maximum of eight channels – that is, a theoretical maximum of around 168 kbps combining several downlink and uplink channels. Typically, a device will in practice accommodate four downlink and two uplink channels, with the bandwidth divided up between data and voice according to the strategy of the operator, delivering in the range 60-80 kbps with bursts up to 115 kbps. In some respects, the building out of GPRS networks is the easy part. Once a GSM network is up and running it is necessary only to add two nodes in the core network (Serving GPRS Support Node and Gateway GPRS Support Node) and to add a circuit board to the radio network to allow packets to be transported [1]. The issues that then need to be resolved include maximising the stability of the new network, making arrangements for roaming and developing applications and services. It is also obviously necessary to have sufficient capacity to meet the demand for GPRS-enabled handsets.

Table 2. Evolution of cellular systems

1G	2G	2.5G	3G
Analogue	Digital	Digital	Digital
Mainly speech	Mainly speech	Mainly speech	Speech & video
Circuit-switched	Circuit-switched	More packet-switched	Mainly packet-switched
Local systems	Global roaming	Global roaming	Global roaming

Source: Compiled by author.

Cdma2000 1xRTT

Reference has already been made above to the first stage in the development of cdma2000, namely 1xRTT, which is being overlaid on virtually all CDMA 2G networks throughout the world. Because the transition to the more advanced 1xEV-DO is relatively simple and inexpensive, it is assumed that the great majority of networks will progress beyond 1xRTT with the added advantage that, unlike with W-CDMA, no additional licence is normally required because no additional spectrum is involved.

I-mode

i-mode, as noted, was developed by DoCoMo in Japan and is, in effect, a Japanese equivalent of GPRS and cdma2000 1xRTT, albeit a rather more sophisticated version [2] That i-mode, a play on the Japanese word for 'anywhere' – the 'i' stands for 'information' – is a success is not generally questioned. Commencing on 22 February 1999, it reached 30 million subscribers at Xmas 2001 and 40 million in October 2003. However, the end-year figure for 2006 was 47.2 million, indicating a considerable slowing of growth during the previous two years.

Although there is general agreement as to the factors underpinning why i-mode has been a success in Japan [Fransman, 2002], there is much less agreement as to whether these factors can be replicated elsewhere. It is worth observing that not everyone is persuaded by the i-mode story. There are those who argue, for example, that i-mode did not enable DoCoMo to increase average revenue per unit (ARPU) but merely to prevent it declining at a time when voice call prices and revenues were in steep decline. Now that KDDI is running a successful cdma2000 1xRTT service based upon flat-rate pricing it makes it impossible for DoCoMo to do anything other than respond in kind, with consequent depressive effects upon ARPU. In addition, in order to tempt i-mode users to switch to its 3G service FOMA, DoCoMo has been obliged to offer relatively low prices for the latter, thereby eroding its i-mode revenues.

What is clear is that European and US telcos have not rushed out to introduce i-mode equivalents into their markets. The main reasons initially put forward were that:

- fixed-wire Internet access was already fairly high and growing;
- a lesser proportion of disposable income was devoted to entertainment;
- less time was spent waiting around to get somewhere or for something to happen;
- there was less enthusiasm for gadgets;
- there was a greater preference for spoken communication.

However, some found these points somewhat unconvincing, arguing for example that:

- mobility was a feature of all Western societies;
- younger mobile users were always keen to utilise new services;
- the growing use of e-mail and text messaging had accustomed people to accessing written communications via a screen;
- everyone spent some time waiting around with time to kill;
- the substantial sum already expended on mobile phones suggested that people would pay for additional services provided they fulfilled some kind of need.

In Europe, the initial adherents of i-mode were KPN Mobile (Belgium, Germany, the Netherlands) – in which DoCoMo took a (now much reduced) stake – Bouygues Télécom (which launched in France in November 2002), Telefónica Móviles (which launched e-moción services in Spain in late June 2003), Wind (which launched in Italy in November 2003), CosmOTE (which launched in Greece in June 2004) and O_2 in Ireland and the UK. Notably, Hutchison 3G UK steadfastly refused to introduce i-mode despite the 20 per cent stake held (but now sold) by DoCoMo. The other main agreements relate to Far EasTone (KG Telecom) in Taiwan which launched an i-mode service in June 2002, to Telstra in Australia, to Mobile TeleSystems in Russia which launched in September 2005 and to StarHub which launched in Singapore in November 2005. DoCoMo also had an agreement with the former AT&T Wireless in the USA and the Caribbean (where it was branded as m-mode). For most other operators, the issue is much more about the route to be taken in the evolution towards 3G, which generally involves GPRS.

By the end of 2006, the prospects for i-mode outside Japan were looking decidedly patchy. In March, Cellcom cancelled i-mode after only six months. In August, MTS suspended development mainly because it did not have available an acceptable range of handsets and those that were available were too expensive at between $200 and $300. Although, at the time, MTS claimed to have 100,000 subscribers, it was claimed that only 43,000 handsets had been sold and that only 25,000 subscribers were actually tapping into i-mode services. For its part, Telstra still provided a service but so little use was being made of it that it was eventually terminated in December 2007. As for E-Plus, it referred to i-mode as a 'niche service' while O_2 failed to launch at all in Germany and, in July 2007, put i-mode on notice of termination in the UK citing poor take-up and a lack of attractive handsets [Turner, 2007].

DoCoMo had licensed i-mode primarily to the third-placed operators in a market who wanted to use it to gain a market advantage, but their limited subscriber numbers were a hindrance to achieving this goal. It also annoyed E-Plus in Germany by giving O_2 the right to launch i-mode there after KPN was assured that it would have exclusive rights.

2.75G

Enhanced Data [Rates] for GSM Evolution (EDGE)

EDGE is now generally regarded, somewhat clumsily perhaps, as equivalent to 2.75G. The situation in relation to EDGE, which offers very good spectral efficiency up to 100 kbps – its likely transmission speed in typical circumstances – and does not require a separate licence, remains somewhat unclear. In essence, the advantage of EDGE is that it offers improvements upon GPRS by utilising more error-tolerant modulation and coding schemes. Each timeslot is theoretically capable of providing 59.3 kbps, and hence bundling eight timeslots provides a theoretical maximum of 473.6 kbps although, as noted, far less is deliverable in practice. Until the delays with UMTS became severe, few expected EDGE to be introduced other than very occasionally. The reason for this is that EDGE needs its own handsets and possibly new hardware in base stations – the latest GSM/GPRS networks need only changes to radio planning and operational parameters via software upgrades. It should also be borne in mind that EDGE cells are smaller than GSM cells which are themselves smaller than TDMA cells [3]. Because EDGE was essentially developed as a means for TDMA operators to provide more sophisticated data services, it was thought that it would not be introduced in Europe other than, perhaps, by operators that either failed to acquire, or chose not to bid for, UMTS licences. The early adopters, not surprisingly, were not European: They included Telecom Personal in Argentina, SingTel Optus in Australia, Batelco in Bahrain, Rogers Wireless in Canada, America Móvil in Brazil, Colombia and Ecuador, HK CSL and People's Telephone in Hong Kong, DiGi.Com in Malaysia, Telcel in Mexico, Globe Telecom and Smart Communications in the Philippines, StarHub in Singapore, TAC in Thailand and AT&T Wireless, Cingular Wireless and T-Mobile USA in the USA. However, TIM, Bouygues Télécom and Orange in France moved quickly to set in hand the facility to use EDGE if necessary, although this mystified some analysts who argued that although EDGE can offer faster data speeds and increase voice capacity, there is no point in using it for the latter purpose when there is a viable alternative – namely taking over an existing operator.

The current position appears to be that EDGE is not particularly viable for an operator with a GSM network built before 2000. Newer networks have been designed to be upgradable to EDGE simply by altering the software at a cost of roughly $1-2 per person, and in future most operators building out their GPRS networks, such as T-Mobile in Hungary, China Mobile (HK) and C&W West Indies, will become EDGE-capable by default. Given the shortage of W-CDMA networks in the USA, it is hardly surprising that the GSM operators there are among the most enthused about EDGE's potential. One of the curiousities of EDGE is that handsets were available prior to most operators' initial launch date for the service simply because adding EDGE to a handset is cheap. This can be compared to the situation in relation to 3G where roll-outs were repeatedly postponed due to the absence of functional handsets.

By early 2006 it had become evident that EDGE would increasingly be seen as a means for covering relatively rural areas that were expensive to cover with 3G and which could be left uncovered for some considerable time because the maximum 3G coverage requirements were, typically, around 70 per cent of the population. This appears to be the strategy of operators such as Orange. This may not be too bad a deal in practice since Nokia is expected to launch a faster version of EDGE called EDGE Evolution or Evolved EDGE – work is proceeding to produce a 3GPP standard for Evolved EDGE via Release 7 [3G Americas, 2006]. In principle, this uses two simultaneous radio channels compared to the one channel currently used for EDGE and hence doubles the data download speed. However, it has the potential to double this again – with a theoretical maximum of 1 Mbps downstream – because it uses an enhanced modulation and coding scheme. As ever, there is a delay while compatible hardware is made available on the requisite scale, so a wide-scale launch is probably not due until 2009. At this point, despite relatively low speeds compared to W-CDMA and HSDPA [4], an Evolved EDGE upgrade will be a very viable alternative where the cost of a 3G network is deemed to be prohibitive.

As a final comment, it may be noted that these intermediate stages between 2G and 3G may be as far as some carriers wish to go – while GPRS, for example, is strictly a best-effort technology, it works perfectly well for the provision of most services up to and including streaming video. Because of this, some potential bidders were expected to be discouraged from seeking 3G licences although an opposing factor was that some operators were becoming desperate for additional spectrum because the number of subscribers to their existing 2G networks was growing very rapidly, and hence they had no option other than to seek 3G licences. In other cases, it was anticipated that non-licensees could apply to licence holders to become mobile virtual network operators (MVNOs) [5]

Evolution of Standards for 3G

In late 1997, the International Telecommunication Union (ITU) published a report – *Framework for Modularity and Radio Commonality within IMT-2000* – which advocated the introduction of a standardised system within the context of International Mobile Telecommunications-2000 (IMT-2000) by the year 2002 [Commission of the European Communities, 1997]. The idea was that both carriers and equipment manufacturers would be provided with a set of common building blocks with which any third-generation system had to be made compatible. This was a critical aspect since it signified that there was not to be a unique standard but rather a 'family of standards' which would allow all existing networks, with their attendant historic heavy investment, to be modified rather than discarded. The actual equipment could take a variety of forms.

The requirements placed upon a 3G network by the ITU were as follows:

- 2 Mbps in fixed or in-building environments;
- 384 kbps in pedestrian or urban environments;
- 144 kbps in wide area mobile environments;
- variable data rates via satellite.

Technically, a 3G network could utilise almost all spectrum lying between 400 MHz and 3 GHz. Despite this potentially vast spectrum range, it had already been determined, as a result of a decision made in 1992 by the World Radiocommunications Conference (WRC) that the common frequencies in all third-generation systems would be restricted to 1885-2025 MHz and 2110-2200 MHz. This would make international roaming between networks a reality, together with the transmission of data at potentially up to 144 kbps on the move and 2 Mbps while stationary.

Because W-CDMA initially needed new spectrum compared to GSM, the issue of roaming was particularly important since whereas the transmission of mobile data was not expected to be relevant to more than 25 per cent of wireless subscribers even by 2010, it was likely to represent at least 50 per cent of carriers' profits. However, the decision on the preferred frequency ranges did lead some commentators to wonder whether the technical advantages derived from a move to new spectrum for mobile communications were greater than the drawbacks arising from the need to design, build and test new networks from scratch, build additional base stations and deal with the need to migrate progressively from 2G networks using lower frequencies.

In June 1997, the European Radiocommunications Committee (ERC) decided [6] to divide up the WRC frequency bands, allocating 1900-1980 MHz, 2010-2025 MHz and 2110-2170 MHz to terrestrial UMTS as well as 1980-2010 MHz (earth-to-space) and 2170-2200 MHz (space-to-earth) to satellite applications which use a Space Division Multiple Access (SDMA) interface [7], but recognised that this would probably be insufficient and that additional spectrum would almost certainly need to be freed or reallocated by 2005 within the 900, 1800 and 1900 MHz bands used most commonly by 2G systems [Commission of the European Communities, 1998a].

In a further move, the ITU divided the new system into two parts based respectively upon the core network architecture and the radio (air) interface – linking base station and handset and known as the Universal Terrestrial Radio Access (UTRA) – and specified common components in each case. The telecoms industry was then left to work out how best to meet the requirements of each. So far as the core architecture was concerned, the choice was fairly limited in practice given the predominance of GSM in Europe, and cdmaOne and TDMA in the USA – the two respective core architectures are generally referred to as GSM-MAP and ANSI-41. The ITU wanted everything to be settled by the end of 1999 so that services could commence in 2002 [The Economist, 1998].

Nokia, Ericsson, Alcatel and Siemens, at the time Europe's leading mobile handset manufacturers, announced that they would be backing GSM-MAP for the core architecture. For the radio interface the choice lay essentially between wideband CDMA (W-CDMA – now called 3GSM by the GSM Association to avoid confusion with cdma2000) and a hybrid wideband TDMA/CDMA (W-TD/CDMA). Whereas Japan opted for W-CDMA throughout – effectively the same as for GSM-MAP in the core architecture – the UMTS approach to the radio interface contained elements of W-TD/CDMA. Work was set in hand to bring these into close compatibility.

The issue of compatibility with standards in the USA presented the most awkward problems because Lucent Technologies and Qualcomm wanted to use cdma2000. The ITU agreed to accept cdma2000 1xEV-DO – known in South Korea as Enhanced Version-Data Only and occasionally as Evolved-Data Optimized – as equivalent to 3G in August 2001. As noted previously, in its original Release 0 variant, 1xEV-DO was capable of delivering downlink at up to 2.4 Mbps (although the actual speed was generally in the 300-600 kbps range when first deployed) while Revision A, approved by the 3GPP in April 2004 and first tested successfully by Nortel Networks and Sierra Wireless in March 2006, upgraded the downlink to a theoretical rate of 3.1 Mbps with a maximum of 1.8 Mbps for the uplink. Although Rev. A only came on stream towards the end of 2006, and only began to become prevalent during 2007, Revision B will in due course upgrade the downlink to a maximum of 4.9 Mbps [8].

Radio Interface Modes

The present position is that there are three main radio interface modes within IMT-2000, known respectively as IMT DS (Direct Sequence or Spread), IMT MC (Multi-Carrier) and IMT TC (Time Code), which between them are compatible with GSM, CDMA and TDMA. These modes are known more commonly as Direct Sequence Frequency Division Duplex (DS-FDD – a type of CDMA (W-CDMA) supported by GSM carriers, which uses paired (separated uplink and downlink) spectrum and is sometimes referred to as UTRA FDD); Multi-Carrier Frequency Division Duplex (MC-FDD – which uses paired spectrum and is based on cdma2000); and Time Division Duplex (TDD – which comes in a version known as UTRA TDD or one harmonised with the new TD-SCDMA standard, both of which use unpaired spectrum for the provision of both uplink and downlink) [9, 10] In addition to these, the ITU has established a specification for IMT SC (Single Carrier), also known as UWC-136 (which, like W-CDMA, is a FDD system related to EDGE), and for IMT FT (Frequency Time), commonly known as DECT [11]. The forecast speed of the new technology is claimed to be 384 kbps in urban areas and up to 2 Mbps in indoor locations, although these speeds are viewed with a certain amount of scepticism.

It is to be noted that FDD/W-CDMA uses a channel spacing of 5 MHz (as compared to 1.25 MHz for cdma2000 1xRTT), and that an operator normally needs either three or four channels to provide an acceptable high-speed service. The standard allocation of spectrum is accordingly 1920-1980 MHz paired with 2110-2170 MHz – that is, 60 MHz paired – providing either 15 MHz paired (four licences) or 20 MHz paired (three licences) per licensee. TDD/SCDMA simply provides 5 MHz of spectrum within the 1900-1920 and 2010-2025 MHz bands. However, it is unusual in practice for seven lots of TDD to be allocated, and TDD spectrum can be paired up if so desired, just as it is possible to get by with a minimum of 10 MHz of paired spectrum (six licences).

WRC 2000

At the World Radiocommunications Conference 2000 (WRC-2000) [12], governments of 150 countries again addressed the issue of spectrum allocation in the hope of agreeing a single global range for further expansion after 2005, essentially consisting of 160 MHz in one continuous band in the 2.5 GHz (2,500 MHz) band. However, the best that could be agreed was a choice of three frequency bands, namely 806-960, 1710-1885 and 2500-2690 MHz – the latter heavily favoured by European organisations because it was the only one not currently occupied for 2G. In addition, the bands 1525-1544, 1545-1559, 1610-1625.5, 1626.5-1645.5, 1646.5-1660.5 and 2483.5-2500 MHz could be used for satellite transmission, as well as 2500-2520 and 2670-2690 MHz should the need arise. For its part, the European Commission issued on 9 March 2001 a fourth Mandate to the Conference of European Post and Telecommunication Administrations (CEPT) to harmonise frequency usage in the additional bands identified in 2000. This resulted in a Report and an ECC Decision in line with the scope and tasks given under Mandate 4. The key points included in the adopted EEC Decision, ECC/DEC/(02)06 of 15 November 2002, were:

- designation of the whole 2500-2690 MHz band to IMT-2000/UMTS systems;
- making the whole so-called 'Expansion Band' available for use by 1 January 2008, subject to market demand and national licensing schemes;
- designation of the frequency band 2520-2670 MHz for terrestrial use;
- detailed channelling arrangements for the whole band as well as utilisation of the bands 2500-2520 MHz and 2670-2690 MHz should form a separate Decision to be adopted by the end of 2004.

On 30 September 2003, the latter point formed part of the Mandate 5 to the CEPT. It was anticipated that its report would be submitted in November 2004 and ratified by a Decision in March 2005. However, all did not go smoothly essentially because member states split into three factions. One group, led by France and Finland (seeking to protect the interests of Nokia), continue to exert pressure to restrict the Expansion Band to IMT-2000 systems, although recently the French have relaxed their position to encompass only the paired spectrum. However, another group, led by the European Commission, want to mandate a free-market, technology-neutral approach to the use of the entire spectrum, arguing that it might be better given over to newly-emerging technologies such as WiMAX (see below). A third group would prefer the Commission not to mandate anything at all but simply to leave matters to the discretion of individual regulators. A vote on the matter was to be taken by the Radio Spectrum Committee in October 2005, but even at a further meeting in March 2006 it was clear that no agreement would be forthcoming from regulators.

For its part, the USA refused to commit itself to any of the original spectrum bands because much of all this bandwidth was already occupied [Curwen, 2005]. This meant that handsets would need to have a built-in capacity to roam between frequencies and would probably be heavier and costlier than was desirable. To ameliorate matters somewhat, the WRC also accepted that 3G networks could be built using existing 2G spectrum although that would either require spare capacity on the 2G network or an upgrade of 2G subscribers to 3G services. In certain cases the spectrum could be in the 450 MHz band (CDMA-450) which would enable large, essentially rural, areas to be covered without building large numbers of base stations. This is because whereas a GSM cell has a maximum 37 km radius, a CDMA-450 cell can 'breathe' and hence cover large but empty territories, a characteristic shared, in practice, by any variant of W-CDMA [13]

It should be noted that because the chosen frequency bands for the W-CDMA version of 3G are much higher than those used for GSM, this means than a W-CDMA network needs far more cells than a GSM network because the signal range decreases as the frequency rises. Furthermore, higher frequency signals travel in straighter lines, which makes little difference in urban areas but a good deal in open rolling countryside. As a rule of thumb, a W-CDMA network requires twice as many cells as a GSM network, but that ratio may prove to be unnecessarily high in practice, especially if 3G facilities are built on top of 2G base stations.

The issue of standardisation of W-CDMA/UMTS is currently being addressed by the 3GPP [14]. The initial standards for W-CDMA were completed in April 1999 and became known as Release 1999 (R'99). A major set of specifications was published in mid-December 2000, with amendments agreed in March 2001, which are known as Release 4 (Rel'4). In March 2002, Release 5 (Rel'5) was completed which, inter alia, defined the High-Speed Downlink Packet Access (HSDPA) channel and introduced the IP Multimedia Subsystem (IMS) architecture designed to enhance the end-user experience for integrated multimedia applications and to offer operators a more efficient means for delivering such services. Rel'5 also introduced the IP UTRAN concept to realize network efficiencies and reduce network costs. The fact that specifications are constantly being updated is somewhat problematic because of the need for networks developed at different times, in different countries and by different equipment makers, to permit international roaming. This, in turn, has some bearing on realistic deadlines for network roll-outs. However, the Release 5 standard was designed to be backwards compatible with R'99.

UMTS

The deadline for Member States of the European Union to introduce an authorisation system for UMTS was initially specified by the European Commission as 1 January 2000 [Commission of the European Communities, 1998a and 1998b], and for harmonised provision of UMTS services as 1 January 2002, although the European Council added an amendment in September 1998 to the effect that Member States which have 'exceptional difficulties in adjusting their national frequency plans ... can obtain an extra period of up to 12 months' beyond these two deadlines [Council of the European Communities, 1998]. In addition, by

February 1999, Member States were expected to devise plans for the provision of such additional spectrum as would be needed. Most countries needed as a consequence to shut down their analogue services and, if necessary, remove the relevant spectrum from organisations such as the armed services that had traditionally used it. A final factor is that the frequencies allocated to UMTS may have been set aside by other parties for the provision of satellite or other services such as PCNs.

At this point it is necessary to make a brief comment on 2.75G since the above discussion indicates that the original expectation was that there would be a single jump from 2.5G to 3G. The problem, at least for GSM-based systems, was the need to invest heavily in new networks so the question increasingly began to be asked whether 3G-equivalent speeds could be obtained via a GSM upgrade from GPRS. The current GSM speed of 9.6-14.4 kbps can be raised in principle through the successive implementation of HSCSD (high-speed circuit-switched data) capable of handling between 38.4 kbps and 57.6 kbps using four bonded GSM circuits; GPRS, theoretically capable of up to 168 kbps using eight 21 kbps GSM timeslots but for now delivering something much nearer 80 kbps in practice, which allows subscribers to remain permanently online and to be billed only for the amount of data received; and EDGE, nominally providing between 384 kbps and 513 kbps through the expedient of simultaneously feeding bundles of spare capacity to a single handset. Although EDGE could thus potentially compete in speed with early manifestations of 3G – since the latter is a shared provision the speed of delivery slows as the amount of data conveyed grows – it generally required additional towers to compensate for its high frequency range as well as new handsets, and was expected to be ignored by most operators given that the time frame for its introduction was similar to that of UMTS. However, the lengthy delays in launching UMTS, together with the development of workable and reasonably cheap EDGE infrastructure, led a good many operators to roll out EDGE networks either as a prelude to, as a complement to or occasionally as a substitute for UMTS.

Towards the end of 2005, there was much discussion as to whether the 900 MHz band, already in heavy use for GSM, could also provide 3G services. The use of this spectrum was first authorised in France where, due to the existence of only three operators, GSM spectrum was relatively plentiful. Spain and Italy also seemed to provide possible test beds for the use of the GSM band but not countries such as the UK and the Netherlands with five GSM operators. For its part, Germany's Federal Network Agency began testing and was considering the possibility of re-launching 900 MHz licences when they expired on a technology-neutral basis. The potential use of the GSM band was generally thought to be limited to rural areas where GSM usage was less heavy, but in any event the regulatory and spectrum management issues were far from full resolution – the EU GSM Directive reserved the 900 MHz band exclusively for 2G services – so the UMTS900 band was not going to be used before 2007 at the earliest, not least because of the need to develop the requisite handsets. There is pressure for the GSM Directive to be amended – the Commission officially proposed this in July 2007 – and if it is sanctioned by the various bodies concerned, there may be significant developments during 2008.

In December 2006, Siemens Networks announced that it had developed, as part of its 'One Network' family of products, a single base station that could be used simultaneously for GSM and W-CDMA purposes. At the same time, Huawei announced that it had completed video phone and HSDPA 3.6 Mbps service trials using UMTS900. In both cases, significant cost savings were said to be forthcoming.

At the beginning of 2007, the CEPT adopted a decision to permit the use of UMTS900 throughout the EU, subject to authorisation by national regulators. The first countries to take advantage of this opportunity were Finland, France and Portugal. In March 2007, the Portuguese regulator decided to open up the 900 MHz band for 3G services without imposing any conditions on technical aspects. However, it was on the Isle of Man that O_2, a subsidiary of Telefónica, completed the first UMTS/HSDPA900 network using equipment supplied by Alcatel-Lucent. It has not yet determined whether the spectrum is to be reserved exclusively for the 2G incumbents. Suitable handsets were expected to appear towards the end of 2007. However, the biggest difficulty will be how to appease those new entrants who only have 3G spectrum and who are not going to be handed GSM spectrum by existing licensees on a voluntary basis.

It is worth noting that 3GPP currently lists nine bands for evolved GSM core networks and the radio access technologies that they support, namely: Band 1 – 2100 MHz; Band 2 – 1900 MHz; Band 3 – 1800 MHz; Band 4 – 2100/1700 MHz; Band 5 – 850 MHz (for the USA); Band 6 – 850 MHz (for Japan); Band 7 – 2500 MHz; Band 8 – 900 MHz; Band 9 – 1700 MHz (for Japan).

The next World Radio Conference (WRC-07) is to take place in October 2007. ITU Working Party 8F (WP8F) has responsibility for the 'overall radio system aspects of IMT-2000 and beyond'. It has been asked by the IEEE and the WiMAX Forum (see below) to consider adding mobile WiMAX – officially known as 802.16e and based upon IP OFDMA techonology – as a complement to existing 3G technologies with a view to offering operators an additional migration path capable of providing wireless broadband services either by way of additional capacity in urban areas or as a stand-alone in rural areas too expensive to provide for by other means. The spectrum bands for which mobile WiMAX was considered to be best suited were 2.3-2.4 GHz and 3.4-3.6 GHz while fixed WiMAX was already well-established in the 3.5 GHz band.

In June 2007, the International Telecommunications Union (ITU) gave its initial approval to plans to include a specific subset of WiMAX known as Orthogonal Frequency Division Multiple Access Time Division Duplexing (OFDMA TDD WMAN) as a terrestrial radio interface under IMT-2000, meaning that spectrum will be reserved for this on an international basis (Global Insight, 2007).

In practice, although the use of IMT-2000 frequency bands for mobile WiMAX is particularly contentious, WRC-07 needs to grapple more broadly with how to find additional spectrum wherever it can for what used to be referred to in the ITU as 'systems beyond IMT-2000' – that is, beyond 3G – but which are now commonly known as IMT-Advanced [Office of Communications, 2007]. The ITU Radiocommunication Sector Working Party 8F is heavily responsible for progressing this project. The full list of bands under consideration is set out in Table 3.

According to Report ITU-R M.2078 produced by WP8F, IMT would require 1,280 MHz of spectrum (as compared to the 580 MHz already identified for IMT-2000) by the year 2020. Hence, in effect, a further 700 MHz of spectrum was being sought in the bands identified in Table 3. In this respect, it is worthy of note that the ITU's Radio Regulations contain a table of Frequency Allocations whereby frequency bands are allocated to services such as fixed-wire or mobile. A 'primary' allocation means that protection from interference by other 'secondary' services can be sought. Furthermore, certain frequency bands are identified as suitable for particular uses such as IMT-2000. Although this provides no specific benefits, it does tend to mean in practice that throughout the world the same bands are favoured for specific uses and hence much increases the probability of agreement about such uses at the WRC. So far as Table 3 is concerned, it includes many instances where no 'primary' allocation yet exists, so such an allocation would need to be assigned if there was sufficient backing at WRC-07.

Getting agreement on a worldwide basis is no mean feat – it can be problematic enough at the level of each regional grouping. For example, within the EU there was no unanimity in the run-up to WRC-07. The full range of preferences is set out in Ofcom (2007) Annex 6. This shows that 18 of the 29 'European' countries listed favoured use of the 3400-3600 MHz and 3600-4200 MHz bands but disagreed on almost everything else. So far as the use of the 470-862 MHz band was concerned, the main disagreement concerned whether this band should be considerd at WRC-07 or WRC-11 in 2011.

Table 3. Candidate bands for WRC-07

Candidate band	Anticipated use	Support	Available
410 to 430 MHz	Coverage extension of existing IMT-2000 networks	Very little	No
450 to 470 MHz	Coverage extension of existing IMT-2000 networks	Limited	No
470 to 862 MHz	Coverage extension and possible IMT-Advanced	Yes, but in 2011	>112 MHz
2300 to 2400 MHz	IMT-Advanced	For mobile WiMAX	No
2700 to 2900 MHz	IMT-Advanced	Very little	No
3400 to 3600 MHz	IMT-Advanced	Considerable	Partly
3600 to 3800 MHz	IMT-Advanced	Considerable	Partly
3800 to 4200 MHz	IMT-Advanced	Limited	Very little
4400 to 4990 MHz	IMT-Advanced	Very little	No

Source: Ofcom (2007) p.2.

As was noted previously, the 2500-2690 MHz band has for some time been set aside as an Expansion Band for 3G. However, as shown in Table 3, the 2300-2400 MHz band is favoured for mobile WiMAX and this technology has also been licensed in some countries in the 2500-2690 MHz band. Led by the GSM Association, there has been much opposition to allowing mobile WiMAX to encroach into the 3G space, but a good many operators would be keen on refarming their 2G networks for 3G use and then using the freed-up 2500-2690 MHz spectrum for mobile WiMAX. Since this would make the latter spectrum considerably more valuable, most regulators are waiting for WRC-07 to (hopefully) resolve the issue.

Cdma2000 1xEV-DO

Because of the complexity involved in shifting from GSM to W-CDMA, the above discussion has necessarily concentrated upon this version of 3G. However, it is also necessary to clarify the situation with respect to the cdma2000 1x family of technologies. The initial design of 1xEV-DO was created by Qualcomm in 1999 to meet IMT-2000 requirements for a stationary communications downlink that would operate at over 2 Mbps. Originally called High Data Rate (HDR), it evolved into EV-DO after ratification by the ITU and was allocated the designation IS-856. Although DO initially stood for ‘Data Only’, this was altered to ‘Data Optimized’, probably because it had more positive connotations. In order to function alongside a voice connection, a further radio channel of 1.25 MHz has to be added.

In practice, there are three main locations for the technology, namely the USA, Asia and South America – it is largely irrelevant in Western Europe. In the former case, the likes of Verizon Wireless, Sprint PCS quickly introduced the technology. Meanwhile, in Canada, Bell Mobility launched a service in February 2002 with Telus following suit in June. In Asia, the second-largest Japanese operator, KDDI, favoured the technology, but was heavily outweighed in aggregate sze by the combined W-CDMA adherents DoCoMo and J-Phone (subsequently Vodafone K.K. and now Softbank). Ironically – given that the two initial 3G licences were for W-CDMA – South Korea initially evolved into a major provider of cdma2000, driven by the need to have something reasonably fast available for the World Cup football tournament in May 2002. By the end of 2001, W-CDMA licensees SK Telecom and KTF were both well on the way to cdma2000 1xEV-DO, with the cdma2000 licensee, the LG Telecom consortium, trailing behind having uniquely opted to develop 1xEV-DV. There was also interest in the likes of Thailand (CAT Wireless Multimedia) and Taiwan where Eastern Broadband acquired a licence to develop a cdma2000 network using spectrum transferred from a previous 1G licence. The first Chinese network was launched by China Unicom in Beijing early in 2003, while in India, Reliance Infocomm was the first operator to contract the construction of a nationwide network in October 2002. Meanwhile, in South America, the relative popularity of CDMA as 2G is developing into a significant reliance upon cdma2000 as in the cases of Brazil, Chile, Ecuador and Venezuela. Given the desire of most East European countries to join the European Union, it was assumed that they would prefer to move via GPRS to W-CDMA.

At some point in the future cdma2000 and W-CDMA were expected to be rendered compatible – Nokia, for example, announced in December 2002 that it hoped to have a compatible technology called Global W-CDMA (G-WCDMA) [15] available for spectrum auctions in the USA in 2004 and Cingular Wireless expressly involved it in its future plans in June 2004 – and in late 2003 Samsung was said to have launched a dual-mode W-CDMA/cdma2000 1xEV-DO device branded the SCH-W110. Meanwhile, Nortel Networks announced in March 2003 that it had completed test calls for cdma2000 1xRTT and 1xEV-DO over the 2.1 GHz band.

It is unlikely that cdma2000 1xEV-DV (which combines circuit-switched voice and packet data in the same radio channel) will ever be needed. Initially, operators with existing voice networks favoured EV-DV because it was backwards-compatible without the need for an overlay, while those without such a network and design engineers favoured EV-DO because it did not need to be backwards compatible and hence could be designed more flexibly within a fully-IP-enabled network, doing away with the likes of mobile switching centres (MSCs) and hence cutting costs. In the event, EV-DV equipment was slow to come to market, and in the meantime operators were able to meet their objectives with 1xEV-DO. Hence, they were inclined to resort to 1xEV-DV only if they expected to run short of spectrum, which was not the norm. Nevertheless, Nokia announced in August 2003 that it had made the world's first 1xEV-DV transmission achieving a peak rate of roughly 3 Mbps, and Samsung became the first manufacturer to offer EV-DV commercially in March 2004 in conjunction with UTStarcom. However, with LG Telecom of South Korea struggling to roll out a network, and little interest being expressed by other operators, Qualcomm eventually decided to pull the plug and effectively consign EV-DV to the technological dustbin.

cdma2000 1xEV-DO, like other technologies, is undergoing progressive development which could in principle be termed 3.5G, 3.75G and so forth. Revision (Rev.) A came on stream in 2006 – Motorola unveiled its prototype solution in November 2005 – and should provide 3.1 Mbps downstream and up to 1.8 Mbps upstream. At the same time, Revision B was nearing completion within standards body Third Generation Partnership Project 2 (3GPP2). Commercial products are expected reach the market nom earlier than 2007Q4. Rev. B is designed to enable dynamic scalable bandwidth capability, potentially enabling operators to combine up to fifteen 1.25 MHz channels capable of 46.5 Mbps downstream although initially two or three channels will probably be used to provide in excess of the maximum with Rev. A.

Rev. C is already under way. In July 2006, 3GPP2 selected from among eleven different proposals one that was loosely backward compatible (LBC), using OFDMA on the forward link and a combination of CDMA and OFDMA on the reverse link. The strictly backward compatible (SBC) approach favoured by many which enabled compatibility between legacy and new handsets was rejected a;though it may be used as an intermediate step. In December 2006, the CDMA Development Group announced that Rev. C would be known as Ultra-Mobile Broadband (UMB) and claimed that it would operate at up to 280 Mbps. A standard would hopefully be published in 2007Q1 with commercial availability in 2009. LBC-based Rev. C will be able to hand-off to the likes of 3G LTE and WiMAX.

One final point worth noting is that it is perfectly possible in principle to overlay a GSM network in order to switch it to the cdma2000 pathway via what is known as GSM 1x, combining an unmodified cdma2000 RAN with an unmodified GSM core network. This route was launched in July 2004 by China Unicom which already had both GSM and CDMA networks, with the prospect of keeping voice traffic on GSM and data traffic on the CDMA overlay. However, there is no sign for now of other operators getting involved, although advocates claim that GSM operators in less-developed countries should consider using GSM 1x as a relatively cheap short-cut to faster data speeds.

3.5G

High-Speed Packet Access (HSPA) [16]

G-WCDMA is put forward by Nokia as doing for W-CDMA what EDGE does for GSM. In that respect it can be referred to as 3.5G. It enhances the benefits of W-CDMA's inherent economies of scale by recourse to High-Speed Downlink (sometimes Data) Packet Access (HSDPA) technology which is a standardised feature in 3GPP Release 5. HSDPA adds a new downlink channel dedicated to carrying data with additional intelligence added to the network to ensure an efficient allocation of channel capacity between users within a cell. The theoretical maximum data transfer speed is 14.2 Mbps – although this is optional while a theoretical maximum of 3.6 Mbps is obligatory – which is sufficient to transfer e-mails with bulky attachments extremely quickly and to allow for high-quality video streaming. A reasonable real-world average will lie (at least initially) between 550 kbps and 1.1 Mbps although some consider that twice as much should be achievable. Bur speed is not the only virtue of HSDPA since it also doubles network capacity and much improves quality of service. Nevertheless, although HSDPA is essentially an overlay and hence requires no more than a software upgrade, this is true only for operators that installed 3G infrastructure with built-in HSDPA capability (essentially since 2002). Earlier 3G networks will need some hardware additions such as channel cards and traffic processing units. Furthermore, hardware additions will be needed at customer premises including additional handset memory.

In January 2005, Siemens claimed to have conducted the first live demonstation with a HSDPA network although Nortel made a similar claim at roughly the same time, and in March, Nortel and LG Electronics completed the first live test calls using a commercial handset solution. O_2 (initially on the Isle of Man where Lucent has been contracted to deploy a converged fixed-mobile network incorporating HSDPA, and subsequently in Ireland, Germany and the UK with Nokia and Nortel as equipment suppliers), mobilkom in Austria (in conjunction with Nortel) and Cingular Wireless (in conjunction with Lucent) were considering a launch before the end of 2005 – in the event, Cingular was the first to launch in December – while KPN pencilled in 2006 as did Telfort in conjunction with Huawei and DoCoMo in conjunction with Siemens. The initial downlink speed of 1.8 Mbps was first raised to 3.6 Mbps by Bité in Lithuania in June 2006. The situation on the Isle of Man is particularly interesting because O_2's Manx Telecom also supplies broadband over ADSL. According to 3G Americas (2007a), there were 97 HSDPA networks in operation in 51 countries at the end of 2006 with as many again expected to come on stream during 2007.

As per usual, the initial launches of HSDPA all involved data cards and were aimed at business users. Needless to say, much depends upon the arrival of decent handsets and these did not put in an appearance until early 2006 with the Samsung SGH-Z560 and BenQ-Siemens EF91 designed for use mainly in Europe, and the Samsung ZX20 designed primarily for the USA.

3GPP Release 5 merely specified that the uplink associated with HSDPA should operate at a minimum of 64 kbps. 3GPP Release 6, finalised in April 2005, also contained specifications for High-Speed Uplink Packet Access (HSUPA) otherwise known as the Enhanced Dedicated Channel (E-DCH). This permits symmetrical two-way high-speed data communications. It was first demonstrated by Ericsson and 3 Scandinavia in May 2005. It was planned that the Ericsson W-CDMA Enhanced Uplink would be introduced in line with terminal and handset availability, commencing in 2006Q2 with data cards and providing 2 Mbps. Meanwhile, in February 2006, Nortel demonstrated uplink data speeds of 1.4 Mbps using commercially available equipment and announced that it would be launching a complete system in 2007. In conjunction with Option, an alleged 'first live' demonstration of a HSUPA data card with uplink at 1.3 Mbps and downlink at 2.7 Mbps was achieved in October 2006. However, mobilkom Austria also claimed to have made the first live transmission in November and to be the first to launch – in Vienna in February 2007 accompanied by a HSDPA upgrade to 7.2 Mbps (3G Americas, 2007b:.7).

4G

The above discussion has already touched upon the issue of 4G in the context of IMT-Advanced. The ITU working definition for 4G is variously reported as a downlink speed of at least 100 Mbps when fully mobile or (according to the ITU which is operating via Working Party 8F) 1 Gbps with limited or local mobilility, although it remains unclear for now exactly how this is to be measured. An alternative definition which would take account of the fact that in practice nothing like 100 Mbps is going to be achieved, no matter what the technical possibilities are, would concentrate upon the existence of an all-IP network capable of delivering in excess of the maximum set for 3G – that is, more than, say, 4 Mbps. In all probability, subscribers will be unable to tell whether data speeds are greater than this in any event. However, 100 Mbps is certainly going to be achievable under highly controlled circumstances, and, indeed, it was first achieved in September 2006 by Samsung in a bus being driven around Jeju Island in South Korea at speeds of up to 38 mph.

The approach based upon a 100 Mbps downlink is also reflected in 3GPP Release 8 (which specifies 50 Mbps on the uplink) although this is not necessarily referred to as 4G even though it is certainly a significant step beyond HSDPA [3G Americas, 2007c]. Networks achieving these kinds of speeds are based upon packet-switched technology combined with a new OFDM air interface and multiple input multiple output antenna (MIMO) technology [17]. The latter means that it is more problematic to step up from HSDPA to what is now generally called Long-Term Evolution (LTE) – but may not continue to be called that in the longer term – than from W-CDMA to HSDPA.

In March 2006, T-Mobile, Orange, KPN and Sprint Nextel set out their joint vision in a White Paper entitled *Next-Generation Mobile Networks. Beyond HSPA & EV-DO*. This led to the formation of the NGMN Forum in June to establish performance targets and deployment scenarios. China Mobile (HK) and DoCoMo and others joined the Forum with a view to creating standards for LTE by the end of 2008 and conducting operator trials during 2009, although it is possible that 3GPP will produce the UTRA-UTRAN standard by the end of 2007 to be followed by the standards for the System Architecture Evolution (SAE) which addresses the core network supporting the radio access network at a later date [*Idem.*:5]. This timetable would place LTE roughly two years behind WiMAX in terms of development. Meanwhile, hopefully, suitable handsets would be developed. DoCoMo was keen to pre-empt the discussion, and to this end issued its proposed pre-standard version of LTE called Super 3G. It had behaved in exactly the same way when W-CDMA was being developed, issuing a pre-standard version of Release '99 technology in 2001 and updating its network once 3GPP members agreed a common specification [*Idem.*: 6]. However, other members of the Forum are not keen on a repetition of this arrangement. As a result, a LTE/SAE Trial Initiative was set up in May 2007 to demonstrate the potential of LTE and SAE through joint tests, including radio transmission performance, early inter-operability, field trials and full customer trials. The Initiative is expected to last up to two years.

LTE is expected to utilise existing GSM spectrum together with the 2.5 GHz band. Although LTE removes the need for circuit-switched voice, it is not expected to bring about its immediate demise. However, this does create some opportunities for economies in constructing networks capable of handling LTE and these will be enhanced if existing spectrum can be utilised although LTE is also optimized for the 2.5 GHz band. This means that LTE and WiMAX may be rendered interoperable – they use the same basic system architecture – although WiMAX is expected to remain largely the province of operators that do not possess 3G spectrum [18].

As noted, the Expansion Band is critical to the 4G road map because it offers the opportunity for worldwide interoperability. Because of, inter alia, the ongoing squabble over the role of WiMAX in 4G, there are indications that the band will be divided into 70 MHz paired of FDD and 50 MHz (less a guard band) of TDD. Because TDD is considered to be more spectrally efficient for asymmetric data, it is favoured for WiMAX networks (see below) although cellular networks are all based on FDD. Originally, WiMAX was simply referred to as IP-OFDMA but, as noted above, it is now being submitted for IMT-2000 approval as OFDMA TDD WMAN, generating fears that the best spectrum will be reserved for FDD networks [Brown, 2007].

There are those who will doubtless wonder why anyone wants a mobile network to operate at 100 Mbps in the first place, especially if it requires investment in new infrastructure. For now, there is no clear answer, although advocates tend to argue that there are so many activities being transferred to mobile devices that by the time higher speeds are available there will be a clear demand for them.

Core Services and Platforms

Short Message Service (SMS)

For those brought up on full 'qwerty' computer keyboards, the idea of tapping out messages on a tiny mobile handset keyboard, where each number also represents several letters and symbols, is little short of purgatory. However, for those so disposed, predominantly teenagers, it is only a matter of practice before messages can be tapped out at great speed. Furthermore, given limitations upon the length of a message that can be sent for a standard rate – usually 160 characters – it has become the custom to use short symbolic abbreviations or representations for quite complex ideas, in effect designing a new form of language.

SMS – sometimes known as short messaging – is not particularly cheap in advanced countries, and does not accordingly seem worthwhile to those with monthly contracts who can often talk free in off-peak periods. However, most teenagers have pre-paid handsets where SMS represents an economic means of communication compared to non-focused conversation which may anyway not be practical in noisy bars and nightclubs. In relatively recent times, mobile-to-PC SMS and vice versa has evolved, which means that information alerts about sports scores, share prices and breaking news can, for example, be sent direct to a handset. SMS is increasingly being used, for example, to send messages on special occasions.

SMS initially became very popular in Europe and in parts of Asia. Because large numbers of messages are carried free there, usage is extraordinarily high in Singapore and the Philippines. However, until very recently SMS barely signified in the USA where 'instant messaging' between computers was commonplace, local calls generally free, mobile operators obliged to offer huge bundles of minutes – up to 5,000 a month – in monthly contracts and data services sometimes charged for separately.

SMS has been a real boon to operators because it is very cheap to set up. Unlike voice, messages can be stored in short message centres and delivered when the networks have spare capacity. As a result, the marginal cost to the operator is close to zero. Furthermore, text messages can be used for premium-rate services – the downloading of ring tones is especially popular. The use of SMS in conjunction with television is growing surprisingly rapidly. For example, the use of SMS to send to broadcasters either the answers to quiz questions or votes as to who should win contests has become widespread. Entering a daily lottery draw by sending a three-digit number has also been a great success in Sweden. This is good news for network owners since they tend to hold most of the aces where the division of spoils is concerned given the need to use a reliable network for any kind of data transmission, to have access to the audience and to provide a means for making micro payments.

In the longer term, some analysts see SMS as being overtaken by instant messaging. This development was, as ever, held up by problems of interoperability between networks despite the creation of the Mobile Instant Message and Presence Services initiative (which has been adopted as an Open Mobile Alliance standard). Nevertheless, in February 2005, Echovox launched the the first cross-network mobile-to-mobile instant messaging application branded 'ZeeWee' which was compatible with all Java-enabled handsets and which was fully interoperable across both Europe and the USA. Subsequently, in June, MSN and Vodafone announced plans to launch the first-ever instant messaging service between PCs and mobile handsets, while in November Swisscom launched an instant messaging handset loaded with MSN Messenger. In October 2006, Telecom Italia announced the November launch of the world's first service running on both handsets and PCs (provided the user has an Alice Mail address), branded the 'Super SMS' service.

Interoperability between GSM networks was boosted with the formation in February 2006 of 'Personal IM', a joint initiative involving China Mobile (HK), Orange, TeliaSonera, Telefónica Móviles, TIM, T-Mobile, Turkcell and Vodafone and the Indian operators Aircel, Bharti, BSNL, Hutchison Essar, Idea, MTNL and Spice. All their IM will be on the caller-pays principle.

Multimedia Messaging Service (MMS)

SMS is evolving into a multimedia messaging service (MMS) – generally defined as the capability to convey audio and video clips, photographs and images (known as m-postcards) although individual operators appear to prefer to define the term to suit themselves – as the capabilities of the networks are upgraded through 2.5G to 3G. MMS must be distinguished from 'instant messaging' which represents a form of real-time communication intended for brief, impromptu communications [19]. It should firstly be borne in mind in the context of MMS that a still image only requires 50 kilobytes of memory whereas a 30-second video requires 3 MB. Handsets with built-in digital cameras – sometimes known as imaging phones – became available in Japan during 2001, and proved to be a major factor in J-Phone's increase in market share during the latter part of the year [20]. In addition to a camera, MMS requires that the handset incorporate a colour LCD for video display and the necessary encoding and decoding features for MPEG-4 video compression and decompression. MMS operates via a Multimedia Messaging Service Centre (MMSC) which stores messages prior to delivery as well as hosting a number of interfaces to connecting networks and an application programming interface (API) to enable deliveries of value-added services and network interconnection to e-mail and the like.

MMS is a premium product requiring relatively expensive handsets. Whereas SMS is predominantly the preserve of those on pre-paid contracts, MMS will for some time remain largely the preserve of those on monthly contracts. For many operators, the route from SMS to MMS passed through an intermediate stage called the enhanced messaging service (EMS).

However, European handset manufacturers were somewhat slow off the mark in getting MMS-capable handsets onto the market. Market leader Nokia, for example, which had chosen to ignore EMS, only began to supply its models with a built-in camera after mid-2002. Nokia nevertheless expected the market to reach critical mass before the year-end – a quite different experience from the relatively slow build-up of SMS.

Insofar that there probably needed to be some kind of trigger to turn MMS into a mass-market phenomenon, the obvious candidate was inter-network roaming – arguably the most important reason for the success of texting although it should be borne in mind that, for example, it took five years in the UK to achieve full interoperability whereupon the market exploded. Sonera (now TeliaSonera) and Radiolinja claimed to have made the world's first MMS interconnection in late-November 2002, and TeliaSonera addressed the interoperability issue on an internal basis by introducing the Nokia MMSC in all four of its Nordic operations in June 2003. The MMS Inter-operability Group, consisting of CMG, Comverse, Ericsson, Logica, Motorola, Nokia, Siemens and SonyEricsson was set up in February 2002 with a view to promoting interoperability between MMS-enabled handsets and servers made by different companies. Nevertheless, as sceptics were keen to point out, the MM4 interface standard – the MMS specification as set by 3GPP – was difficult to translate across the great variety of MMS handsets deployed in Europe: somewhat ironically, the problem was not the lack of a set of standards but the fact that there were too many options within the specifications thereby permitting manufacturers to pick and choose those to incorporate in their equipment. Furthermore, it was less straightforward to receive an MMS picture than operators claimed it to be. For this reason, even in the Far East those who bought camera phones tended to use the handsets to store images for later viewing and this still largely remains the case. However, their popularity in that region is something of a revelation. DoCoMo, for example, which was struggling with its more sophisticated FOMA 3G service, nevertheless managed to sign up two million subscribers to its i-shot service within four months of its launch in June 2002. More recently, US operators, acting via the CTIA's Inter Carrier MMS Working Group, agreed a set of guidelines enabling MMS interoperability in November 2004.

Further work was done by a collaborative venture involving the Open Mobile Alliance (OMA), the GSM Association, the CDMA Development Group, 3GPP and 3GPP2 which initially produced OMA MMS version 1.2. This specified a minimum set of requirements and conformance to enable end-to-end interoperability of MMS applications, MMS-capable handsets and servers and content provisioning. In effect, the solution was dependent upon a ruthless 'no options' approach. A very promising initiative embodying such an approach took the form of trials based on a multilateral hub concept whereby operators would only have to connect once in order to reach all of the networks connected to that hub and, through interconnection arrangements between hubs, operators connected to other hubs. In the trials, Belgacom and C&W added switching and billing functionality to their global roaming exchange (GRX) networks to provide the hub function and cascade billing support while Vodafone connected several of its subsidiaries into the hubs.

Mobile E-mail

MMS is a fairly sophisticated type of service for those who wish to venture beyond the realms of SMS. Yet, it is well-known that sophistication does not rank particularly high on the wish-list of large numbers of mobile handset users. What they want is connectivity on the move as provided by voice telephony and SMS. Fairly recently, it has begun to dawn on equipment manufacturers that there is a third form of connectivity on which vast numbers of people are reliant in the form of e-mail. Furthermore, there is clear evidence to indicate that if a simple choice had to be made between e-mail and browsing the Web, the latter would barely get a look in. One great advantage of e-mail is its simplicity combined with familiarity, which suggested that if it were to be offered as a standard feature on handsets then it could be a 'killer application'.

Traditionally, e-mail is a facility provided only on smart-phones designed for business users. The 'Blackberry' made by Research in Motion (RIM) is arguably the iconic device [21], although it is aimed at 'high-end' users. However, there is no particular reason why a different type of handset could not be marketed that concentrated upon voice telphony, texting and e-mail – something occasionally referred to as a 'dumb smart-phone'. And, not surprisingly, this is now available. It is equally unsurprising that it is powered by Microsoft which, on 10 May 2005, unveiled a new version of its Windows operating system designed to lie at the heart of mobile handsets. This system will allow independent software companies to run programmes involving e-mail, and, to this end, Visto of the USA concluded deals with Vodafone and Rogers Wireless to start rolling out mobile e-mail services. However, they were beaten to the post by Cingular Wireless which, in May 2005, launched its GoodLink e-mail service for corporate customers in conjunction with Good Technology. For its part, Sprint Nextel incorporated Visto's e-mail client into the majority of its hand-held devices, potentially making the service accessible to seven million former Nextel subscribers.

In November, Vodafone Ireland launched its Business E-Mail portfolio in Ireland, in conjunction with Visto, with the addition of Vodafone's 'push e-mail' solution providing business customers with real-time access to their e-mail, calendars, contacts and corporate files. On offer was an 'Internet Solution' designed for customers with ISP-based e-mail accounts, a 'Small Office Solution' for small companies with a server-based e-mail such as Microdoft Exchange and an 'Enterprise Solution' for larger companies. In February 2006, Vodafone launched the new Windows Mobile direct Push e-mail solution based upon Windows Mobile ® 5.0 with the first countries to be served being France, Germany and the UK. This was also launched by Orange Business Solutions in April using the Orange SPV C600, SPV M600 and SPV M5000 devices.

It may be observed that the smart-phone market is currently quite small [22], but dropping an e-mail facility into the mass consumer market may eventually have a much wider effect. Typical of the operators that have announced the launch of mobile e-mail during 2006 are the Ecuadorian subsidiary of Telefónica which introduced 'Correo MóviStar' in February 2006, while LMT launched Latvia's first service in April followed by Sun Cellular in the Philippines.

Push-to-Talk (PTT)

Push-to-talk is a bit of a curiousity. Readers will doubtless be familiar with so-called 'walkie-talkies' whereby it is possible to connect instantly with another party by pressing a button on a handset followed by something along the lines of 'over and out', whereupon the other party can respond in turn. It enables point-to-multipoint communication, and is a very useful device when short bursts of speech are the norm and hence is well suited for the likes of taxi companies and emergency services (although speech does sound a little discordant which is why cellular phones may still be preferred), and historically was purely a business service which was seen as irrelevant for the consumer market. However, it was an established feature of one network, that of Nextel in the USA, precisely because so many of their customers were business rather than mass consumer.

And there it might well have rested – after all, operators hardly want to encourage succinct and terse exchanges – but for not altogether obvious reasons PTT suddenly became the 'in-thing' in the consumer market during the latter part of 2003. Perhaps the typical mobile user's desire for instant gratification was partly to blame, together with the sheer simplicity of the technology – no sophisticated 3G services here, just a speedier way to convey voice calls involving one-push button access, a high-audio microphone, a speaker phone and an intuitive design. However, even that does rather depend upon what version of PTT is being referred to. One variant, for example, involves the speaker sending a voice message which is stored until such time as the intended recipient is in a position to listen to it, but the variant that is all the rage is that involving real-time communications, known as push-to-talk over cellular (PoC), which streams voice calls over any kind of data network once the sender has ascertained that the intended call recipient is available – a similar concept to instant messaging between PCs.

Not surprisingly, even though Nextel's service was based upon iDEN technology used by no other operator, all parties were agreed that interoperability was the crucial issue to be addressed. The Open Mobile Alliance (OMA) accordingly took it upon itself to develop agreed standards, and produced a draft in August 2003. This was promptly picked up by Nokia whereas most other developers including Ericsson, Motorola and Siemens elected to support a different set of specifications even though these were also based upon the August draft [23]. Nokia claimed that it had merely set out to remedy various deficiencies inherent in the draft, that it had not developed a proprietary version of the technology and that other companies had chosen a different path because they were in a less advanced position; they begged to differ. The situation in mid-2004 was that such implementations as already existed either used the Nokia solution or a proprietary one developed by Kodiak.

By the middle of 2006, only about 40 PoC launches had taken place, and all had in varying degrees been tarred with failure due to proprietary technology rejected elsewhere, poor targeting of handsets and/or poor performance of the technology. Publication of the OMA PoC 2.0 standard in June 2006, which cost $500 million to develop and three years to reach agreement, finally reinvigorated the technology, and version 3.0 followed by interoperability testing was set in hand with a view to full commercial availability by mid-2008..

Pending this, some tendency to hold back is evident. Inevitably, however, the unstoppable progress of technology means that much more sophisticated versions of PoC are becomingy available. For example, in March 2007, ZTE Corporation of China launched (on behalf of Telstra) the world's first HSDPA-enabled handset with PTT functionality, the F159.

Java versus BREW versus Linux

In essence, all wireless applications must run on a platform – that is, a software system that interprets the application's instructions and instructs a device how to execute them. Java, a Sun Microsystems platform, made the running with its Java 2 Platform, Micro Edition (J2ME) version which was supported by virtually all major handset manufacturers and most major operators such as Vodafone. As of the end of June 2003, J2ME was installed in one hundred million handsets, ten times the figure for Binary Runtime Environment for Wireless (BREW). Its greatest virtue in principle was the ability of a Java application to run on all Java-enabled devices with little or no tweaking. In practice, such tweaking is necessary, but the standard is to have a 'common code base' covering at least 90 per cent of an application's code, meaning that relatively little tweaking is needed to adjust for the peculiarities – such as different screen sizes – of the multitude of devices on the market.

One issue is that carriers tend to demand proprietary versions of the platform because, understandably, they want to differentiate their products. DoCoMo, for example, launched its original Java service, called iAppli, in January 2001 using DoCoMo Java (DoJa), whereas Vodafone added special improvements to its game platform launched in October 2002. In mid-2003, Sun also formed an alliance with China Unicom to promote UniJa.

However, in August 2004, Nokia and Vodafone jointly announced the formation of a mobile service architecture initiative designed to simplify mobile Java standards by defining next-generation open standards-based mobile Java services architecture specifications. The initiative would function and create its specifications within the Java Community Process and would seek to harmonise the licensing terms of the component Java Specification Requests to be open, fair and predictable.

Java's rival BREW, developed by Qualcomm and embedded in its chipsets, is for now largely restricted to CDMA networks and the Americas and Asia – the first contract with a GSM/W-CDMA operator in Western Europe was with (Telefónica's) O_2 in November 2005 (using own-branded handsets first launched in September 2006 with the 'Ice' and where the intention is ultimately to have 20-30 per cent of subscribers using BREW-enabled devices) and it was also adopted by Telecom Italia (like O_2 using the uiOne interface platform) for its fixed-mobile convergence programme in 2006 and by 3 UK in June 2007. Its virtue is that it provides a turnkey system for managing application downloads including hosting, authentication and billing services, and generally uses less processing power and runs faster than Java-based rivals, but in return Qualcomm's royalties are relatively high. By August 2003, nine operators had deployed BREW, of which four were US-based including Verizon Wireless and the others in Japan, South Korea, China, Australia and Brazil. By September 2006, the total was said to be 70. Given that South Korea is the only CDMA-exclusive market of consequence, it was expected that it would be the primary market for BREW. However, in April 2004, KTF announced that it would be adopting Korea's home-grown standard called the Wireless Internet Platform for Interoperability (WIPI) even though BREW had already established a foothold with 5.7 million users. This stemmed in good part from a desire by the government for the three incumbents to share the same platform. The platform was to be opened up to all developers, so BREW was likely to be adapted to become compatible with WIPI. However, the Korean government shut down that option in April 2005, at which point there were 2.5 million WIPI-enabled handsets in use compared to six million that were BREW-enabled. With KTF, its main supporter in South Korea, turning away from BREW, WIPI is likely to become dominant there. Its prospects elsewhere in the world are another matter. It is worth noting that WIPI version 2 incorporates the Java 2 Micro Edition platform.

In an interesting development in May 2004, Zapp Mobile of Romania was the first European operator to introduce BREW on a CDMA-450 network. The 'Zapp Me' service was introduced using a Zapp Series Z710i handset. However, even though it was also technically possible to run BREW over GSM, little interest was shown in so doing. By forming a co-operative agreement with UK mobile content delivery company Elata in June 2005, Qualcomm hopes that Elata's operator customers will look more kindly on the technology.

It is also possible to write applications specifically for Symbian or another operating system used in a particular device. However, while this enhances the device's capabilities, the software is relatively unwieldy and hence difficult to download. This suggests that devices capable of handling more than one platform is the way of the future, and one such example is, as ever, SK Telecom in Korea which uses both Java for big games applications and a proprietary platform, GVM, for downloading simple games.

It is finally worth noting that there are those who eschew both Java and BREW in favour of Linux, developed in a modular manner as 'open source' software by a community of programmers. However, its many variants are a cause of confusion and mobile devices have struggled to incorporate the memory, processor power and battery life to make the most of its potential. Nevertheless, the likes of Samsung and Motorola began work to develop Linux as an alternative to Symbian and Windows Mobile, and in June 2006 they were joined by NEC, NTT DoCoMo, Panasonic Mobile and Vodafone which agreed to set up an independent foundation dedicated to creating an agreed specification for a Linux-based platform. In August 2006, Orange agreed to incorporate Linux, developed by Japan's Access (owner of PalmSource), in its range of Signature Devices.

Wireless in the Local Loop (WLL)

WLL is a variant of what is generically known as broadband wireless access (BWA) and which also incldes WiMAX. WLL is known by a number of synonyms, but for now it is probably most commonly referred to as 'fixed wireless' and sometimes as 'limited mobility'. The reason for this is that it represents a 'last mile' connection via wireless instead of copper wire or fibre-optics. However, it does not form part of a conventional cellular network. Usually, WLL is introduced into major conurbations where it is possible to set up a dense network of equipment with limited range, and is self-contained within the conurbation. Hence, it is ideal for potential subscribers who stay fairly continuously within the conurbation and simply want to use their handsets while, for example, at work, wandering the streets or pottering about in traffic (as against travelling at high speed).

Because cellular networks have a long history in Western countries, have almost complete population coverage and use PCN/PCS spectrum as well as that in lower spectrum bands, there was never a good economic case for creating a significant WLL presence. Attempts were certainly made to licence it (with the fees set necessarily low), but it never took off. However, the situation in less-developed countries was another matter, Since cellular networks there were often either poorly developed in terms of population coverage or simply unable to cope with demand, WLL had a much brighter future. The countries where this was most obviously the case were China and India. In China, despite the extraordinary growth of GSM/CDMA networks, penetration remained low and much of the population simply wanted something cheap and cheerful for use within their own locality. Technically, only two companies are licensed to supply mobile connectivity of any kind, but this has not prevented fixed-wire operators China Telecom and China Netcom from proceeding unilaterally to introduce the Xiolingtong or 'Little Smart' service which rapidly gained over 30 million subscribers and had achieved 85 million by the end of 2005, at which point the majority of new fixed-wire connections were via this service. The government has recently authorised inter-connection with China Mobile (HK) and China Unicom, so the service is now competing directly with their cellular networks.

Retrospective acknowledgement of the status of fixed wireless has also been a characteristic of provision in India. Here, despite the existence of a range of cellular operators and a rapidly rising penetration level, the major opportunity arguably exists in providing a service to those on very low incomes. At the turn of the decade, a number of small, privately-owned operators such as Shyam Telecom and HFCL began to provide services limited to short-distance charging areas (SDCAs). The government promoted this development by licensing on a first-come-first-served basis with a much reduced level of licence fee compared to that exacted on cellular licensees which had been licensed via an auction. This initially made little difference, but when a subsidiary of industrial conglomerate Reliance as well as Tata Teleservices entered the fray, the number of subscribers rose very sharply. The cellular operators predictably protested fiercely that they were unfairly disadvantaged in relation to such matters as licence fees, and the government responded at the end of 2003 by offering limited mobility operators the opportunity to upgrade to a 'unified access' licence by paying the difference between the original licence fee and that paid by the fourth cellular operator in the relevant local circle. Those operators without a full cellular licence (all bar BSNL and MTNL) predictably paid up.

It may be argued that the Indian experience is somewhat untypical because the real competition to cellular operators was forthcoming from heavily resourced subsidiaries of conglomerates with wide footprints and purchasing power in relation to handsets. Indeed, much the same has been true in China. But does that imply that the model can be transferred to, say, Africa or even Eastern Europe? One central issue would appear to stand out, which is whether the incumbent cellular operators are already providing, or even interested in providing, the kind of cheap and cheerful (essentially voice + SMS) service provided by limited mobility operators but with a full roaming service thrown in for good measure? In practice, there is surprisingly little evidence of new limited mobility networks, but this may be because cellular networks are leaving too little room for them to operate profitably.

The obvious country to test this hypothesis is Nigeria where there are a number of established limited mobility operators in a market with a much lower cellular penetration rate than might be expected of a country with oil riches. Limited mobility services have been available since 1998 whereas GSM came on stream only in 2001. Significantly, although GSM operators cannot provide national coverage, they are most prominent in the main urban areas which is precisely where WLL has its natural market, so the networks compete head-on. Technically, no new WLL operators are permitted to enter the market, although this has not prevented the likes of Reliance (not the Indian company) from joining in regardless, and it is clear that WLL operators can easily undercut their cellular rivals given that voice/SMS is the core service being provided by all. Nevertheless, WLL operators operate at a severe disadvantage in relation to handsets since these are much more expensive than those available to GSM subscribers, and hence the advantage of cheap calls is largely negated.

Alternative Wireless Technologies

Wireless Local Area Networks (W-LANS/Wi-Fi)

What is known technically as 802.11, a wireless local area network (W-LAN) standard approved by the Institute of Electrical and Electronics Engineers (IEEE) – uses the 2.4 GHz spectrum, which is free and unlicensed, in its 802.11b variant now commonly known as Wi-Fi, for which there is a competing technical standard in the form of Bluetooth [24]. 802.11b uses Direct Sequence Spread Spectrum (DSSS) to transmit signals at up to 11 Mbps and this provides some protection against interference originating from microwave ovens or cordless handsets. The 802.11b band is typically less than 100 Hz wide, and other radio allocations on either side mean that it is not practical to create a wider band to ensure international consistency.

The cost of fixing a base-station transmitter to a building – thereby creating a 'hotspot' – is often less than $1,000 and individuals need pay no more than $60 for a data card to enable a laptop – the introduction of the Intel Centrino into laptops is increasingly doing away with the need for the card – and hence one which is being rapidly adopted in big cities and especially in airports, hotels and cafes as well as on school and university campuses. The crucial aspect of 802.11 is that it recognises that people on the move tend to become stationary before accessing the Internet. However, although the number of users already runs into the millions, it was set up as a data-only system which cannot therefore (at least for now) supplant 3G but rather can be expected to act as a complement to it [25]. Furthermore, although suitable for laptops, it was originally designed to operate with desk-top computers and for now tends to require more power than can be handled comfortably by smaller battery-powered equipment such as PDAs and mobile handsets, especially in its 802.11a variant which has become available using spectrum in the 5 GHz band (5-6 GHz) (and was embedded in the second-generation Centrino introduced in January 2005). 802.11a is theoretically capable of data transmission at up to 54 Mbps although the average speed in practice is roughly 22 Mbps. A competing technology known as High-performance Radio Local Area Network Type 2 (HiperLAN2) has been introduced in Western Europe but the two are expected to converge in due course. Compared to 2.4 GHz, 5 GHz provides more channels and a higher quality of service, but for the time being issues relating to lack of equipment, regulation and business models are in the developmental stage. The ITU has sanctioned 5.150-5.250 GHz (for indoor use only), 5.250-5.350 GHz (for 'predominantly' indoor use) and 5.470-5.725 GHz (for outdoor use provided 'interference mitigation' techniques are employed) as suitable for this purpose [26]. Similar speeds to 802.11a are becoming possible on the 2.4 GHz band by switching to 802.11g which is interoperable with 802.11b and was approved by the IEEE in June 2003, but for the moment the slower technology is sufficient for most purposes. 802.11n, which uses Multiple Input Multiple Output (MIMO) technology, is currently in development, backed by the Enhanced Wireless Consortium (EWC), and received the approval of the IEEE in January 2006. Capable, in principle, of speeds up to 540 Mbps, it is supported by the likes of Broadcom. However, the initial chipsets were rejected by the IEEE in May 2006 and the standard is not now expected to obtain final approvals until 2007 or even later [27].

For the time being there are a variety of technical problems to overcome – especially concerning interference and security – before 802.11's role in relation to 3G is clarified [28]. Furthermore, while the use of unlicensed spectrum for private purposes is regarded as acceptable, annexing it for public applications such as LANs and charging for services provided may well instigate regulatory interference – for example, it is illegal to use W-LAN frequencies for profit in some countries – possibly prompted by complaints from 3G licensees. Problems of congestion may anyway induce self-imposed limits on use of the spectrum, and suppliers of broadband services are not enthusiastic about multiple access to their networks. In any event, until cheap and ubiquitous access to broadband is available the widespread use of Wi-Fi will be constrained by the need to link into broadband fixed-wire networks. The issue of backhaul is often glossed over – advertisements show Wi-Fi-enabled laptops being used on Mt. Everest, but are a little vague about backhaul. It is all very well having a super-high-speed Wi-Fi connection but it can only travel over the backhaul network at whatever speed the latter permits – and it is likely to be DSL in most cases. Once this is taken into account and joint use of a hotspot is factored in, the service provided may be indistinguishable from 3G.

However, even assuming away many of these difficulties, there appears to be a difference in approach between Western Europe and the USA since the dominance of mobile operators in Europe tends to favour their rolling up the development of W-LANs into the provision of GPRS/UMTS. Arguably, the development of W-LANs in Europe is being held back by the need to roll out 3G networks, but operators do appear to be aware that they cannot afford to cede control over W-LANs to other parties. Nevertheless, it is important not to get carried away with hotspot numbers running into the thousands. Wi-Fi operates over a short-range at a relatively high frequency. Hence, it was calculated that 700 hotspots would be needed in mid-2003 to approximate one mobile base station. Even allowing that the ratio will improve somewhat over time, it must be borne in mind that a typical mobile network involves several thousand base stations, so the equivalent number of hotspots is extremely large.

The European Union has chosen to intervene. In August 2002, armed with the knowledge that it would, in certain circumstances, be able to over-rule national anti-trust authorities as from July 2003, the Commission warned Member States that any unfair restriction on the use of Wi-Fi would potentially be in breach of EU law. However, operators appear to be uncertain as to whether Wi-Fi is an extension of a fixed-wire or mobile network. In the UK, for example, it is BT that is leading the way, whereas in Switzerland and Denmark it is the incumbent's mobile arm. In March 2003, the Commission accordingly issued a Recommendation (formal guidelines) calling on Member States to subject radio spectrum usage for LANs to the least onerous authorisation system. The Commission intends at a later date to identify frequency usage restrictions to prevent neighbouring LANs using similar spectrum from interfering with each others' signals.

More recently, the Commission has taken action to prepeare for what it sees as a coming 'explosion' of Wi-Fi users' by allocating 5150-5350 MHz and 5470-5725 MHz specifically to meet that demand. It required all Member States to allocate the spectrum by 31 October 2005.

So far, companies seeking to create viable businesses based on the provision of W-LAN services have not been successful. At heart, the primary difficulty is that prospective customers will only pay fees to access a network with sufficient coverage, yet different locations are generally controlled by different operators and this forces users to set up multiple accounts and passwords while travelling around. As ever, some 'amateur' enthusiasts are willing to provide access to all-comers via the simple expedient of erecting a Wi-Fi antenna and entering an availability listing on an online directory, but whereas this has the virtue of providing free access to passers-by it lacks any kind of controlling mechanism and hence is open to abuse. The latest thinking is to provide an aggregation service which can provide access to a large number of networks via a single account, with the aggregator sharing revenue with the network operators. Alternatively, any Wi-Fi-enabled PC can have free software installed that turns it into a hotspot linked up to other hotspots on the same network. Achieving the necessary scale to attract subscriptions is then the difficult bit if commercial viability is to be achieved. Logic suggests that Wi-Fi will evolve in tandem with high-speed mobile Internet access under the control of a single operator able to switch subscribers between different access modes as appropriate [29].

International roaming between hotspots is underway. A company such as Boingo is very active in both selling subscriptions directly to consumers as well as helping operators to provide roaming facilities [TelecomWeb, 2007b]. A divergent approach is being adopted by RoamPoint which specialises in providing interconnection agreements between service providers and by other brokers that also negotiate commercial roaming agreements.

Recent evidence that the international aggregation of hotspots is an unstoppable force is the roaming agreement signed by iPass and European provider The Cloud in April 2006 such that the latter's 6,000+ hotspots would be added to its Global Broadband Roaming network, raising its total above the 50,000 mark, once they had passed the iPass Enterprise Ready certification programme which aims to ensure inter-operability with common policy and security tools.

The really crucial issue, needless to say, revolves not around data cards in laptops, which is necessarily a limited market, but upon the possibility of handsets capable of making voice calls over a W-LAN. Handling data is not a particular problem from a technical viewpoint, and operators are well aware that it will be to their advantage to offer such handsets, not only because this will discourage the development of Wi-Fi-only devices but also because it will generate revenue and encourage the use of their GPRS/EDGE/3G networks. What is far less desirable for them is to have handsets equipped with a Voice over Internet Protocol (VoIP) facility that will circumvent the need to pay high charges for voice calls.

From a marketing perspective, one issue that will increasingly arise is that the popular hotspots will need to be covered by all of the 3G licensees. While that is reasonable in relation to a large airport, it is likely to lead to excessive amounts of equipment being located in, say, city centre locations. Clearly, the efficient solution is for, say, a hotel to provide a common service on behalf of all the operators, but that would mean that each operator would end up with an identical provision, thereby preventing them from gaining any marketing advantage.

WiMAX

Another way forward is encapsulated in the Worldwide Interoperability for Microwave Access (WiMAX) Forum which is behind the IEEE 802.16 standard first published in April 2002 [30]. This is a data-only standard – not strictly a technology as such – that initially came in variants 802.16a and 802.16c of the common 802.16 *fixed* standard – the former covering the 2 GHz to 11 GHz band, making it suitable for connection to Wi-Fi hotspots, and the latter covering the 56 GHz to 66 GHz band – as well as the original variant designed for the intervening bandwidth which tends to be already occupied and hence of limited practical use. 802.16 has evolved into 802.16-2004 which was previously known as 802.16d [OECD, 2006].

With a peak range of at least 50 kilometres – and hence suitable for metropolitan area networks (MANs) – and a peak shared data rate of 70 Mbps, 802.16 certainly appears to offer an attractive prospect for areas currently not covered at all, or poorly covered, by other types of network. However, these sort of outstanding numbers reflect the optimum channel size, power and best-case modulation, and take no account of the network overheads. In fact, roughly thirty per cent of the bandwidth is needed for error correction and encoding, and what is left – roughly 50 Mbps – covers both directions of data transfer given that WiMAX is a TDD system. Hence, with equal directional flows, each could carry only 25 Mbps, and a further ten per cent is needed for other purposes. What is left – 20 Mbps upstream and downstream – is accordingly rather less than the original amount claimed, but on the other hand it is a lot faster than 3G.

Significantly, this standard defines neither the frequency bands nor channel width, and if the channel is as wide as 28 MHz then speeds of up to 130 Mbps are possible. However, being a multi-point system, the capacity must be shared between however many customers as are logged on at any one time. One of the earliest networks was that constructed by Telabria in south-east England, commencing in November 2004 and going live in mid-2005. A *mobile* variant with a 5 km radius from a base station known as Wireless Broadband Korea (WiBRO or 802.16e) – but occasionally as Wide Broad Internet – has been developed by Samsung (initially launching its H1000 and M1000 handsets in November 2005) and LG Electronics in South Korea using a 10 MHz channel in the 2.3 GHz band – the first European network is planned by TIM of Italy to launch in 2007 but as ever there are significant differences in the approach taken by different Member States [Ballon, 2007; Taafe, 2007]. Originally seen as a regional competitor to WiMAX – because it permits handover between cells it can, unlike WiMAX, be used in vehicles moving at up to 120 kph – the IEEE 802.16e Task Group approved the specifications for worldwide 802.16e – often referred to as mobile WiMAX – in December 2005. It is expected to be capable of delivering 30 Mbps but this translates into a real-world speed of up to 10 Mbps under normal operating conditions which is not a huge step forward relative to the speeds expected from HSPA in the reasonably near future. Unfortunately, there is no agreement as yet on a single spectrum band to be used on a world-wide basis; the USA and Asia-Pacific countries favour 2.5 GHz while Europe and Latin America favour 3.4-3.6 GHz. Should the disparity continue, business travellers would need to have their laptops configured to deal with both bands.

At the end of the day, mobile WiMAX was seen as a real contender to take on 3G in good part because of its relative speed, but its real-world performance is now thought to be comparable to that of HSDPA which accordingly is likely to be preferred due to its lower cost [TelecomWeb, 2007a]. WiBro has certainly not been a success. When launched in South Korea, it was expected to attract 3.5 million subscribers worldwide by the end of 2006. In practice, SK Telecom had achieved 151 subscribers by that date at a cost of $186 million, KT Corp. had achieved 906 subscribers and Hanaro had returned its licence.

The fixed WiMAX market is dominated by Alvarion and is largely the preserve of proprietary broadband service providers taking on incumbents. Many major equipment manufacturers such as Alcatel, Motorola and Nokia, are developing mobile WiMAX, which will not be inter-operable – a mobile WiMAX modem will not be able to connect to a fixed WiMAX base station – but there are doubts as to whether the market will ever achieve sufficient scale to warrant the investment needed. It must be borne in mind that the great bulk of potential usage is clustered geographically and hence that a decent integrated set of hotspots will not be much different in effect from a WiMAX network.

It is also appropriate to refer to an alternative to WiMAX known as Flash-OFDM, a technology linked to IEEE 802.20 developed by Flarion Technologies (acquired by Qualcomm in August 2005) to operate within a pure IP-based network. So far, little has been heard of this – the technical aspects can be found at www.flarion.com – although it was put on trial by Nextel in North Carolina during 2004. Subsequently, Nextel was taken over by Sprint, a company that was allegedly not interested in the technology but which announced that it was considering both WiMAX and Flash-OFDM in November 2005 only to come down in favour of WiMAX. However, the first significant launch was by T-Mobile in Slovakia in November 2005 – to be followed by others in Europe although T-Mobile's strategy is to use whatever spare spectrum is available which is why it is using unpaired UMTS TDD spectrum in, for example, the Czech Republic. In Slovakia, T-Mobile is using the 450 MHz band for which it claims Flash-OFDM is optimised – it is initially operating at comfortably above 1 Mbps downstream and 500 kbps upstream – and when analogue is switched off across Europe in this band is probably going to provide the earliest opportunity to demonstrate what the technology can achieve. The recent upgrade to Flexband has potentially boosted the average downstream speed from 1.25 Mbps to 6 Mbps within a cell, with roughly 700 kbps at the edges, which compares favourably with cdma2000 1xEV-DO.

In October 2006, Nortel launched a new mobile WiMAX range of equipment based upon OFDM and MIMO as well as a collaboration deal with Runcom to develop mobile WiMAX devices and chipsets. MIMO-enabled devices are expected to appear in mid-2007.

Table 4. High-speed networks

System	Standard	GHz frequency	Maximum Mbps	Range
EDGE	2.75G	0.85, 0.9, 1.8, 1.9	0.473	Low mileage
W-CDMA	3G	0.8, 0.9, 1.7, 1.8, 1.9, 2.1, 2.6	2.1	Low mileage
cdma1xEV-DO	3G	0.4, 0.8, 0.9, 1.7, 1.8, 1.9, 2.1	2.4	Low mileage
HSDPA	3.5G	0.9, 1.8, 1.9, 2.1	14.2	Low mileage
UWB	802.15.3a	7.5	480	Very short
Wi-Fi	802.11a	5.0	54	Short
Wi-Fi	802.11b	2.4	11	Short
Wi-Fi	802.11g	2.4	54	Short
Mesh Wi-Fi	802.11b/g	2.4, 5.0	54	Short
WiMAX	802.16d	<11	70	Low mileage
WiMAX	802.16e	2.6	30	Low mileage

A final issue to address is that of hand-off. Until very recently, there was no way of using a handset to connect up with a hotspot. However, there is now a new standard known as Unlicensed Mobile Access (UMA) that has been created to deal with the hand-off of phone calls between base stations and Wi-Fi access points. This was accepted for GSM in February 2005. There is also work going on to adapt IP Multimedia Subsystem to the same end.

Mesh Networks

Given that broadband connections can only be provided on a large scale by either adding DSL to copper wire or via cable, the issue of whether a connection to either can be effected becomes of great consequence. This accordingly appears to present an opportunity to install a fixed wireless broadband service which blankets a neighbourhood without the need for physical connections such as routers. The core of the mesh is the standard arrangement of a 'neighbourhood access point' (NAP) which is a base station connected to the fibre-optic ring. The homes/offices within range of the NAP then install antennas to provide the last-mile link. However, the economics of this arrangement are predicated upon each connected antenna acting as a relay for other antennas, thereby considerably widening the geographic scope of the mesh. Normally, signals travel point-to-multipoint, a process that needs a lot of power because the signals often do not have line of sight and must circumvent obstacles. However, it is only necessary for any antenna in a mesh to have an unencumbered access to another antenna – it does not matter whether the shortest route is taken – so power levels can be much reduced. Siting antennas on rooftops is ideal for this purpose.

A great virtue of mesh networks is that they are spectrum-agnostic and hence can be cheaply located in unlicensed spectrum. Given that they also do not need an expensive fibre-optic backbone, the inherent attraction to new entrants is evident. However, there is nothing to prevent incumbents from setting up mesh networks for their own benefit.

Although it is widely accepted that the same radio band should not be used for both access and backhaul, what is as yet unclear is whether mesh architecture will consist of dual-band Wi-Fi or Wi-Fi in the 2.4 GHz band for direct access to customers combined with WiMAX in the 5.8 GHz band for the backbone in metropolitan areas. The driving force is the need to keep the cost of delivering broadband down to an acceptable level, particularly as many customers will need to be subsidised by municipalities..

Ad hoc Networks

Ad hoc networks are often treated in tandem with mesh networks (Ballon, 2007, p.199). A device equipped with, say, Bluetooth, is already capable of a single leap between pieces of equipment. By linking a whole series of, say, computers, with 'one-hop' links one can envision the creation of an entire network which is accordingly created on an 'ad hoc' basis as needed. This idea has already been given consideration in relation to Wi-Fi such that, assuming the relevant software is installed, one person's Wi-Fi-equipped laptop can act as a connection to that of another person and so on down the line. One obvious advantage of this approach is that it can extend the scope of a network, with those outside the immediate range of a base station able to hop in via a device, or series of devices, already within range. Also on the plus side is the fact that service quality overall is much improved, while the negative aspect is that strangers are draining a device's battery as they use it en passage. A system of micro-payments might possibly address this problem, but it would introduce complications.

Ultra-Wideband (UWB)

UWB, or IEEE 802.15.3a, involves the simultaneous transmission of a series of very short pulses, typically lasting less than one billionth of a second, at low power on a wide range of frequencies. The speed of the pulse is such that it remains undetected by standard receivers but can be picked by a UWB receiver. Information needs to be encoded into the stream of pulses so that it can be read by the receiver. The FCC authorised the use of UWB on a restricted basis in February 2002, with a maximum distance for transmission fixed at ten metres in order not to create problems of interference with other signals, but the potential speed of more than 100 Mbps over such distances certainly opens up opportunities even if, for the time being, these are not particularly relevant for mobile telephony as such given the low power involved. One problem is that UWB does not work well in sending signals around corners or through material such as brick whereas Wi-Fi can deal comfortably with corners and most materials. Furthermore, its range is roughly 80 feet compared to Wi-Fi's 300 feet. In any event, the process of establishing standards and introducing a range of practical devices is still some way off. As ever, there are competing versions of the technology, with the direct-sequence (DS-UWB) camp led by Freescale in competition with the Multiband OFDM Alliance (MB-UWB) led by Texas Instruments.

However, the fact that UWB is legally established primarily in the USA – although the EU outlined the mandatory conditions for its use in a harmonising decision published in February 2007 [31] and the UK regulator has permitted its unlicensed use since August 2007 [32] – is for now a major obstacle in the path of its widespread introduction. In Spring 2006, the European Communications Commission announced that it would support transmissions in the 6.5-8.5 GHz band – the harmonising Decision merely refers to 'exclusively above 6 GHz in the longer term' – while Japan looks to prefer the 7-10 GHz band. This suggests that there may be a common worldwide band lying between 7 GHz and 8.5 GHz but as ever, no regional grouping wants to cede control over the technology to any other.

xGTM [33]

In February 2006, xGTM Technology was approved by the FCC to launch a novel wide-area wireless network transmitter. The xGTM system is based upon a new low-frequency wireless networking protocol called xMAX. Unlike WiMAX or Flash-OFDM, xGTM has a range of up to thirteen miles and provides a signal that penetrates more deeply into buildings. Furthermore, it does not need dedicated spectrum because it transmits at power levels that are sufficiently low as not to interfere with other signals. In a test, a xMAX device transmitted 100 feet using only 300 nanowatts of power compared to roughly three million times as much needed for 802.11.

xMAX is not a compression technique but it has the potential to boost data transfer rates considerably. What it does is to combine elements of traditional narrowband carrier systems with non-interfering elements found in low-power wideband systems. xMAX is accordingly particularly suitable for broadband wireless access (BWA) services. It is possible that in dense metropolitan areas, only one xMAX base station will be needed every 35 miles. This would allow one tower to provide backhaul for hotspots across an entire city. Furthermore, the ability to pass through dense obstacles means that in the UHF and VHF bands the problems customarily associated with the need for line-of-sight are eliminated.

Smart Antennas

A simple base station broadcasts in all directions simultaneously. This is not particularly efficient, especially when viewed in terms of use of scarce bandwidth. The initial solution is to use multiple antennas, each covering one sector spanning, say, 120°. However, a much more efficient approach is to pinpoint the position of a handset by comparing the signal strength at different antennas located around the base station. Although in theory capacity is enhanced by the same multiple as applies to the number of antennas used, this is hard to achieve in practice. From an economic viewpoint, adding antennas is now cheaper than adding base stations, but the differential has not been large enough to induce much substitution in the context of rapidly falling equipment prices. Smart antenna manufacturers understandably believe that companies making base stations are trying to lock them out. Because W-CDMA networks require the construction of new base stations on a significant scale, this may tilt the playing field in favour of smart antennas during the rest of the decade. ArrayComm uses this approach, and claims to boost capacity by a factor of between three and seven with its iBurst technology. So far, iBurst has been introduced by Personal Broadband Australia using 3G spectrum acquired in Australia by the consortium in which ArrayComm was an investor. It launched in Sydney in March 2004, while Wireless Business Solutions launched in South Africa in 2005.

Notes

[1] See www.cellular-news.com/gprs

[2] There are some potentially crucial differences between the USA and elsewhere. In particular, whereas the Japanese and Europeans often commute by public transport, leaving them free to play with their mobile phones en route, Americans prefer where possible to go by car. On the other hand, Americans spend far more time hanging around airports where they want to connect up their laptop computers.

[3] See Parker, T. (2003). A brief history of EDGE is to be found at www.cellular.news.com of 18 July 2003. See also www.gsacom.com for a list of adopters; Pearson and Rojas (2003); and *Wireless Business Review: 3GSM World Congress 2004 Edition* pp.60-62 at www.gsmworld.com.

[4] For further information see section on HSDPA, *Data Capabilities: GPRS to HSDPA*, a report prepared by PYSAVY for 3gamericas in September 2004, and ‘Siemens brings end-to-end HSDPA solution to market’, available at www.the3g portal.com of 22 October 2004.

[5] MVNOs do not need to buy a licence or build a network. However, to be successful they probably need to have an existing relationship with a customer base that they can exploit, a well-known brand, a distinct set of charges, and their own distribution channels and billing systems.

[6] See ERC/DEC/(97)07.

[7] 'Commission decision of 14 February 2007 on the harmonised use of radio spectrum in the 2 GHz frequency bands for the implementation of systems providing mobile satellite services', *Official Journal of the European Union*, L 43/32, 15.2.2007 stipulated that, because of the threat of harmfiul interference, the spectrum should be allocated to MSS 'on a primary basis' and that Member States should allocate the spectrum on that basis as from 1 July 2007.

[8] See entry for Evolution-Data Optimized at http://en.wikipedia.org/wiki/EVDO.

[9] For a concise discussion of the history of GSM see Pelkmans, J. (2001). There is also a discussion of technology at www.itu.int/newsarchive/press/releases, at www.ida.gov.sg/Website/IDAConten, at www.cellular-news.com/3G, at www.umtsworld.com and in Federal Communications Commission (2001). FDD is a technique whereby the uplink and downlink are at different frequencies, whereas with TDD they are on the same frequency. Although the paired FDD spectrum has found a ready use in 3G networks, a situation that was initially puzzling in some respects since whereas voice needed roughly equal bandwidth in both directions, data was heavily biased towards download (although that is changing gradually with the likes of interactive gaming), the 5 MHz unpaired TDD slots have been more problematic. However, the Global UMTS TDD Alliance was launched in February 2004. The current view is that three technical options are available. The first is to put it to the use originally envisaged for it, namely for high-capacity, short-range 'femtocells' – smaller and less expensive than enterprise-focused 'picocells' – in locations such as railway stations and airports – 3Way Networks launched the 3Way Home Base Station, the world's first commercially available 3G femtocell, in February 2007. The second is to use it for wireless broadband as is the case for T-Mobile CZ and others. The third is to use it for tdTV as developed by IPWireless.

[10] For a discussion of the component parts of a mobile network see the Nokia White Paper *Mobile Network Transmission*, available at www.totaltele.com. For a history of the development of mobile communications in Japan where somewhat different technology has been introduced see Srivastava, L. (2001). Whether any other standards will be authorised for 3G is a moot point, but there is some support for LAS-CDMA (Large Area Synchronized CDMA) given that LinkAir has developed an initial version of TD-LAS that is considered to be a rival for TD-SCDMA in China [Tanner, 2002]. For its part, TD-SCDMA currently comes in two versions: TSM, which uses the same base network as GSM and is supported by Siemens; and TDD-LCR, which is supported by Datang. For the more technically minded, an explanation of CDMA in the context of W-CDMA can be found at http://imtdesign.com/hottop.

[11] Unlike DECT, PHS – also a TDD-based TDMA cordless access technology – was not submitted for consideration as 3G. The idea that one can pick up a fixed-wire handset and wander around within a limited range from the cradle is nothing new, although reception may still be patchy. However, this concept has now been turned on its head with a large number of operators seeking to introduce a means whereby a mobile handset effectively reconfigures itself into a fixed-wire handset once within range of a box within domestic or other premises, the objective being to provide fixed-wire QoS and to charge at fixed-wire prices once the connection to the box is established The first operator to launch such a service was BT in June 2005. In this case, the mobile handset switched to a BT Broadband line once in the home, and hence necessitated a

BT connection as well as a broadband subscription. Thereafter, BT Fusion subscribers were entitled to standard BT fixed-wire rates – see 'BT Group launches combined fixed, mobile phone service', available at www.cellular-news.com of 15 June 2005 and Curwen (2006).

[12] The WRC meets every two or three years – it met, for example, in 1995, 1997, 2000 and 2003. Its conferences are organised by the International Telecommunication Union, a specialised agency of the UN – see www.itu.int/ITU-R/conferences/wrc. The WRC's job is to review and, where necessary, revise the Radio Regulations – the international treaty obligations governing the global use of the radio spectrum and of satellite orbits.

[13] The obvious justification for exploring the use of CDMA-450 (often CDMA450) is that one base station provides roughly the same coverage as 25 UMTS base stations, and a CDMA-450 network is roughly 30 per cent cheaper than a CDMA-800 network and up to 70 per cent cheaper than a CDMA-1900 network. However, the absence of a facility to roam out of the 450 MHz band has been a problem. Opinion as to the value of CDMA-450 in Europe is somewhat divided. As analogue networks are shut down, the vacated 400-450 MHz spectrum band becomes available for other purposes, and although Europe is something of a desert for CDMA in the other spectrum bands, it could potentially be introduced to occupy the vacated space, especially for use in public mobile radio (PMR). Because the typical spectrum available is less than the 5 MHz paired required by W-CDMA, the 450 MHz band can in general only be developed along the cdma2000 route. Indeed some former Eastern European countries are intent upon so doing such as EuroTel Praha in the Czech Republic and Latvia's Telekom Baltija. Qualcomm already has a 1.8 MHz paired CDMA-450 licence in Denmark and a licence for cdma2000 in the 872.5-876 + 917.5-921 MHz bands in Norway. The spectrum vacated by Telenor's analogue service was licensed, as from 1 January 2005, to Nordisk Mobiltelefon for a probable CDMA-450 service. The main objections to the introduction of CDMA-450 were that it took no account of the overriding need for interoperability, and that with the number of subscribers heavily circumscribed by the limited spectrum, costs would be relatively high. Furthermore, there would be insufficient spectrum for a range of operators to compete, and the choice of terminals and services would be limited. Finally, there would be issues about licensing fees, especially if it was intended that 3G services would be provided using the spectrum. Nevertheless, if digital public mobile radio (TETRA) systems grow over time, they will need to be allocated additional spectrum. The CEPT voted in favour of permitting CDMA-450 in March 2004 and also supported the allocation of 1.25 MHz in the GSM band for CDMA even though many operators would need to re-farm their spectrum to allow this. However, the vote related to the use of the spectrum for PMR rather than mass-market mobile. At the end of the day, national regulators will do their own thing.

[14] 3GPP is an international standardization initiative concerned with GSM and itrs derivatives that was originally founded by ETSI (Europe), Association of Radio Industries and Businesses (ARIB - Japan) and the Alliance for Telecommunications Industry Solutions (USA), with the latest of its six Organizational Partners, the China Communications Standards Association (CCSA) joining in April 2004 The others are the Telecommunications Technology Association (TTA) of South Korea and the

Telecommunication Technology Committee (TTC) of Japan. 3GPP periodically produces a complete current list of its specifications in a sequence of Releases. In late August 2002, 3GPP published Release 5, which it claimed to represent a major step forward in enabling IP multimedia capabilities to be provided over GPRS and 3G networks. The central problem being addressed by 3GPP's core network technical specification group, TSG-CN, is that Internet protocols allow largely free services to be provided over 'best effort' quality of service (QoS) networks, whereas 3G operators require a very high QoS if subscribers are to be willing to pay much higher fees than for 2G. The other two groups are concerned with the radio access network (RAN), terminals and service and systems aspects (see www.3GPP.org). Much of the history can be traced via publications of 3G Americas, notably *The Evolution of UMTS – 3GPP Release 5 and Beyond*, a White Paper first published in June 2004 and updated in November 2004, and *The Evolution of UMTS/HSDPA – 3GPP Release 6 and Beyond*, a White Paper published in July 2005.

3GPP2 is the equivalent of 3GPP but applied to derivatives of CDMA. Its members comprise ARIB, CCSA, TIA, TTC and the Telecommunications Industry Association of the USA. Its working groups are concerned with access network interfaces, cdma2000, services and systems aspects and intersystem operations (see www.3GPP2.org). 3GPP does not publlish a complete list of its specifications but announces each one as it is approved.

3GPP should not be confused with 3G3P – the 3G Patent Platform Partnership which is concerned with intellectual property rights – see Taaffe (2001). Over 100 companies are thought to own patented technologies which might be 'essential' to the realisation of 3G. 3G3P has the job of sifting through the thousands of patents claimed to be 'essential', selecting several hundred which are worthy of that epithet and finding a way to speed up the process of issuing licences and of controlling the amounts of licence payments. For a history and discussion of the innovatory features of the 3G3P see Ewing (2003). There are also other bodies concerned with standardisation. For example, the Open Mobile Alliance was formed in June 2002 with 200 members including Microsoft, and the Open Base Station Architecture Initiative was set up by LG Electronics, NEC, Nokia and Samsung in October 2002. The issue of standardisation forms the theme of Sehier, Gabriagues and Urie (2001).

[15] See the preceding section on i-mode. It may be noted that i-mode is not sold on the basis of its technological prowess but its ability to provide a wide range of services, and hence DoCoMo is providing a range of 3G handsets to suit the different levels of service sophistication desired by customers. For a recent view, see Peltokorpi, Nonata and Kodama (2007).

[16] For a recent review of HSPA see, for example, Harrowell (2006).

[17] In practice, while the downlink is based on OFDMA, the 3GPP specifies SC-FDMA for the uplink. For a discussion of the complexities of introducing LTE technology see 'LTE: a testing environment – fast forward', available at www.mobileeurope.co.uk of 31 January 2007.

[18] See Morris (2006). A brief review of LTE is to be found on the www.gsacom.com website.

[19] See www.radicati.com. For a general discussion of MMS see 'What is MMS?', available at www.ameinfo.com of 19 May 2002, and for a more recent review see

Kattan (2004) and *Wireless Business Review: 3GSM World Congress 2004 Edition*, pp.57-59 at www.gsmworld.com.

[20] The former Vodafone K.K.'s MMS service was known as Sha-mail. In March 2002, a Movie Sha-mail service, providing video messaging of clips lasting up to five seconds together with text, was also introduced. DoCoMo was initially a laggard, having started late in June 2002. Somewhat curiously, it sought to market handsets such as the 7650 as 'imaging devices' rather than mobile phones. By 2010, nearly 90 per cent of all handsets are expected to contain a camera [Tilak, 2006].

[21] For an analysis of the pros and cons of various platforms and devices see 'Mobile email – the kit', available at www.telecoms.com of 10 August 2006.

[22] See 'Worldwide handheld device market continues downhill slide', available at www.cellular-news.com of 16 May 2007.

[23] Push-to-talk has been discussed on a number of occasions in *Global Mobile* – see, for example, 10 March 2004, p.12 and 7 April 2004, p.11. The Open Mobile Alliance puvblished its first open industry standard for PoC in June 2005 called the OMA PoC 1.0 Candidate enabler – see www.telecoms.com of 13 June 2005.

[24] Bluetooth is technically IEEE standard 802.15.1 with a range of 10 metres. A single Bluetooth 'master' can communicate with 'slaves' thereby creating a so-called 'piconet', and piconets can be joined together to form 'scatternets' – see www.Bluetooth.com. If Wi-Fi and Bluetooth are used in proximity their slightly different modes of operation can cause the Wi-Fi signal to degrade. It must also be borne in mind that the 2.4 GHz band not only has to accommodate W-LANs and Bluetooth, but the likes of microwave ovens and cordless phones. In October 2006, Nokia announced that it had launched a new radio technology branded as 'Wibree' which it hoped to establish as a new industry standard for short-range wireless connections between small devices. Wibree is compatible with Bluetooth but is lower-powered and hence can, for example, be used in connectivity to watches. The first commercial version was expected in mid-2007 – see 'Nokia launches new short-range standard' at www.bwcs.com of 3 October 2006. However, Wibree has been downplayed as 'Bluetooth lite' and the suspicion raised that it is simply Nokia sniping at Ericsson which first developed Bluetooth – see 'Why Wibree?', available at www.telecoms.com of 12 October.

[25] For this reason, equipment manufacturers are concentrating upon effecting seamless hand-off between W-LANs and 3G networks. A crude comparison between Wi-Fi with 3G would suggest the following, inter alia, is true for the time being:

- both are wireless technologies
- both need to be connected to a fixed-wire backbone
- both provide broadband connections
- 3G is normally licensed; Wi-Fi is unlicensed
- 3G is designed to carry voice and data; Wi-Fi to carry data only
- 3G is introduced as a cohesive, monolithic system; Wi-Fi grows piecemeal
- 3G is designed to cover large areas; Wi-Fi to cover small areas
- 3G requires huge expenditures; Wi-Fi is relatively cheap
- 3G heavily favours wireless incumbents; Wi-Fi is more entrepreneurial
- a much higher QoS can be guaranteed for 3G (e.g., less interference, more security).

Not unnaturally, much of the debate about 3G versus W-LANs is in practice couched in terms of data transmission speeds since early forms of 3G struggled to get past the 384 kbps barrier. HSDPA has improved this situation significantly, and 3G achieved 1,000 Mbps in laboratory conditions as early as December 2004 using so-called Orthogonal Frequency Division Multiplexing (OFDM) although this required 100 MHz channels in the 5 GHz band. Looking forward to 2015 and beyond, OFDM is likely to become commonplace as a form of so-called Multiple Input, Multiple Output (MIMO) system which uses multiple antennas simultaneously to transmit different flows of data over a single radio channel and frequency band.

[26] The 2003 World Radiocommunication Conference (WRC-03) was principally concerned with the allocation of spectrum in the 5 GHz band. In total, 455 MHz was allocated in the band even though, since the spectrum is normally unlicensed, international allocations are not strictly necessary. Nevertheless, most advanced countries were anxious that the spectrum bands be harmonised to ensure inter-operability and also that developing countries with no history of unlicensed cspectrum should legitimise the unlicensed use of the bands as agreed at the WRC (Sung, 2003/2004).

[27] See Clark (2006), 'New Wi-Fi IEEE 802.11n standard suffers further delay', available at www.telecomdirectnews.com of 16 August 2006 and 'Wi-Fi Alliance rushes in with 'pre' 802.11n certification', available at www.telecomdirectnews.com of 30 August.

[28] 28. The drawback that Wi-Fi has a range of a few hundred metres, and that it uses a shared pool of bandwidth, appears to have been addressed by the iBurst technology developed by ArrayComm of the USA. This technology is not shared and has a range of up to 10 kilometres providing 1 Mbps of data transfer. iBurst is being introduced by the likes of Telecom New Zealand. The issue of insecurity, caused by signals bleeding outside the local network, has yet to be fully addressed because the Wired Equivalent Privacy Standard proved to be weak but 802.11g is expected to be much more secure than 802.11a and b. The 802.11i standard promises improved security, but probably at a price since hardware will probably have to be replaced. Recent products are based upon Multiple Input, Multiple Output (MIMO) technology from Airgo Networks. The specification involving MIMO is known as 802.11n, is backwards compatible and is intended as a replacement for 802.11b and 802.11g [Mossberg, 2005].

[29] Information dates quickly in this area, but a useful review is to be found in Rao and Parikh (2003). This compares a number of business models, ranging from the top-down WISP arrangement associated with Boingo (which had a worldwide network of 18,000 hotspots in September 2005 incorporating 100 operators in 50 countries) and Sputnik to community-based networks such as NYC Wireless. The advantages of the top-down model are seen as reliability, stability and predictable pricing whereas the bottom-up model has the advantage in terms of innovation and experimentation. Obviously, control in the top-down model has to be exercised by the likes of a WISP which views provision as a commercial exercise, whereas in the bottom-up model community leaders and volunteers are responsible which can lead to fragmentation partly because the objective is not to make money.

[30] Although no official WiMAX certification was forthcoming until 2005, certain operators such as Millicom International in Argentina had already contracted to have

WiMAX overlaid on an existing Wi-Fi network. The role of the WiMAX Forum is discussed in Fitchard (2005). For a recent overview see Judge (2005).

[31] The Radio Spectrum Committee of the EU issued a Working Document on 18 February 2004 (RSCOM04-08 EN) entitled 'Mandate to CEPT to harmonise radio spectrum use for Ultra-wideband Systems in the European Union'. Eventually, the EU published 'Commission Decision of 21 February 2007 on allowing the use of the radio spectrum for equipment using ultra-wideband technology in a harmonised manner in the Community', *Official Journal of the European Union*, L 55/53, 23.2.2007. This stipulated that equipment would be used indoors and that any transmission would cease after ten seconds if not acknowledged.

[32] See 'Digital home technology given OK', available at www.bbc.co.uk of 9 August 2007.

[33] See '"Better than WiMax" device gets FCC nod', available at www.telecomdirectnews.com of 8 February 2006.

References

Ballon, P. (2007). Changing business models for Europe's telecommunications industry: The impact of alternative wireless technologies. *Telematics and Informatics*, **24**, 192-205.

Brown, G. (2007). WiMax gets ITU standards boost. Available at www.unstrung.com of 4 June.

Clark, D. (2006). New WiFi devices get faster as gear improves. Available at www.totaltele.com of 13 June.

Commission of the European Communities (1997). *Communication from the Commission to the Council, the European Parliament, the Economic and Social Committee and the Committee of the Regions: strategy and policy orientations with regard to the further development of mobile and wireless communications (UMTS)*, COM(97) 513 final. Brussels: CEC.

Commission of the European Communities (1998a). *Proposal for a European Parliament and Council Decision on the co-ordinated introduction of mobile and wireless communications (UMTS) in the Community*, COM(1998) 58 final. Brussels: CEC.

Commission of the European Communities (1998b). Amended proposal for a European Parliament and Council Decision on the co-ordinated introduction of mobile and wireless communications in the Community. *Official Journal of the European Communities*, **C276/4**.

Council of the European Communities (1998). Common position (EC) No 52/98 adopted by the Council on 24 September 1998 with a view to adopting European Parliament and Council Decision No/98/EC, of … on the co-ordinated introduction of a third-generation mobile and wireless communications system (UMTS) in the Community. *Official Journal of the European Communities*, **C333/56**. Annex 1.

Council of the European Communities (1999). Decision No 128/1999/EC of the European Parliament and of the Council of 14 December 1998 on the co-ordinated introduction of a third-generation mobile and wireless communications system (UMTS) in the Community. *Official Journal of the European Communities*, L17/1. Annex 1.

Curwen, P. (2005). *The Prospects for 3G in the USA: the view from Europe. info*, **7**, 16-32.

Curwen, P. (2006). *Fixed-Mobile convergence. info*, **8**, 3-11.

Ewing, K. Jnr (2003). EC and DOJ approval of the 3G Patent Platform. Available at www.wirelessweek.com of 17 March

Faulhaber, G. (2006). The future of wireless telecommunications: Spectrum as a critical resource. *Information Economics and Policy*, **18**, 256-271.

Federal Communications Commission (2001). *Final report. Spectrum study of the 2500-2690 MHz band.* Available at www.fcc.gov/3G.

Fitchard, K. (2005). Confusion keeps WiMAX up in the air. Available at www. printthis.clickability.com of 5 September.

Fransman, M. (2002). Explaining the success of NTT DoCoMo's i-mode wireless internet service. A paper available from M.Fransman@ed.ac.uk.

Global Insight (2007). WiMAX gains initial approval as IMT-2000 3G standard. Available at www.communicationsdirectnews.com of 5 June.

Gruber, H. (2004). *Is spectrum still a constraint for mobile telecommunications? info*, **6**, 289-297.

Harrowel, A. (2006). More Faster 3G Now. Available at http://84.233.149.205/ ActiveMagazine.

Hausman, J. (2002). Mobile Telephone. In M. Cave, S. Majumbdar & I. Vogelsang (Eds.), *Handbook of Telecommunications,* Volume 1. Amsterdam: Elsevier.

Judge, P. (2005). WiMax – a pantomine horse going in two directions? Available at www.techworld.com of 29 July.

Kattan, F. (2004). Accelerating MMS adoption and usage. Available at http://telephonyonline. com of 31 March

Morris, A. (2006). Nokia sees commercial LTE solution from 2009. Available at www. totaltele.com of 5 October.

Mossberg, W. (2005). New WiFi products expand range, speed of networks. Available at www. totaltele.com of 6 January.

OECD (2006). The implications of WiMAX for competition and regulation. OECD Working Party on Telecommunication and Information Services Policies, DSTI/ICCP/TISP (2005)4/FINAL. 2 March.

Office of Communications (2007). *WRC-07 agenda item 1.4: consultation on candidate bands under consideration at WRC-07 for IMT.* London: Ofcom. 27 February. See also *Statement* of 7 June.

Parker, T. (2003). EDGE has arrived, but does anyone care? Available at www.telecoms.com of 2 September.

Pearson, C., & Rojas, E. (2003). *Wireless Trends in the Americas: The Proliferation of GSM 850 MHz and EDGE.* 3gamericas. Available at www.3gamericas.org.

Pelkmans, J. (2001). The GSM standard: explaining a success story. *Journal of European Public Policy*, *8*. 432-453.

Peltokorpi, V., Nonaka, I., & Kodama, M. (2007). NTT DoCoMo's launch of I-mode in the Japanese mobile phone market: A knowledge creation perspective. *Journal of Management Studies*, **44**.

Rao, B., & Parikh, M. (2003). Wireless broadband drivers and their social implications. *Technology in Society*, **25**,477-489.

Sehier, P., Gabriagues, J-M., & Urie, A. (2001). Standardisation of 3G mobile systems. *Alcatel Telecommunications Review*, *1st quarter*, 11-18.

Srivastava, L. (2001). *3G mobile policy: the case of Japan. info*, **3,** 461-484.

Sung, L. (2003/2004). Observations from the WRC-03. *International Journal of Communications Law and Policy*, **8.**

Taafe, J. (2001). 3G devices – plotting a course for 3G. *Communications Week International*, No. **265**.

Taafe, J. (2007). Mobile disruption. *Total Telecom Magazine*, April, 16-19.

Tanner, J. (2002). The other 3G. Available at www.telecomasia.net of 26 September.

TelecomWeb (2007a). Yet another study dismisses mobile WiMAX. Available at www. communicationsdirectnews.com of 9 February.

TelecomWeb (2007b). Long distance goes Boingo over Wi-Fi. Available at www.communicationsdirectnews.com of 16 February.

The Economist (1998). *Public Network Europe 1998 Yearbook.* London: The Economist.

Tilak, J. (2006). Camera phones to account for 87% of handsets by 2010. Available at www.dmeurope.com of 19 January.

Turner, D. (2007). DoCoMo loses two foreign mobile carriers. Available at www.ft.com of 18 July.

3G Americas (2006). *Mobile Broadband: The global evolution of UMTS/HSPA Release 7 and beyond.* July. Available at www.3gamericas.com.

3G Americas (2007a). *World Digital Wireless Subscribers by Technology.* Available at www.3gamericas.org/English/Statistics.

3G Americas (2007b) 'mobilkom austria first European mobile operator to go live with HSDPA 7.2 and HSUPA', available at www.3gamericas.com of 22 February.

3G Americas (2007c]. *UMTS Evolution from 3GPP Release 7 to Release 8 HSPA and SAE/LTE.* July. Available at www.3gamericas.com.

In: Mobile Telephones...
Editors: A.C. Harper and R.V. Buress, pp. 77-105
ISBN: 978-1-60456-436-5

Chapter 2

European Union Mobile Telecommunications in the Context of Further Enlargement

Peter Curwen and Jason Whalley
University of Strathclyde, Glasgow, Scotland, UK

Abstract

On 1 May 2004, the European Union (EU) witnessed its single largest expansion when ten countries joined. The accession of these ten countries – Cyprus (South), Czech Republic, Estonia, Hungary, Latvia, Lithuania, Malta, Poland, Slovakia and Slovenia was followed by that of two further countries, Bulgaria and Romania, on 1 January 2007. These countries may well not be the last to join the EU, but their accession represents a useful point at which to assess just how big an effect EU expansion has had upon its industrial structure and the strategies of its companies, in this case with respect to mobile telecommunications.

The purpose of this paper is to explore two particular issues in the context of the mobile telecommunications market: Firstly, whether as a consequence of EU expansion, the accession countries will indeed be the recipients of inward investment from those companies that have, until now, largely ignored them; and secondly, whether those companies already active in the accession countries will seek to reinforce their existing positions.

To this end, the paper is structured as follows. The initial section charts the evolution of telecommunications regulation within the EU and addresses the accession process. In the main section, the ownership of mobile communication licences across the enlarged EU is described. In addition, the geographical footprint of operators is established, with a distinction being made between those operators that have invested in the accession countries and those that have not. In the second main section the focus shifts onto the accession countries, with the analysis being driven by the issues outlined above. Conclusions are drawn in the final section.

Introduction

On 1 May 2004, the European Union (EU) witnessed its single largest expansion when ten countries joined. The accession of these ten countries – Cyprus (South), Czech Republic, Estonia, Hungary, Latvia, Lithuania, Malta, Poland, Slovakia and Slovenia – irrevocably changed the EU [Rachman, 2001, Cottrell, 2003]. Although the absorption of so many new

Member States at the same time inevitably created problems [Dinan, 2006; Peet, 2007; Sedelmeier and Young, 2006], the EU was already committed to welcoming two additional members into the fold on 1 January 2007, namely Bulgaria and Romania. These countries may well not be the last to join the EU – there is still a lengthy queue of prospective entrants – but their accession represents a useful point at which to assess just how big an effect EU expansion has had upon its industrial structure and the strategies of its companies, in this case with respect to mobile telecommunications.

The purpose of this paper is to explore two particular issues in the context of the mobile telecommunications market: Firstly, whether as a consequence of EU expansion, the accession countries will indeed be the recipients of inward investment from those companies that have, until now, largely ignored them; and secondly, whether those companies already active in the accession countries will seek to reinforce their existing positions.

To this end, the paper is structured as follows. The initial section charts the evolution of telecommunications regulation within the EU and addresses the accession process. In the main section that follows, the ownership of mobile communication licences across the enlarged EU will be described. In addition, the geographical footprint of operators will be established, with a distinction being made between those operators that have invested in the accession countries and those that have not. In the second main section the focus shifts onto the accession countries, with the analysis being driven by the issues outlined above. Conclusions will be drawn in the final section.

Issues Pertaining to Accession

Until the mid-1980s the EU had no role in the regulation of the telecommunications industry [Thatcher, 2001: 562]. Since then, however, the role of the EU has been transformed to such an extent that it is now a key actor in the industry's regulation and development. Drawing on Thatcher (2001) and Walden (2005), among others it is possible to divide this transformation into three phases. The first phase is associated with the 1987 Green Paper that sought to create a telecommunications single market across the EU. This set out three principles to guide the subsequent creation of the regulatory framework: liberalisation, harmonisation and the application of competition rules (*Ibid.*: 110) [1].

The second phase commenced in late 1994 when the EU committed itself to full liberalisation by 1 January 1998 [2]. To this end, Directives were issued covering those parts of the telecommunications industry hitherto excluded from liberalisation by the EU, namely satellite services, mobile communications, voice telephony and infrastructure [Thatcher, 2001: 568]. In addition, the EU also issued Directives covering licensing, numbering, interconnection and universal service. The first three of these are central to the development and functioning of competitive telecommunication markets, while the latter focuses on the provision of telecommunication services to all citizens throughout the EU [3]. According to Walden (2005: 112), the third phase emerged from the 1999 review of communications that sought to address external pressures such as conformity with the rules of the World Trade Organisation, diverging practices between Member States [4] and a desire to simplify regulation to reflect the widening prevalence of competition within the industry. The EU proposed six Directives in July 2000 for implementation by July 2003 5]. The new framework took into account convergence, and distinguished between ex-ante regulation and competition law [6].

When countries join the EU they are required to adopt its existing laws and institutions [7]. As a consequence, the accession countries joining in 2004 and 2007 were required to incorporate the considerable and wide-ranging body of EU telecommunications regulation outlined above into national legislation. This is by no means an insignificant task. Not only is the body of telecommunications regulation substantial [8], but it is just one of thirty-one areas where accession negotiations are undertaken [9]. Having said this, benefits begin to accrue to accession countries before they become actual members. Grabbe and Hughes (1998: 24) note how the 2004 accession countries benefited from foreign direct investment, preferential access to the EU and increased competition before they became full members of the EU. While foreign investors are likely to be attracted to accession countries for a variety of reasons, they are also likely to be reassured by the country's negotiation of membership as this necessitates a functioning market economy, efficient administration and the rule of law [10]. Hitran (2006) illustrates how Romania adopted the 2003 framework and liberalised its telecommunications market before it formally joined the EU, accruing substantial benefits in the process. Thus, the date at which the EU agrees to begin accession negotiations is as important as the actual accession date.

Mobile Licence Distribution across the Enlarged EU

At the heart of the analysis of the implications of EU expansion on the strategies of mobile communications companies is the information contained in Table 1. This table depicts networks using second-generation (2G) technology – known generically in the EU as the Global System for Mobile (GSM) – and third-generation (3G) technology – known in the EU as the Universal Mobile Telecommunications System (UMTS) [11]. The table encompasses mobile licence ownership across the twenty-seven countries that were Member States of the EU on 1 January 2007. In essence, 2G represents a digital technology whose main purpose is to carry voice telephony while also accommodating low-speed data transfer as exemplified by the short message service (SMS), while 3G is capable of much higher speeds of data transfer suitable for large data files and still and video photography. UMTS normally requires the licensing of new spectrum, but there are also two intermediate technologies, known as the General Packet Radio Service (GPRS) and Enhanced Data rates for GSM Evolution (EDGE) that can operate at higher speeds than GSM while using the same spectrum, as well as a recently launched technology, High-Speed Downlink Packet Access (HSDPA), capable of even higher speeds than UMTS over the same spectrum band.

Table 1 builds on Whalley & Curwen (2003) in several ways. In the first place, the table differentiates between the two bandwidths used for what is generically called GSM, namely GSM900 and GSM1800 (PCNs). Secondly, the table identifies when each mobile service was launched. Thirdly, the table details the number of subscribers that each company had as of 1 January 2007. By detailing the number of subscribers that a company has in each country, the table begins to differentiate between a simple presence in a country where the company is not a significant player and a presence where the company is actually (one of) the largest in the market in terms of number of subscribers. By combining the service launch date and the number of subscribers, the table also provides an impression as

Table 1. Continued

Country	GSM[1]			PCNs[1]			UMTS[2,3]		
Ireland	-	-	-	-	-	-	H3G*	07/05	87
	Meteor	02/01	803	Meteor	-	-	-	-	-
	O_2	03/97	1,632	O_2	-	-	O_2	03/05	-
	Vodafone	06/93	2,178	Vodafone	-	-	Vodafone*	07/04	-
Italy	-	-	-	-	-	-	H3G*	03/03	7,080
	-	-	-	-	-	-	IPSE 2000	-	-
	TIM	11/92	34,245	TIM	-	-	TIM*	05/04	-
	-	-	-	Wind	03/99	14,634	Wind	10/04	-
	Vodafone	12/95	26,175	Vodafone	-	-	Vodafone*	02/04	-
Latvia[4]	-	-	-	Bité	05/05	204	Bité*	09/06	-
	LMT	01/95	803	LMT	-	-	LMT*	12/04	-
	Tele2	03/97	1,025	-	-	-	Tele2	12/04	-
Lithuania	Bité	08/95	1,839	Bité	01/99	-	Bité*	04/06	-
	Omnitel	03/95	2,074	Omnitel	01/99	-	Omnitel*	05/06	-
	Tele2	12/99	1,708	Tele2	06/00	-	Tele2	-	-
Luxembourg	EPT	06/93	303	EPT	03/99	-	EPT	06/03	14
	LuXcom	05/05	102	-	-	-	LuXcom	05/05	35
	-	-	-	-	-	-	Orange	-	-
	Tele2	05/98	224	Tele2	03/99	-	Tele2	07/04	11
Malta	-	-	-	Go Mobile	12/00	160	Go Mobile*	09/06	-
	Vodafone	09/97	188	-	-	-	Vodafone*	08/06	-
	-			-	-	-	3G Telecoms	-	-
Netherlands	KPN	07/94	8,642	KPN (Telfort)	10/98	-	KPN*	07/04	-
	-	-	-	Orange	01/99	2,115	Orange	11/06	-
	-	-	-	T-Mobile	02/99	2,552	T-Mobile*	01/06	-
	Vodafone	09/95	3,817	Vodafone	-	-	Vodafone*	02/04	-
Poland	Centertel	07/99	12,521	Centertel	08/97	-	Centertel*	01/06	-
	-	-	-	-	-	-	Netia P4	-	-
	Polkomtel	02/96	11,855	Polkomtel	09/99	-	Polkomtel*	09/04	-
	PTC	02//96		PTC	08/99	-	PTC*	08/04	-
Portugal	Optimus	08/98	2,023	Optimus	08/98	-	Optimus*	06/04	-
	TMN	10/92	5,700	TMN	-	-	TMN*	04/04	-
	Vodafone	10/92	4,618	Vodafone	-	-	Vodafone*	02/04	-
Romania[5]	-	-	-	CosmOTE	04/00	1,226	-	-	-
	Orange	06/97	7,760	-	-	-	Orange	06/06	383
	Vodafone	03/97	7,419	-	-	-	Vodafone*	04/05	298
Slovakia	Orange	01/97	2,690	Orange	-	-	Orange*	08/06	-
	T-Mobile	02/97	2,201	-	-	-	T-Mobile*	01/06	-
Slovenia	Mobitel	06/96	1,374	Mobitel	-	-	Mobitel*	12/03	-
	Si.mobil	03/99	421	Si.mobil	-	-	Si.mobil	-	-
	-	-	-	Tus mobil	12/01	-	-	-	-
	-	-	-	-	-	-	T-2	-	-
Spain	-	-	-	Orange	04/99	10,692	Orange*	10/04	422
	Telefónica	07/95	21,446	Telefónica	07/98	-	Telefónica*	02/04	-
	Vodafone	10/95	14,464	Vodafone	07/98	-	Vodafone*	02/04	-
	-	-	-	-	-	-	Xfera	12/06	24

Table 1. Continued

Country	GSM[1]			PCNs[1]			UMTS[2,3]		
Sweden	-	-	-	-	-	-	Hi3G*	04/03	518
	Tele2	09/92	3,508	Tele2	-	-	Tele2	06/04	-
	Telenor	09/92	1,733	Telenor	-	-	Telenor	01/04	-
	TeliaSonera	11/92	4,603	TeliaSonera	-	-	TeliaSonera	03/04	-
UK	O_2	01/94	17,633	O_2	-	-	O_2	09/04	-
	-	-	-	Orange	04/94	14,404	Orange*	07/04	931
	-	-	-	T-Mobile	09/93	16,905	T-Mobile*	02/04	-
	Vodafone	07/92	16,939	Vodafone	-	-	Vodafone*	04/04	-
	-	-	-	-	-	-	3 UK*	03/03	3,863

[1] The entries consist of name of operator, the month when its service was first launched and the number of subscribers in thousands on the first of January 2007. Where an operator provides both GSM (900 MHz band) and PCN (1800 MHz band) services, the subscriber data are generally provided for both services together in the GSM column. Subscriber data often differ depending upon the source, but such differences tend not to be significant in the context of EU countries. There is, however, some controversy over the counting of 'inactive' customers which may be done differently by individual operators. For example, in Germany Vodafone and E-plus release both registered and active subscriber numbers while T-Mobile only releases registered subscribers and O_2 only active subscribers.

[2] UMTS is the name used for W-CDMA technology in the EU. It is also known as 3GSM. The licensees on 1 January 2007 are listed but not all have yet launched. The term 'launch' in the context of UMTS can mean many things, but normally refers to the launch of a service for corporate customers via data cards inserted in laptops. A consumer service via handsets – sometimes referred to as a 'commercial' launch – usually follows months later but may be simultaneous or even come first. Where subscriber data are known they are included in this column and the total deducted from the total number of subscribers to derive the total for GSM.

[3] Operators marked with an asterisk have launched High-Speed Downlink Access (HSDPA).

[4] Zapp Mobile launched cdma2000 1xEV-DO (approximating to somewhere between UMTS and HSDPA) in October 2004. It had 370,000 subscribers.

[5] Telekom Baltija operated a cdma2000 1xRTT network with roughly 30,000 subscribers.

[6] Formerly known as tele.ring.

[7] Including the former tele.ring.

Source: Details obtained from a wide variety of regulators' websites, company websites and media and Internet websites.

Market Structure

Across the enlarged EU in 2004, the incumbent fixed-wire operator owned the largest mobile operator as measured in terms of subscribers in nineteen of the twenty-five Member States: Austria, Belgium, Cyprus (South), Czech Republic, Denmark, Estonia, Finland, France, Germany, Greece, Hungary, Italy, Latvia (minority stake), Luxembourg, Netherlands, Portugal, Slovenia, Spain and Sweden. By 1 January 2007, the Czech Republic had disappeared from the list and both of the two new entrants qualified, albeit via a minority stake in Romania, so it was now a case of twenty-one among twenty-seven. This observation can also be expressed informatively in a slightly different fashion; that is, in just two of the fifteen 'old' member states of the EU the incumbent operator did *not* own the largest mobile operator either in 2004 or subsequently. The exceptions were Ireland and the UK. In the case of Ireland, Eircom, the incumbent fixed-wire operator, divested its mobile subsidiary, Eircell, in May 2001 and Vodafone subsequently acquired Eircell for €4.5 billion in December 2001. BT also divested its mobile arm, known at the time as mmO_2 (which has recently become part of Telefónica). In November 2001, BT spun off mmO_2 in order to ease the financial problems that it was facing in the aftermath of acquiring 3G licences and buying out its partners in its British, Irish, Dutch and German mobile businesses. Moreover, mmO_2 was not at that time the largest mobile operator in the UK and had not been for many years. This accolade had alternated between Orange, a subsidiary of France Télécom, and Vodafone. As of December 2003, all four GSM network operators had at least 13 million subscribers and only 897,000 subscribers separated the largest company, Vodafone, from the smallest, mmO_2. No other Member State has ever had anything like such consistent equality between so many operators although Lithuania in 2006 produced virtual equality between three much smaller operators.

In 2004 this left four Member States – all accession countries – where the incumbent fixed-wire operator did not own the largest mobile operator. In all of these – Lithuania, Malta, Poland and Slovakia – the largest mobile operator was partially owned by foreign investors. In Lithuania, the incumbent fixed-wire operator did not own a stake in any mobile operator, while in the case of the other three countries the incumbent owned a stake in the second-largest mobile operator. Subsequently, Ceský Telecom sold its mobile subsidiary, EuroTel Praha, to what is now Telefónica O_2 Czech Republic. Meanwhile, in Bulgaria, BTC eventually got around to launching a GSM network in November 2004 and, in Romania, RomTelecom continues to own 30 per cent of CosmOTE Romania.

Related to the above is the observation that those mobile operators with multiple licences across the EU were usually the second- or third-largest operators in the market. It is possible to determine the market position of operators in EU Member States if we assume that an operator is present provided it has a stake in any operator outside its home market. The precise size of most of these stakes is set out in the discussion that follows, and it may be observed that some are controlling stakes – which is not necessarily the same thing as majority stakes – while others are not, and that in certain cases there is multiple ownership of the same operator by companies listed in the table. With two exceptions – Tele2 and Telefónica subsidiary O_2 (the former mmO_2) – each company identified below was the largest operator in its home market denoted in bold. Tele2 was the second-largest operator in Sweden after TeliaSonera while O_2 was the third-largest (albeit not by much) of the four second-generation network operators in the UK.

If we had turned our attention to those mobile operators that were the largest operators in a foreign country just three years previously, then a common trait was that those foreign markets where they were the largest were comparatively small. However, this was no longer the case at the end of 2006. It remained true, for example, that TeliaSonera was the largest mobile operator in two of the three Baltic States which were numbered among the smaller EU markets, while Vodafone was dominant in Ireland and Malta. In contrast, however, Deutsche Telekom was the largest operator in the Czech Republic and Hungary, and France Télécom in Poland and Romania.

Market Concentration

Drawing on the subscriber information contained in Table 1, it is possible to calculate the percentage of the mobile market controlled by the largest (two/three) mobile operator(s) as of

31 December 2006. As can be observed from Table 3 below, it was commonly the case – basically reflecting the advantage of GSM incumbency – that the largest mobile operator controlled at least 40 per cent (but generally less than 50 per cent) of the market although, clearly, the figure always exceeded 50 per cent in the case of duopolies. Nevertheless, in five countries – Austria, Germany, Greece, Lithuania and Poland – the largest operator accounted for only between 30 per cent and 40 per cent of all subscribers while in the UK the figure was a mere one-quarter, indicating fierce competition.

Table 2. Market position, 1 January 2007

	Deutsche Telekom	France Télécom	KPN	mobilkom	OTE	TDC	Telefónica incl. O_2	Telenor[1]	Tele2[2]	Telia Sonera	Vodafone[3]
Austria	2	3		**1**		3		3			
Belgium		2	3								
Bulgaria				1	2						
Cyprus (S)											
Czech Rep.	1						2				3
Denmark						**1**		2		3	
Estonia									2	1	
Finland										1	
France		**1**									2
Germany	**1**		3				4				2
Greece					**1**						2
Hungary	1							2			3
Ireland							2				1
Italy											2
Latvia						3			1	3	
Lithuania						2			3	1	
Luxembourg									2		
Malta											1
Netherlands	3	4	**1**								2
Poland	3	1				2					2
Portugal		3					1				2
Romania		1			3						2
Slovakia	2	1									
Slovenia				2							
Spain		3					**1**			4	2
Sweden								3	**2**	**1**	
UK	2	4					3				**1**

[1] Telenor is the largest mobile operator in its home market of Norway, a non-EU Member State.

[2] Also present through MVNO arrangements in Austria, Denmark, France and The Netherlands.

[3] Present through a Partner Network Agreement in Austria, Belgium, Bulgaria, Cyprus (South), Denmark, Estonia, Finland, Latvia, Lithuania, Luxembourg, Slovenia and Sweden.

Source: Calculated by authors from data in Table 1.

Table 3. Mobile market concentration, 1 January 2007

Country	Total subscribers (000s)	% market share: largest operator	% market share: largest 2 operators	% market share: largest 3 operators
Austria	9,289,000	39.0	73.1	95.3
Belgium	9,744,000	43.6	75.8	100
Bulgaria	8,233,000	51.9	91.6	100
Cyprus (South)	861,000	91.2	100	n/a
Czech Republic	12,327,000	41.0	80.4	100
Denmark	5,139.000	46.8	74.0	95.8
Estonia	1,590,000	47.7	81.4	100
Finland	5,584,000	43.1	82.4	100
France	50,589,000	47.7	83.0	100
Germany	85,699,000	36.6	72.4	87.2
Greece	14,081,000	37.0	72.3	92.4
Hungary	9,718,000	45.6	78.0	100
Ireland	4,701,000	46.3	81.1	98.1
Italy	82,134,000	41.7	73.6	91.4
Latvia	2,032,000	50.4	90.0	100
Lithuania	5,621,000	36.9	69.6	100
Luxembourg	689,000	46.0	80.1	100
Malta	348,000	52.0	100	n/a
Netherlands	17,126,000	50.5	72.7	87.6
Poland	36,334,000	33.7	67.4	100
Portugal	12,341,000	46.2	83.6	100
Romania	17,371,000	46.3	90.7	97.8
Slovakia	4,891,000	55.0	100	n/a
Slovenia	1,795,000	76.5	100	n/a
Spain	47,048,000	45.6	76.4	100
Sweden	10,362,000	44.4	78.3	95.0
UK	70,675,000	24.9	48.9	72.8

Source: Calculated by authors from data in Table 1.

One explanatory factor is clearly the existence of a fourth operator, although only in Germany, the Netherlands and the UK do these control more than 10 per cent of the market, and it has to be said that even attaining 10 million subscribes in the largest markets may not prove to be economic in the medium term.

If the calculation is extended to include the second-largest mobile operator in each market, then in a majority of Member States the mobile market was, to all intents and purposes, a duopoly with the percentage standing in excess of 80 per cent. Elsewhere, the two largest mobile operators still tended to control over 70 per cent of the market, with the only exceptions being Lithuania and Poland (both narrowly) and the UK. Where three or more mobile operators had been licensed, a considerable gap often existed between the number of mobile subscribers controlled by the second-largest operator and the number of subscribers controlled by the third-largest operator. In eleven Member States, the subscriber base of the third-largest mobile operator was more than half the size of that of the second-largest, but in a further eight cases it was less than half the size – notably so in the case of recent entrants Bulgaria and Romania. In the Czech Republic and Ireland the third-largest was almost exactly half the size of the second-largest, but the real anomaly was the UK with four almost equal-sized operators plus a respectably-sized 3G operator (3 UK).

It may also be noted that in the majority of Member States the most recent mobile operator to launch its service was also the one with the fewest subscribers. Although this was true in fifteen cases, it is surprising that only four of these could be found among the accession countries. In other words, the date when a mobile operator launched its services was more important in the EU15 than in the accession countries. For five Member States – Austria, Finland, Latvia, Slovenia and the UK – the gap between the smallest and the next-largest operator was less than one million subscribers, while for the remaining ten countries the gap was greater, sometimes considerably greater, than one million. For example, in Italy there was a gap of almost ten million subscribers between Wind, the last of the three 2G operators to launch its service, and Vodafone, the second-largest operator in the market. Gaps of more than two million subscribers could also be found in France (8.1 million subscribers), Germany (2.6 million), the Czech Republic (2.3 million) and Greece (2.1 million). In Belgium, Hungary, Ireland, Portugal and Spain the gap was less than one million subscribers.

Mobile Communication Markets in Accesson Countries

The first issue to address at this point is the extent to which mobile operators were EU-centric in respect of their geographical footprints, distinguishing between operators with a heavy presence in the EU-15 and those with a presence in the 12 accession countries.

Table 4 is drawn up so as to include those operators with licences in at least two accession countries. This is a modest enough total, but reflects the fact that only one operator, Vodafone, was present in more than four of the twelve. Even here, however, there is a need to distinguish carefully between operators with licences and those companies operating under other arrangements. For example, it is possible for an operator to act as a mobile virtual network operator (MVNO) by leasing spare capacity on an incumbent's network [Shin & Bartolacci, 2007]. Technically, the definition of an MVNO requires an operator to own its own switches and sell under its own brand, although there are also less rigorous ways to operate such as an enhanced service provider or simply as a reseller of another operator's branded service. The primary advocate of the MVNO approach has been Tele2 [Curwen & Whalley, 2007] although, as Table 4 shows, it has tended to prefer direct investment in networks in accession countries while operating as a MVNO in more established markets. For its part, Vodafone has preferred to negotiate Partner Network Agreements involving no investment [12], whereby the network in question is usually re-branded with the original operator's name hyphenated to that of Vodafone. By this means, Vodafone has enjoyed brand recognition without needing to lay out huge sums of money, and has been able to introduce its Vodafone live! portal with associated roaming benefits, while the network owner has enjoyed improved subscriber numbers and reduced churn because the Vodafone brand is more attractive than its own. In practice, therefore, Vodafone has a much greater brand recognition factor in accession countries than any other operator and is absent only in Slovakia.

The next in line is Deutsche Telekom's wholly-owned subsidiary T-Mobile. This is not surprising since the geographical position of Germany clearly lends itself to investment in countries close to its borders, many of which are accession countries (with possibly even more to come). It may also be noted that T-Mobile is present in the relatively large accession markets of Hungary, Poland and the Czech Republic. This is an important point because it is immediately noticeable that three of the big five EU incumbent mobile operators, Telefónica Móviles, O_2 (its sister company) [13] and Telecom Italia Mobile (TIM – which became a wholly-owned subsidiary of Telecom Italia in June 2005) do not appear in Table 4. For Telefónica Móviles in particular, this is ultimately a question of history, culture and language. Telefónica Móviles (and/or occasionally its parent although there is no longer any difference since it is currently a wholly-owned subsidiary) has long operated overseas, primarily in Latin America with the only exception prior to the end of 2004 being Morocco, its immediate southern neighbour. In other words, apart from a minor reciprocal stake in Portugal Telecom and some toying with 3G licences that had so far resulted in nothing other than fairly substantial write-offs, the company had zero interest in the EU let alone in accession countries. This strategy, it must be said, had served it well up to that point in time.

However, the Brazilian operation, Vivo, half-owned by Portugal Telecom, has struggled to show a profit [14] (and is currently being switched from CDMA to GSM at great expense), so Telefónica has decided to diversify back into Europe. It began with the acquisition of an initial 51.1 per cent stake in EuroTel of the Czech Republic in April 2005 [Roman & Rousek, 2005] – subsequently raised to 69.4 per cent. In March 2006, it completed the purchase of O_2 (which also has an independent network in the Isle of Man) [Telefónica, 2006] and thereby brought its proportionate subscriber numbers – that is, total subscribers multiplied by the ownership stake expressed as a percentage – within and outside Latin America into rough equivalence. Subsequently, it won a combined GSM/UMTS licence in Slovakia [15].

As for O_2 (changed from mmO_2 in March 2005), both prior to and after its divestment from what is now the BT Group it had spent a period of retrenchment involving the shedding of minority interests such that it remained operational in only Germany, the Netherlands and the UK (plus the Isle of Man). Even so, it has to be said that in its earlier life it was never really interested in the accession countries, preferring to get involved in South-East Asia and North America.

Telecom Italia and its then 56.1 per cent subsidiary TIM also had a fairly significant presence in Latin America at the end of 2004 although the great majority of proportionate subscribers were to be found in Brazil. Elsewhere, its presence in a single accession country, the Czech Republic, merely represented a tiny stake in the operator controlled by T-Mobile, and was the least significant of its overseas holdings bar Cuba. The stake was accordingly sold in March 2005 and the sale of its Greek operation shortly thereafter means that Telecom Italia is currently present in Europe only in its home market.

It is also be useful for the purposes of clarification to examine briefly the operations of mobile companies in what used to be termed Eastern Europe since only some of its constituent countries had become accession countries. As of 31 December 2003, four EU incumbents had a significant presence involving investment in Eastern Europe, namely Telenor, OTE, T-Mobile and TeliaSonera. OTE, interestingly, had stakes in Albania, Armenia, Bulgaria, Macedonia, Romania and Serbia, so it had not profited so far from accession nor was it about to do so during 2004 although it proved to be the main beneficiary in 2007. TeliaSonera's accession stakes were in practice entirely in the Baltic countries but its stakes to the east, in Azerbaijan, Georgia, Kazakhstan, Moldova and Russia were also in line to miss the accession boat. For its part, Telenor had twelve overseas interests but, interestingly, it was not focussed upon the Nordic/Baltic area being present in only Denmark, Norway and Sweden (as a MVNO), whereas it had stakes in Albania, Montenegro, Russia and the Ukraine in respect of which it would miss out on accession. T-Mobile accordingly stood out because it had stakes in four prospective accession countries, of which three (Czech Republic, Hungary and Poland) generated more than two million proportionate subscribers during 2003. In addition, it owned stakes in Belarus, Bosnia, Croatia, Macedonia and Russia. Given its limited overseas stakes within the EU at the time (consisting of Austria, the Netherlands and the UK), accession would serve to enlarge significantly its EU coverage even if the USA comfortably generated the third-largest number of proportionate subscribers after Germany and the UK (with Russia in fourth place).

Table 4. Operators present in at least two accession countries, 1 January 2007

	Vodafone	T-Mobile	Tele2	Telia Sonera	TDC	Orange	mobilkom	OTE
Austria	1	■	■[3]		■	■	■	
Belgium	1					■		
Bulgaria	1						■	■
Cyprus	1							
Czech R	■	■						
Denmark	1		■[3]	■	■			
Estonia	1		■	■				
Finland	1			■				
France	■		■[3]			■		
Germany	■	■			■[3]			
Greece	■							■
Hungary	■	■						
Ireland	■							
Italy	■			2				
Latvia	1		■	■	■			
Lithuania	1		■	■	■			
Luxembourg	1		■			2		
Malta	■							
Netherlands	■	■	■[3]			■		
Poland	■	■			■	■		
Portugal	■					■		
Romania	■					■		■
Slovakia		■				■		
Slovenia	1						■	
Spain	■			■		■		
Sweden	1		■	■				
UK	■	■				■		
Total Accession	5 + 6[1]	4	3	3	3	2	2	2
Total	14 + 12[1]	8	9	7	6	10	3	3

[1] Via Partner Network Agreement not involving direct investment.

[2] 3G licence only. No network.

[3] Trading as a MVNO. Sold provisionally in June 2007.

Source: Calculated by authors from data in Table 1.

What the above suggests is that there is a useful distinction to be made between the Baltic and Eastern European aspects of the 2004 accession – Cyprus (South) and Malta are of little significance because of their size and lack of potential for the entry of major operators. Taking the three Baltic accession countries as a whole, the eight operators listed in Table 4 generated eight entries but only three operators among them accounted for this – Tele2, TeliaSonera and TDC – although account should also be taken of Vodafone's three Partner Network Agreements. In contrast, the five broadly Eastern European countries generated ten entries. This was not a significant difference, so it is worth asking whether it resulted from the companies sampled. To answer this, we can return to the data in Whalley & Curwen (2003) which encompassed thirteen major European operators, and these reveal that increasing the sample size makes almost no difference when compared to Table 4. Of the ten accession countries, only two are affected at all by the altered size of the sample, namely Hungary where Telenor had a substantial stake and the Czech Republic where TIM had a very small stake. It is also possible to establish whether any significance can be attributed to the fact that two Nordic countries – Iceland and Norway – were not members of the EU. In practice, Iceland was not significant since the only EU operator there was Vodafone via a Partner Network Agreement, but in Norway we find (predictably) both Telenor and TeliaSonera (trading as NetCom GSM) as incumbents with Tele2 as a MVNO (although it had returned its 3G licence).

In summary, accordingly, the situation was as follows at the time of the 2004 accession: Vodafone had invested in accession countries in the former Eastern Europe (Group A) but had been keen to extend its footprint to the Baltic accession countries (Group B) without investing heavily. T-Mobile had heavily invested in Group A but was wholly disinterested in Group B. Orange was less involved in Group B but equally indifferent to Group A. TDC was slightly interested in both, while both TeliaSonera and Tele2 were heavily invested in Group B while wholly disinterested in Group A. Curiously, Telenor (not in Table 4) was the only Nordic operator acting in a wholly non-Nordic manner where accession was concerned.

The 2007 accession added one large operation in Romania to the EU holdings of both Orange and Vodafone but these operators were already EU-centric in terms of subscribers. OTE, as noted, gained two EU operations but one, CosmOTE in Romania, was very small, while mobilkom gained the largest operator in Bulgaria, MobilTel.

Expansion and Consolidation

Given the aforementioned differences in the countries in which the mobile operators identified in Table 4 have chosen to invest prior to the 2004 accession, an inevitable question to ask is whether the accession of twelve new Members States either has resulted so far, or is expected to result, in changes in their strategic priorities. In the first sub-section the focus is on T-Mobile, Orange and Vodafone whose ability to expand further is to a degree interlinked with one another, while the second sub-section concentrates on those other mobile companies with a presence in the accession countries.

Table 5. T-Mobile, European stakes, 31 December 2006

Country	Stake %	Proportionate subscribers
Austria	100	3,180,000
Bosnia	29.9	93,000
Croatia	51.0	1,101,000
Czech Rep.	60.8	3,070,000
Germany	100	31,398,000
Hungary	59.4	2,632,000
Macedonia	30.3	320,000
Montenegro	39.9	131,000
Netherlands	100	2,552,000
Poland	97.0	11,861,000
Slovakia	51.2	1,127,000
UK	100	16,905,000
Total world		99,476,000
Accession		*18,690,000*

Source: Compiled by authors from operator and other websites.

T-Mobile, Orange and Vodafone

If we begin with T-Mobile, then the strategic importance of the Eastern European countries to the company is clear for all to see in Table 5. Indeed, the CEO Kai-Uwe Ricke, basking in predictions of massive cash inflows during 2004, stated in May 2004 that 'Taking into account the EU's enlargement towards the east, we are placing a special focus on this region'. It is possible to calculate the importance of this region to T-Mobile as at 31 December 2003 when it had in total 68.7 million proportionate subscribers. Of these, 26.3 million were in Germany and 43.9 million in total in the pre-accession EU. Accession transferred a further 7.7 million to that total, yielding 51.6 million in total in the post-accession EU. The rest were largely accounted for by the USA (12.8 million) and Russia (3.4 million) with Croatia and Macedonia adding 0.8 million between them.

T-Mobile had a choice between moving into new countries and expanding into existing ones. In both cases, much depended upon the identity of existing shareholders and their willingness to sell. Faced with a cash offer above the market price, many shareholders might have been expected to succumb, but Deutsche Telekom's own shareholders were unlikely to sanction using up cash reserves to support a move into the likes of Moldova. Predictably, T-Mobile was not willing to fight for the 2G licence issued in Bulgaria in May

Hence, the probability was that T-Mobile would prefer to increase its existing stakes. In some cases, the purchase of additional equity would consolidate its existing control over the operator while in other cases the purchase could allow T-Mobile to take control of the operator for the first time. T-Mobile was particularly keen to acquire the 51 per cent of PTC it did not own in Poland, if only to keep one step ahead of Vodafone in a country with a modest penetration rate. Thus, its existing stakes provided T-Mobile with ample incentives and opportunities to continue its Eastern European-focused investment strategy. However, one intriguing prospect lay in the Czech Republic where, despite its majority stake in an incumbent, T-Mobile was alleged to be interested in acquiring EuroTel Praha via a bid for parent Ceský Telecom. Presumably, if it did so it would be forced to dispose of its existing network which was almost the same size, but this would get around the problem of trying to obtain full ownership of T-Mobile CZ.

It is fair to say that, by 2005, many major European operators had decided that non-controlling stakes were often more trouble than they were worth. This would be particularly true of stakes in countries such as Russia where the rule of law could best be described as shaky. Hence, Deutsche Telekom's decision to sell its 10.2 per cent stake in Russia's MTS – and with it its indirect stakes in the likes of Belarus, the Ukraine and Uzbekistan – which was completed in September 2005, was not unexpected. Its travails in Poland, where it embarked upon an immensely long-winded trawl through the law courts of Europe in order to determine whether Vivendi Universal or itself could lay claim to the other half of the shareholding in PTC, were not entirely of its own choosing but did, however, appear to have turned out well in the end since it was seemingly able to take control of the shares at well below their market value in November 2006 – although Vivendi has yet to accept this apparent *fait accompli* [16].

2005 saw other activity in the EU and prospective accession countries, but to less good effect. For example, talks about a joint bid for O_2 with KPN fell through and Deutsche Telekom declined to make a bid for Turk Telekom. In August, the CEO stated that the operator was not actively seeking new acquisitions, but would continue to evaluate any opportunities arising within its existing footprint. These proved to be fairly plentiful during 2006 when Deutsche Telekom failed to win Serbia's Mobi 63 and expressed an interest in a licence in Romania while its Hungarian subsidiary expressed an interest in acquiring stakes in Bosnia, Romania, Serbia and the Ukraine. Very recently, Deutsche Telekom has indicated that it may bid for a 20 per cent stake in OTE which is being privatized during 2007 although it will not be prepared to over-pay for a minority stake. This is especially the case given that, in late January, Deutsche Telekom announced its second profit warning in six months and embarked upon a massive cost-cutting exercise [Williamson, 2007].

But would any of the other mobile operators identified in Table 4 follow T-Mobile and respond to accession by increasing their geographical coverage? Vodafone had a presence of one kind or another in all but three of the ten accession countries at the beginning of 2004, but given that it had chosen to use Partner Network Agreements as a substitute for direct investments in many cases, the scope for it to invest in more of these markets was actually quite limited. Of the markets where Vodafone was not active as an investor, the most significant was the Czech Republic. Of the three Czech GSM operators, two – Ceský Mobil and EuroTel Praha – were potential acquisitions. The third operator, T-Mobile CZ, was majority-owned by Deutsche Telekom and thus unavailable unless, as noted above, T-Mobile was forced to sell it. In principle, EuroTel Praha could also be dismissed as an acquisition target of Vodafone since it was possibly being targeted by T-Mobile and, in any event, was a subsidiary of the incumbent PTO which was most unlikely to want to be split off from it mobile operations. However, Vodafone had recently indicated that it was willing, and had the financial resources, to acquire the entire operation.

Ceský Mobil was owned by Telesystem International Wireless, which was possibly prepared to sell its 96.4 per cent stake if the price was sufficiently attractive. Nevertheless, it remained to be seen whether Vodafone's shareholders would be prepared to countenance the comparatively expensive acquisition of the smallest of the Czech Republic's GSM operators. Given the existing investment, and the 16 per cent market share of Ceský Mobil, a more attractive course of action looked to be to enter into a Partner Network Agreement. However, Vodafone duly bought 99.9 per cent of Ceský Mobil in May 2005 which it has re-branded as Vodafone CZ.

As for Latvia and Slovakia, both were comparatively small. As a consequence, it was more likely that Vodafone would enter these markets through the use of Partner Network Agreements rather than an equity investment. However, this assumed that the existing GSM operators would enter into such an arrangement. So far as Latvia was concerned, this was highly unlikely given who owned the two existing operators; Baltkom was owned by Tele2 while LMT was jointly owned by the Latvian state (51 per cent) and TeliaSonera (49 per cent). It seemed inconceivable that either Tele2 or TeliaSonera would sign a Partner Network Agreement with Vodafone as this would expand the brand recognition of their main competitor in the Baltic States. In practice, Vodafone achieved its aim indirectly when Bité GSM, the Lithuanian subsidiary of TDC and a Partner Network of Vodafone, launched a network in Latvia in 2005 [Vodafone, 2006a].

The situation in Slovakia was a little more complicated, not least because Vodafone's ability to enter this market was also dependent on the strategic priorities and intentions of Orange. Orange had only a limited exposure to the mobile markets of the ten accession countries, with a presence across the EU that was increasingly skewed in favour of Western Europe. Orange already possessed two accession country mobile investments; in Poland, where it – or strictly its parent France Télécom – was a majority shareholder in PKT Centertel, and Slovakia where it owned 63.9 per cent of Orange Slovensko, the largest operator. These two investments were, however, somewhat detached from the other investments that Orange had made. Their relative peripherality was further reinforced when subscriber numbers are taken into account; Poland and Slovakia accounted for less than 10 per cent of the wider European subscriber base of Orange. Orange had also invested in Romania, a country that expected (correctly) to be among the next wave of accession countries, though when these subscribers were also included the three countries still only accounted for roughly 17 per cent of the European subscriber base. In contrast, France and the UK, which were the two largest mobile markets of Orange, accounted for over 75 per cent of its European subscriber base as shown in Table 6.

However, given that Orange was committed to consolidation based upon countries where it held majority stakes and hence control, preferably in conjunction with a top-two ranking, and that parent France Télécom's short-term need for cash had abated somewhat, a wholesale withdrawal from the former Eastern Europe no longer seemed likely. Indeed, Orange moved to mop up the minorities in Romania [Middleton, 2005] and Slovakia[17] during 2005 although the situation in Poland was much harder to disentangle despite a re-branding to Orange in August [France Télécom, 2005; Halaba, 2005; McQuaid, 2005]. Nevertheless, Orange was not successful in building up any further presence in accession/potential accession countries: In March 2005, it was part of a consortium that bid unsuccessfully for 51.1 per cent of EuroTel Praha; it withdrew from the bidding for Turkey's Telsim in December 2005; and it failed to win Mobi 63 of Serbia in July 2006.

Table 6. Orange, European stakes, 31 December 2006

Country	Stake %	Proportionate Subscribers
Austria	17.5	362,000
Belgium	50.2	1,588,000
France	100	24,109,000
Liechtenstein	100	neg.
Netherlands	100	2,115,000
Poland	43.9	5,497,000
Portugal	19.2	2,023,000
Romania	96.8	7,786,000
Slovakia	100	2,690,000
Spain	77.7	8,636,000
Switzerland	100	1,367,000
UK	100	15,333,000
Total world		80,580,000
Accession		*15,973,000*

Source: Compiled by authors from operator and other websites.

Table 7. Vodafone, European stakes, 31 December 2006

Country	Stake %	Proportionate Subscribers
Albania	99.9	919,000
Czech Rep.	100	2,413,000
France	43.9	7,851,000
Germany	100	30,622,000
Greece	99.8	4,955,000
Hungary	100	2,134,000
Ireland	100	2,178,000
Italy	76.9	20,129,000
Malta	100	188,000
Netherlands	99.9	3,817,000
Poland	19.6	2,324,000
Portugal	100	4,618,000
Romania	100	7,717,000
Spain	100	14,464,000
UK	100	16,939,000
Total world		198,584,000
Accession		14,776,000

[1] The Nordic region – Denmark, Norway and Sweden – is credited with 4,292,000 subscribers.

Source: Compiled by authors from operator and other websites.

Vodafone would potentially have been interested in acquiring in the majority of Orange Slovensko owned by Orange had Orange been forced to sell it, but Vodafone's shareholders were unlikely to be willing to support an acquisition that would add a comparatively small number of subscribers in a market where growth expectations were limited – Slovakia's population was only 5.4 million and already heavily penetrated – but there were attractive investments with more potential elsewhere. One such was Romania where Vodafone was already a minority shareholder in Connex and whose population was four times that of Slovakia. Telesystem International Wireless (TIW), holder of a 63.6 per cent stake, agreed in July 2004 to buy all or part of the 14.4 per cent stake held by Deraso Holdings. In May 2005, Vodafone acquired this entire stake [Krosnar, 2005a], raising its total holding to 99.1 per cent and subsequently to 100 per cent before re-branding as Vodafone Romania in May 2006. Outside of such an acquisition, Vodafone was unlikely to expand into new markets other than through Partner Network Agreements.

The remaining accession country of interest was Poland where Vodafone already held a 19.6 per cent stake in Polkomtel (Plus GSM) as did TDC and various local partners. Vodafone attempted to acquire a further 23 per cent stake in July 2005 [18] but without success, and also expressed interest in the TDC stake when it was put on the market in December 2005 although this has degenerated into yet another legal battle over who has what rights in relation to the sale [19].

In September 2005, Vodafone CEO Arun Sarin stated that he was now broadly happy with the company's European profile but would be bidding to buy Turkey's Telsim at the year-end [Krosnar, 2005b]. However, in October, Vodafone agreed to sell its 100 per cent stake (yielding 1.5 million subscribers) in Sweden to Telenor (replacing it with a Partner Network Agreement) [Vodafone, 2005], and it followed up with the sale of its 25 per cent stake in Belgacom and its 25 per cent stake in Swisscom to their majority owners in November 2006 and December 2006 respectively [Smith & Simonian, 2006; Vodafone, 2006b]. Neither produced a lot of subscribers (Table 7), but it is of interest that the overall effect was to re-balance Vodafone's European holdings towards the accession countries, a factor enhanced by the agreed purchase of Telsim in December [Boland & Edgecliffe-Johnson, 2005] [20] given Turkey's pressure to become an accession country.

Other Operators

A further company with a presence in the accession countries is Tele2 (see Table 8). Although Tele2 operated in nine EU Member States at the end of 2006 – it subsequently sold its mobile operation in the Netherlands to Versatel in May 2007 and is negotiating the sale of the assets in France – it has made just three investments in the accession countries, namely in Estonia, Latvia and Lithuania, and all prior to the 2004 accession. In other words, as a factor in its basic strategy rather than on account of accession per se, Tele2 has invested in the Baltic States that complement geographically its presence in the nearby Nordic States although none of these Baltic networks yields significant numbers of subscribers. Table 8 also draws attention to a second characteristic of Tele2's investment strategy; namely its tendency to use MVNO arrangements to enter new markets. Of the nine mobile investments in the post-accession EU set out in Table 6, almost half were as a MVNO. Of the five networks owned by Tele2, only one, in Luxembourg, could be found outside of Sweden and the Baltic States. Thus, the geographical preference in terms of ownership is marked as is the preference for control – only in Sweden, where it has an 87.3 per cent stake, does Tele2 not own the entire company although it only has a 51 per cent stake in Croatia, a potential accession country. In addition to the above, it may be noted that Tele2 failed to acquire 3G licences in Hungary (2004) and Bulgaria (2005) and to buy Serbia's Mobi 63 in July 2006.

What of the other companies identified in Table 4? For different reasons, neither TDC nor TeliaSonera was likely to expand its geographical footprint as a result of EU expansion. At the end of 2003, TDC had only two remaining investments in accession countries – in Bité in Lithuania and Polkomtel in Poland – having sold in the course of that year its holding in the Czech Republic (as well as in the Ukraine) Hence, it did not appear to see the former Eastern Europe as other than providing opportunities for financial investments. In any event, any additional investments by TDC in accession countries could be ruled out until the uncertainty over its own future was resolved. In mid-2004, SBC Communications Inc. sold 32.1 per cent of the 41.6 per cent of TDC that it owned, but to financial institutions rather than to another operator. Pending the completion of this sale, TDC stated that it would not enter into any negotiations regarding potential partnerships or strategic transactions at Group level. Moreover, these would only resume once the new board had been able to conduct a strategic review of the company.

Table 8. Tele2, European stakes, 31 December 2006

Country	Stake %	Proportionate Subscribers
Austria	100	174,000
Croatia	51.0	158,000
Denmark	100	n/a[1]
Estonia	100	536,000
France	100	417,000
Latvia	100	1,025,000
Liechtenstein	100	15,000
Lithuania	100	1,708,000
Luxembourg	100	235,000
Netherlands	100	577,000
Norway	100	n/a[1]
Sweden	87.3	3,062,000
Switzerland	100	27,000
Total world		18,679,000
Accession		*3,269,000*

[1] The Nordic region – Denmark, Norway and Sweden – is credited with 4,292,000 subscribers.

Source: Compiled by authors from operator and other websites.

Table 9. TDC, European stakes, 31 December 2006

Country	Stake %	Proportionate subscribers
Austria	15.0	312,000
Denmark	100	1,202,000
Germany	100	3,742,000
Latvia	100	204,000
Lithuania	100	1,839,000
Norway	80.0	11,000
Poland	19.6	2,324,000
Switzerland	100	1,361,000
Total world		11,074,000
Accession		*7,333,000*

Source: Compiled by authors from operator and other websites.

Its subsequent strategy has been less than clear-cut. For example, in May 2005 Lithuania's Bité acquired a licence in Latvia and launched its network in September, but TDC also failed to win the 3G licence in Hungary in December 2004 and withdrew from the bidding for EuroTel Praha in 2005, so it clearly has no particular interest in the accession countries per se. Indeed, Bité was provisionally sold to Mid Europa Partners in January 2007 in line with the view expressed by the CEO in November 2006 that 'everything outside of Denmark isn't core business...if somebody comes with a price, and the price is right, it will be sold'.[21] Furthermore, the (so far frustrated) desire to sell its stake in Poland has been mentioned previously, so taken with its strong interest in expanding in Finland, Norway and Sweden, it is fair to say that TDC is if anything retreating back to its Nordic roots as shown in Table 9.

Table 10. TeliaSonera, European stakes, 31 December 2006

Country	Stake %	Proportionate Subscribers
Denmark	100	1,123,000
Estonia	51.6	408,000
Finland	100	2,407,000
Latvia	60.3	484,000
Lithuania	100	2,074,000
Norway	100	1,641,000
Spain	76.6	24,000
Sweden	100	4,603,000
Total world		42,110,000
Accession		*2,966,000*

Source: Compiled by authors from operator and other websites.

For its part, TeliaSonera has for many years been one of the most widespread operators in the mobile world, deriving more proportionate subscribers from Russia and Turkey than from Sweden at the time of the 2004 accession. With the exception of the three Baltic States, none of which understandably produced large numbers of subscribers, TeliaSonera had no other mobile investments in accession countries in 2004, nor did it gain any in 2007, although with its stake in Turkcell and, via jointly-held Fintur Holdings, in countries to the east of the former Eastern Europe, it might end up with further EU holdings during the next stages of accession.

Telia and Sonera, prior to their merger, did take advantage of the 3G licensing process to enter Germany, Italy and Spain, three of the largest Western European markets. However, there followed a period of post-merger repentance involving the writing off of the investments in all three markets and these barely signify in Table 10. Interestingly, despite its widespread empire, TeliaSonera tended to think of itself as the Nordic and Baltic telecommunications leader, but although this might simply have been an appropriate description of its market position in these two regions, and lacking a 'local' partner to offset the risk inherent in investing in the non-Baltic 2004 accession countries, it became possible that TeliaSonera would sell some of its overseas investments, leaving it predominantly as a Nordic/Baltic operator. The April 2004 offer by TeliaSonera to take outright control of Eesti Telekom, although unsuccessful [Brown-Humes, 2004], together with the acquisition of the outstanding 10 per cent of Lithuania's Omnitel in August 2004 [22], the purchase of Orange Denmark in October 2004 and the subsequent sales of stakes in Hong Kong and Namibia, reinforced the feeling that its strategic priorities lay in the Nordic/Baltic Member States and not elsewhere. Nevertheless, TeliaSonera was set to be debt-free by the end of 2004, and had a substantial war chest for acquisitions, so a contraction of its international footprint was not a foregone conclusion [George, 2004].

It is worth noting that the Finnish government appeared to have agreed to the effective takeover of Sonera by Telia on the understanding that TeliaSonera would pursue a strategy of growth. Ultimately, because TeliaSonera stated in June 2004 that its ambition was to take majority control of its foreign investments, and given the size of the proportionate subscribers involved, its strategy was dependent primarily upon its relationship with its main partners. For example, the relationship between TeliaSonera and Turkcell's largest shareholder, Çukurova, had at times been fraught [23] – Çukurova's stake was confiscated by the government in 2003 as collateral against debts and was about to be returned in stages commencing in July 2004 – and the situation in Russia was permanently unsettled. Such problems are usually addressed either via a takeover or a withdrawal. It is significant that, in late June 2004, the Finnish deputy CEO of TeliaSonera, with responsibility for pursuing the purchase of majority stakes in Turkcell and MegaFon, was dismissed by the Swedish CEO (George, 2004). At the very least, this indicated that TeliaSonera would not 'overpay' to take control, but to remain a permanent minority investor hardly seemed an attractive proposition as TeliaSonera was prepared to acknowledge.

In the event, TeliaSonera has acted in a somewhat conservative manner of late. It is hopeful of raising its stake in Latvia to 100 per cent [24], but failed to win the third GSM licence in Slovakia in August 2006 and it withdrew from the bidding for Mobi 63 in July 2006. Against all the odds, not to mention analysts' reservations, TeliaSonera finally decided to go ahead with its launch of Xfera, its 3G licensee in Spain, in June 2006 [Roman & Hanson, 2006]. The CEO of TeliaSonera explained that this was really not a dramatic change from before when TeliaSonera was exclusively looking for acquisitions in, or close to, its existing markets in the Nordic, Baltic and Eurasian regions [Hansson, 2006]. Rather, with an extension into continental Europe, the company was 'expanding our thinking a bit'. However, it is unlikely that this will turn out to mean that TeliaSonera intends to buy stakes in the non-Baltic accession countries, although there is currently much discussion surrounding a possible merger with Telenor which has networks in Hungary, Montenegro, Norway, Serbia and Sweden.

As shown in Table 11, mobilkom Austria has always tended to look to the east for overseas acquisitions, of which it had three at the end of 2003 among which only Croatia provided more than a modest number of proportionate subscribers. However, since the 2004 accession its record of expansion has been patchy. On a positive note, parent Telekom Austria exercised an option to buy the whole of Bulgaria's MobilTel in July 2005 [25], and won the third GSM licence in Serbia in November 2006 [McDonald, 2006]. In addition, in Slovenia it was able to raise its stake from 75 per cent to 92.2 per cent and then to 100 per cent in May 2006. In contrast, mobilkom made an unsuccessful bid for 51 per cent of Telekom Montenegro (owner of MoNet) in March 2005 and, although mobilkom then provisionally agreed to acquire a 49 per cent stake in Serbia's Mobi 63, the deal fell through, and mobilkom was also unsuccessful when the stake was re-offered in July 2006. Furthermore, mobilkom failed to win a 65 per cent stake in Telekom Srpske in Bosnia in 2006 and to win a licence in Slovakia in July of that year.

Despite these setbacks, mobilkom continues to aspire to become the leading player in south-eastern Europe [Lenningham, 2005; Simonian & Yuk, 2006] and to that end managed to win the third GSM licence in Macedonia in February 2007 [26] – albeit by virtue of being the sole bidder – and is pursuing licences in Kosovo and Romania as well as a stake in OTE. Overall, therefore, while a minor player during the 2004 accession and a relatively significant one in 2007, it may be that mobilkom will play a more important role in future accessions.

Finally, OTE has always been strategically focused on the Balkans, as shown in Table 12. Hence, although the 2007 accession coincidentally meant that OTE was now present in two accession countries and three Member States, this was not the result of strategic decisions related to accession. Given its profile, OTE may end up with further accession candidates.

Table 11. Mobilkom, European stakes, 31 December 2006

Country	Stake %	Proportionate subscribers
Austria	100	3,630,000
Bulgaria	100	4,270,000
Croatia	100	1,910,000
Liechtenstein	100	5,000
Slovenia	100	421,000
Total world		10,234,000
Accession		*4,691,000*

Source: Compiled by authors from operator and other websites.

Table 12. OTE, European stakes, 31 December 2006

Country	Stake %	Proportionate subscribers
Albania	55.0	545,000
Bulgaria	66.7	2,182,000
Greece	66.7	3,480,000
Macedonia	66.7	315,000
Romania	62.9	771,000
Serbia	20.0	806,000
Total world		8,100,000
Accession		*2,953,000*

Source: Compiled by authors from operator and other websites.

Conclusion

The above discussion has focused on the ownership of mobile licences in the enlarged EU. In the course of this a distinction has been made between the original 15 Member States and the twelve accession countries that joined in May 2004 and January 2007. Drawing such a distinction allowed those mobile operators with a presence in the accession countries to be differentiated from those that did not.

The first conclusion that can be drawn is that the largest multiple owners of mobile licences identified by Whalley & Curwen (2003) have, with the exception of Vodafone, only a limited presence in the mobile markets of the twelve accession countries. Both Tele2 and TeliaSonera have focused on the Baltic States while Deutsche Telekom has concentrated its attention on those Eastern European markets that either border, or are close to, its home market. This is not particularly surprising since liberalization offered so many opportunities to expand into the other Member States of the pre-accession EU, and the costs of licence acquisition plus network roll-out were extremely burdensome.

Ten of the 12 accession countries applied to join the EU between 1994 and 1996, with Cyprus and the Czech Republic being the two exceptions that applied much earlier in 1990 [27]. The adoption of EU legislation by the applicant countries liberalized their telecommunications markets so that on the one hand foreign investment was possible while on the other they instituted a regulatory regime that would reassure investors that they would not be discriminated against. As a consequence, the formal accession date did not trigger a wave of foreign investment by mobile operators as they had already responded to inward investment opportunities as they arose. Thus, the second conclusion that can be drawn is that the expansion of mobile operators into accession countries began before they formally joined the EU.

That said, it was reasonable to expect that once the formal accession dates had been announced, the strategic interest of operators could have been refreshed. However, not only were opportunities comparatively limited due to the aforementioned pre-accession inward investment but many operators were struggling with the fall-out from the telecommunications-media-technology (TMT) collapse that became very marked in 2002.

Few accordingly had the wherewithal, let alone the will, to expand into the accession countries. One possible expansionary candidate was Vodafone, given its resources and strategy based on its international footprint, while alternative contender Orange was forced to retrench to the point that it became, to all intents and purpose, a Western European-focused operator with a presence in an increasingly scattered set of markets.

While the need to raise capital for its parent has abated, Orange, like TeliaSonera, is no longer interested in playing bit parts and wants to be either a major player or to exit. Exit is nevertheless easier said than done because of the shortage of buyers. Insofar as stake-building is concerned, it does appear to be far more likely that operators will seek to consolidate their positions in existing markets through purchasing additional equity in companies where they already own a stake. Taken together, these points suggest a third conclusion, namely that little change can be expected in accession countries as those operators wishing for whatever reason to exit are unlikely to find buyers while those wishing to consolidate their position in existing markets will be unwilling to pay the premium sought.

The limited future scope for structural change has implications for the competitiveness of mobile telecommunications markets. Most accession markets are effective duopolies, with the largest two operators controlling 70 per cent or more of the market between them. One consequence of this is that the ability of the third or fourth operators in the market to bring about the competition-derived benefits associated with liberalization is limited, while another is that any would-be investor into the market is stuck with the choice either of investing in one of the smaller operators or to become part of the duopoly. Neither is an attractive option, though for different reasons: The revenue and growth potential of the former is likely to be limited while the latter will involve paying a premium.

The unattractiveness of the investment decision may also be compounded by the relatively small size of many accession countries and, by extension, their mobile telecommunication markets. In this respect, it is significant that although Vodafone is present in seven accession countries, it does not own a network in all seven markets. Indeed, Vodafone owns a network in just two markets – Malta and Poland – and is present in the other five through the use of Partner Network Agreements. Those markets where Vodafone has used these Agreements are all characterised by their small size. The Agreements have enabled Vodafone to offer its services in new markets independently of owning a network and to retain a presence in those markets like Sweden that it has exited. As such, they are a key component of Vodafone's pan-EU footprint. When this observation is combined with the propensity of Tele2 to use MVNO arrangements to enter markets, a final conclusion is that multiple licence owners are using a wider variety of entry modes than was previously the case.

Notes

[1] See, for instance, Bartle (2005: 68-71) or Thatcher (2001: 563ff) for an overview of the Directives introduced as a consequence of the Green Paper.

[2] Five out of the EU's 15 Member States were granted derogations from this. Luxembourg was given until January 2000 due to the small size of its network while the other four Member States - Greece, Ireland, Portugal and Spain – were given until January 2003 (Curwen, 1997: 57). Having said this, all of the Member States

liberalised their markets before the extended deadline with Greece, in January 2001, being the final country to do so (Walden, 2005: 122).

[3] For example, Member States were required to ensure that the terms and conditions of interconnection were published and that number portability was introduced - see, Thatcher (2001: 568-570) for a detailed discussion of the actions taken by the EU.

[4] Eliassen and Sjøvaag (1999) highlight some of the regulatory differences between Member States in the EU.

[5] For details of the six Directives see, for example, Bartle (2005: 72f).

[6] See, Stumpf (2006) or Walden (2005) for a discussion of the new regulatory framework.

[7] The 1993 Copenhagen European Council identified criteria for EU membership. Applicant countries must possess stable institutions, have functioning market economies with the capacity to withstand competitive pressures from elsewhere within the EU and be able to assume the obligations of membership (Blair, 2006: 156). To these three criteria Friis (2003: 182) adds a fourth, namely, that the EU must be able to absorb the Member State without interrupting its continued integration.

[8] Walden (2005: 107) estimates that since 1984 more than 100 different pieces of legislation have been adopted across the telecommunications industry.

[9] See Friis (2003: 183) for a full list of the areas where negotiations occur or Barnes and Barnes (2007) for a description of the access process. Friis implies that the potential for telecommunications-centred disagreements and delays is less than for other areas such as agriculture, energy or regional policy.

[10] This should not, however, be taken as implying that there are no differences between the accession countries and original 15 Member States of the EU. Bak and Szczesniak (2007) identify two areas where differences remain even though many barriers to foreign investment were removed when they joined the EU.

[11] GSM is a generic term encompassing an uplink at 890-915 MHz combined with a downlink at 935-960 MHz and an uplink at 1710-1785 combined with a downlink at 1805-1880 MHz, where the latter are also known as PCNs. PCNs were typically licensed after 1997 either to ameliorate spectrum shortages in GSM900 or to permit the entry of new operators. Fortunately, what was then Eastern Europe chose almost universally to adopt the same spectrum bands and technologies as were being enforced across the EU. Eastern European countries also chose UMTS for 3G, which initially required the use of 1885-2025 MHz combined with 2110-2200 MHz, although other spectrum bands have subsequently been added.

[12] See, for example, the newsletter released by Vodafone on 22 February 2006 relating to its new Agreement with Bulgaria's Mobiltel.

[13] BT was certainly the mobile incumbent in respect of analogue telephony but, somewhat surprisingly, Vodafone was the first operator to launch a GSM service (in July 2002) and hence has a claim to be known as the GSM incumbent. mmO_2, as it was known after being hived off from BT, launched only in January 2004 and the close proximity of the other three launches after that of Vodafone explains why, uniquely, the UK has four mobile operators of roughly the same size.

[14] See, for example, 'Vivo shares fall 39% since merger', available at www.cellular-news.com of 10 August 2006.

[15] See 'Telefónica O_2 Slovakia wins mobile licence with lowest bid', available at www.telegeography.com of 3 August 2006.

[16] See, for example, 'Elektrim says Deutsche Telekom now controls PTC', available at www.cellular-news.com of 19 June 2006; 'Vivendi offers €2.5 billion for Deutsche Tel PTC stake', available at www.cellular-news.com of 26 September 2006; 'Polish court backs Vivendi appeal in PTC row', available at www.cellular-news.com of 19 January 2007 and 'Vivendi gains ground against Deutsche Telekom in PTC case', available at www.communicationsdirectnews.com of 15 August 2007.

[17] See 'France Telecom buys Orange Slovakia minority holders', available at www.cellular-news.com of 20 October 2006.

[18] See 'Vodafone, TDC currently hold 19.61% in Polkomtel', available at www.totaltele.com of 25 July 2005.

[19] See, for example, 'Vodafone files to block TDC sale of Polkomtel shares', available at www.telegeography.com of 28 February 2006.

[20] The purchase was completed in May 2006.

[21] See 'TDC CEO said telco may sell operations outside Denmark', available at www.totaltele.com of 1 November 2006.

[22] See 'TeliaSonera pays $63.5 million for rest of Lithuanian Omnitel', available at www.totaltele.com of 17 August.

[23] And remained so – see, for example, 'TeliaSonera launches new suit against Çukurova', available at www.telegeography.com of 22 August 2005, 'TeliaSonera initiates legal action against Russian Alfa Group', available at www.telegeography.com of 28 November 2005, and Ostrovsky & Bergstrom (2005).

[24] See 'TeliaSonera given permission to buy LMT, but not Lattelekom', available at www.telegeography.com of 11 May 2006, and Kaza & Hansson (2006).

[25] See 'Telekom Austria closes purchase of Mobiltel', available at www. telecomdirectnews.com of 14 July 2006.

[26] See 'Macedonia selects Austrian Mobilkom as third mobile operator', available at www.telegeography.com of 5 February 2007.

[27] For details of the application and subsequent progress of the twelve accession members states see, for example, ec.europa.eu/enlargement.

References

Bak, M., Kulawczuk, P., & Szczesniak, A. (2007). Institutional and regulatory barriers to Foreign M&A in Central Europe. In, K. Meyer. & S. Estrin (Eds.), *Acquisition Strategies in European Emerging Markets.* Basingstoke: Palgrave.

Barnes, I., & Barnes, P. (2007). Enlargement. In M. Cini (Ed.), *European Union Politics*, Second Edition. Oxford: OUP.

Bartle, I. (2005). *Globalisation and EU Policy-making. The neo-liberal transformation of telecommunications and electricity.* Manchester: Manchester University Press.

Blair, A. (2006) *Companion to the European Union,* Routledge, England.

Boland, V., & Edgecliffe-Johnson, A. (2005). Win in Telsim auction hurts Vodafone stock. Available at http://news.ft.com of 13 December 2005.

Brown-Humes, C. (2004). TeliaSonera in cash offer for Eesti Telekom. Available at http://news.ft.com of 14 April.

Cottrell, R. (2003). A survey of European enlargement. When east meets west. *The Economist*, 22 November.

Curwen, P (1997). *Restructuring Telecommunications*, Basingstoke: Palgrave.

Curwen, P. (2002). *The Future of Mobile Communications: Awaiting the third generation.* Basingstoke: Palgrave.

Curwen, P., & Whalley, J. (2007). Tele2 and the strategic role of virtual operations, *info*, **9**, 54-69.

Dinan, D. (2006). Governance and institutional developments: In the shadow of the constitutional treaty. *Journal of Common Market Studies*, **44**, 63-80.

Eliassen, K., & Solvang, M. (1999). *European Telecommunications Liberalisation.* London: Routledge.

France Télécom (2005). France Telecom to take full ownership of 47.5% of TPSA. Press release of July 27.

Friis, L. (2003). EU enlargement...and then there were 28". In E. Bomberg & A. Stubb (Eds.), *The European Union: How does it work?* Oxford: OUP.

George, N. (2004). Shadows over an unhappy Nordic marriage. Available at http://news.ft.com of 29 June.

Grabbe, H., & Hughes, K. (1998). *Enlarging the EU Eastwards.* London: The Royal Institute of International Affairs.

Halaba, M. (2005). Poland's TPSA launching Orange brand on 19 September. Available at www.totaltele.com of 15 September.

Hansson, M. (2006). TeliaSonera extends acquisition strategy to Europe. Available at www.totaltele.com of 25 April.

Hitran, A. (2006). Romania – a vibrant market on the doorstep to the European Union. In J. Kittl, E. Lichtenberger, E-O. Ruhle & C. Schleps (Eds.), *Telecommunications Markets in Central and South Eastern Europe. Market Developments and Regulatory Frameworks*, Germany: EUL Verlag, Köln.

Kaza, J., & Hansson, M. (2006). TeliaSonera aims to end impasse on Latvian ownership. Available at www.totaltele.com of 8 March.

Krosnar, K. (2005a). TIW to complete Oskar, MobiFon sales by end June. Available at www.totaltele.com of 20 May.

Krosnar, K. (2005b). Vodafone looking to expand further in Europe, Asia – CEO. Available at www.totaltele.com of 30 September.

Lenningham, M. (2005). Mobilkom plans €2bn trip down Danube. Available at www.totaltele.com of 16 February.

McDonald, N. (2006). Telekom Austria wins Serbian licence. Available a www.ft.com of 8 November.

McQuaid, D. (2005). TPSA to start rebranding mobile unit to Orange 1 September. Available at www.totaltele.com of 30 August.

Middleton, J. (2005). France Telecom increases Romanian holding for US$523 million. Available at www.telecoms.com of 14 April.

Ostrovsky, A., & Bergstrom, R. (2005). Alfa jumps in with Turkcell bid. Available at http://news.ft.com of 31 March.

Peet, J. (2007). Fit at 50? A special report on the European Union. *The Economist.* 17 March.

Rachman, G. (2001). A survey of European enlargement. Europe's magnetic attraction. *The Economist.* 19 May.
Roman, D., & Hansom, M (2006). TeliaSonera move to Spain rattles local operators. Available at www.totaltele.com of 16 June.
Roman, D., & Rousek, L. (2005). Telefonica outbids rivals with Cesky Telecom offer. Available at www.totaltele.com of 31 March.
Sedelmeier, U., & Young, A. (2006). Crisis, what crisis? Continuity and normality in the European Union in 2005. *Journal of Common Market Studies*, **44**, 1-5.
Shin, D., & Bartolacci, M. (2007). A study of MVNO diffusion and market structure in the EU, US, Hong Kong and Singapore. *Telematics and Infomatics*, **24**, 86-100.
Simonian, H., and Yuk, P. (2006). Telekom Austria vows to continue Balkan search. Available at www.ft.com of 24 August.
Smith, G., & Ionian, H. (2006). Vodafone sells Swisscom Mobile stake. Available at www.ft.com of 19 December.
Stumpf, U. (2006). Markets susceptible to ex ante regulation: methodology and commission recommendations. *Communications & Strategies*, **64**, 41-60.
TDC (2004). SBC intends to sell TDC shares. Available at http://tdc.com/about/press/releases.
Tele2 (2003). *Annual Report 2003.* Stockholm: Tele2.
Telefónica (2006). Nota de Prensa: Telefónica successfully completes the acquisition of O_2. 23 January.
TeliaSonera (2003). *Annual Report 2002.* Stockholm: TeliaSonera..
Thatcher, M. (2001). The Commission and national governments as partners: EC regulatory expansion in telecommunications 1979-2000. *Journal of European Public Policy*, **8,** 558-584.
Vodafone (2005). Sale of Vodafone Sweden. Available at www.vodafone.com on 31 October 2005.
Vodafone (2006a). Vodafone welcomes Latvia to its global community. Available at www.vodafone.com of 11 April.
Vodafone (2006b). Sale of 25% interest in Proximus. Available at www.vodafone.com of 25 August.
Walden, I. (2005). European Union Communications Law. In I. Walden & J. Angel (Eds.), *Telecommunications Law and Regulation.* Oxford: OUP.
Whalley, J., & Curwen, P. (2003). Licence acquisition strategy in the European mobile communications industry. *info*, **5**, 45-57.
Whalley, J., & Curwen, P. (2006). Third generation new entrants in the European mobile telecommunications industry. *Telecommunications Policy*, **30** , 622-632.
Williamson, H. (2007). Deutsche Telekom hit by profits warning. Available at www.ft.com of 28 January.

In: Mobile Telephones…
Editors: A.C. Harper and R.V. Buress, pp. 107-149
ISBN: 978-1-60456-436-5

Chapter 3

MOBILE TELEPHONY RADIATION EFFECTS ON LIVING ORGANISMS

Dimitris J. Panagopoulos* and Lukas H. Margaritis
Department of Cell Biology and Biophysics, Faculty of Biology, University of Athens, Panepistimiopolis, 15784, Athens, Greece

Abstract

A number of serious non thermal biological effects, ranging from changes in cellular function like proliferation rate changes or gene expression changes to cell death induction, decrease in the rate of melatonin production and changes in electroencephalogram patterns in humans, population declinations of birds and insects, and small but statistically significant increases of certain types of cancer, are attributed in our days to the radiations emitted by mobile telephony antennas of both handsets and base stations. This chapter reviews briefly the most important experimental, clinical and statistical findings and presents more extensively a series of experiments, concerning cell death induction on a model biological system. Mobile telephony radiation is found to decrease significantly and non thermally insect reproduction by up to 60%, after a few minutes daily exposure for only few days. Both sexes were found to be affected. The effect is due to DNA fragmentation in the gonads caused by both types of digital mobile telephony radiation used in Europe, GSM 900MHz, (Global System for Mobile telecommunications), and DCS 1800MHz, (Digital Cellular System). GSM was found to be even more bioactive than DCS, due to its higher intensity under equal conditions. The decrease in reproductive capacity seems to be non-linearly depended on radiation intensity, exhibiting a peak for intensities higher than 200 $\mu W/cm^2$ and an intensity "window" around $10\mu W/cm^2$ were it becomes maximum. In terms of the distance from a mobile phone antenna, the intensity of this "window"corresponds under usual conditions to a distance of 20-30 cm. The importance of different parameters of the radiation like intensity, carrier frequency and pulse repetition frequency, in relation to the recorded effects are discussed. Finally, this chapter describes a plausible biophysical and biochemical mechanism which can explain the recorded effects of mobile telephony radiations on living organisms.

Keywords: mobile telephony radiation, GSM, DCS, RF, ELF, electromagnetic fields, non-ionizing electromagnetic radiation, biological effects, health effects, *Drosophila*, reproductive capacity, cell death, intensity windows.

* E-mail address: dpanagop@biol.uoa.gr. Fax: +30210 7274742, Phone: +30210 7274117.

Introduction

As mobile telephony becomes more and more a necessary tool in our daily life enabling modern man to communicate easily with everyone at any place and any moment, serious threats arise from the exposure of all living organisms and the environment to a type of radiation unknown until now. Man made electromagnetic fields and radiations differ substantially from natural electromagnetic radiations like natural light, mainly because artificial ones are polarised, able to induce coherent forced vibrations to any electric charge in their space. All living organisms are made of cells and all cellular functions are of electrical nature, involving movements of electrical charges like clouds of free ions or charged macromolecules. Certain movements of certain type of charges within the cells induce or interrupt corresponding cellular functions. Any wrong, synchronized net movement of charge within the cell, would induce a wrong cellular function. The cell as a highly organized unit of life, has protective mechanisms against wrong cellular function, for example by activating certain genes and consequently producing certain proteins like the "heat shock" ones, made to protect the cell from excessive heat. But if the cell fails to protect itself from an external disturbance, a malfunction may start which can be transferred to a whole tissue or the whole organism. Electromagnetic fields (EMFs) are perceived by the cells as external disturbances or external stress but the cells don't seem to have special genes to be activated for protection against electromagnetic stress. This might be the reason why in response to electromagnetic stress, cells activate heat shock genes and produce heat shock proteins very rapidly (within minutes) and at a much higher rate than for heat itself, (Weisbrot et al, 2003). It seems to be for the same reason why electromagnetic stress from mobile telephony radiation induces cell death to the reproductive cells much more than other types of external stress examined before like food deprivation or chemicals, (Panagopoulos et al 2007a). Thus it seems that cells are much more sensitive to man-made electromagnetic fields (EMFs) than to other types of stress previously known. This is probably due to the fact that man-made EMFs constitute a new and perhaps more intense type of external stress, against which, cells have not developed defensive mechanisms. If cells activate heat shock genes to protect themselves from electromagnetic stress and this happens at a much higher rate than for heat itself, this might be dangerous, since repetitive stress leading to continuous expression of heat shock genes may result to cancer induction, (French et al, 2001).

A number of biological effects induced by man-made (EMFs) and radiations of different frequencies including digital mobile telephony and microwave radiations, have already been reported and documented by many research groups. These include changes in intracellular ionic concentrations, changes in the synthesis rate of different biomolecules, changes in cell proliferation rates, changes in the reproductive capacity of animals, changes in gene expression and even DNA damage and cell death,, (Aitken et al 2005; Bawin and Adey 1976; Bawin et al. 1975; 1978; Barteri et al 2005; Belyaev et al 2005; Blackman et al 1980; 1989; Caraglia et al 2005; Diem et al 2005; Dutta et al 1984; Kwee and Raskmark 1998; Velizarov et al 1999; Magras and Xenos 2001; Xenos and Magras 2003; Panagopoulos et al 2004; 2007a; 2007b; Lai and Singh 1995; 1996; 1997; 2004; Remondini et al 2006; Nylund and Leszczynski 2006; Diem et al 2005; Salford et al 2003). At the same time, some epidemiological studies are starting more and more to indicate a connection between the use of cellular mobile phones and certain types of cancer, (Hardell et al 2007a; Hardell et al 2006; Hardell and Hansson-Mild, 2006; Kundi 2004).

In several cases, melatonin, a hormone which controls the daily biological cycle and has an oncostatic action, produced by the epiphysis (pineal gland) in mammals, mainly during the night, is found to reduce the action of EMR exposure, but the synthesis of melatonin itself seems to be reduced by EMR, (Burch et al, 2002; Ozguner et al, 2006; Oktem et al, 2005).

Technical Characteristics of Digital Mobile Telephony Radiation

Both systems of Digital Mobile Telephony Radiation used in Europe, GSM 900 MHz and DCS 1800 MHz and also the system used in USA, GSM 1900 MHz, use different carrier frequencies, (900, 1800, and 1900 MHz respectively), but the same pulse repetition frequency of 217 Hz, (Hillebrand 2002; Clark 2001; Hyland 2000; Hamnerius and Uddmar 2000; Tisal 1998). As is obvious, the signals of Digital Mobile Telephony Radiation, combine "radio frequencies" (RF) and "extremely low frequencies" (ELF). All three systems use the "Time Division Multiple Access" (TDMA) code to increase the number of people that can simultaneously communicate with a base station. The radiation is emitted in frames of 4.615 msec duration, at a repetition rate of 217 Hz. Each frame consists of eight "time slots" and each user occupies one of them. Within each time slot the microwave radiation uses a type of phase modulation called "Gausian Minimum Shift Keying" modulation (GMSK) to carry the information, (Tisal 1998; Hamnerius and Uddmar 2000). The transmitted frames by both handsets and base stations are grouped into multi-frames of 25 by the absence of every 26th frame. This results to an additional multi-frame repetition frequency of 8.34 Hz. Finally, handsets emit an even lower frequency at 2 Hz whenever the user is not speaking, for energy saving reasons, ("non-modulated" or "non-speaking" emission or "discontinuous transmission mode"- DTX), (Hyland 2000). Of course, when the handsets operate at DTX mode, the average emitted power is much less (about one tenth of the emitted power when they operate at "speaking" mode, (Panagopoulos et al, 2000a, 2004).

Except of the carrier frequency, another important difference between the three systems of digital mobile telephony radiation is that GSM 900MHz antennas of both mobile phones and base stations operate with double the output power than the corresponding DCS 1800MHz ones or the GSM 1900 MHz ones. GSM 900 MHz handsets operate with 2 W peak power output, while DCS 1800 MHz and GSM 1900 MHz ones operate with 1 W peak power output.

Radiation from base station antennas is almost identical to that from mobile phones of the same system (GSM or DCS), except that it is about 100 times more powerful, or to be more accurate, from several tens up to several hundred times more powerful. Thereby, effects produced by mobile phones at certain distances, can be extrapolated to represent effects from base station antennas at about 100 times longer distances. Another difference is that handset signals include one pulse per frame occupying one time slot, whereas base station signals include again one pulse per frame but this pulse may occupy 1-8 time slots depending on the number of subscribers each moment. In other words the ratio between pulse peak power and time-averaged power is usually higher for the handset signals compared to the base station signals, (Hillebrand 2002; Clark 2001; Hyland 2000; Hamnerius and Uddmar 2000; Tisal 1998).

Established Exposure Criteria for Mobile Telephony Radiations

The most stringent international exposure limits in the western world for RF radiation used by digital mobile telephony were set by the International Radiation Protection Association (IRPA) and the International Commission on Non-Ionizing Radiation Protection (ICNIRP). These criteria were established to protect biological tissue from temperature increases, (thermal effects).

The ICNIRP exposure limits are given either in terms of Radiation Intensity (Power Density) usually in mW/cm^2, either in terms of Specific Absorption Rate (SAR) which is defined as the radiation power, absorbed by the unit mass of tissue, in W/kg. Only the radiation intensity in air outside the body can be readily and objectively measured in exposed individuals. The SAR is difficult to be determined for every single tissue as is different for different tissues and radiations. The best way for determining SAR is by computational approximate methods like the Finite Difference Time Domain (FDTP) method, the Finite Element Method (FEM), or the Method of Moments (MoM), (Meyer and Jacobus, 2003).

According to the ICNIRP exposure criteria, the maximum permitted radiation intensity (in mW/cm^2) for the general population exposure, is given according to radiation frequency and it is $f/2$ (f in GHz). Therefore, at 900MHz, the intensity limit according to these criteria is $0.45mW/cm^2$. At 1800 MHz the corresponding limit is 0.9 mW/cm^2, e.t.c). In terms of SAR the ICNIRP limits for the general population are 0.08 W/Kg (for whole-body average absorbed power) and 2 W/Kg (for the head and trunk). All the above values are to be averaged over any 6min period during the 24-h day. (IRPA 1988; ICNIRP 1998).

For the frequency 25-800 Hz, the IRPA-ICNIRP limits for the general population are for electric field intensity E, the value $250/f$ and for magnetic induction B, the value $50/f$, (E in kV/m, B in G, f in Hz). Therefore, at 217 Hz, (the pulse repetition frequency of digital mobile telephony radiations), the ICNIRP limits are 1.15kV/m and 0.23 G for up to 24h exposure during the day, (IRPA 1990; ICNIRP 1998).

As we shall see, during the years after the establishment of the IRPA-ICNIRP exposure criteria, it has been shown that the vast majority of health effects of digital mobile telephony radiations are non-thermal and a lot of biological effects were recorded at radiation intensities much lower than the values of these criteria. This is the reason why several countries in Europe have established much more stringent national exposure criteria, like Italy, Poland, Russia (10 $\mu W/cm^2$), or Salzburg (Austria), (0.1 $\mu W/cm^2$), ("EMF World Wide Standards").

A Review of Biological, Clinical and Epidemiological Data

There is already a very large number of published studies regarding research on possible health risks from cellular mobile telephony radiations. While a large and increasing number of studies (biological, clinical and epidemiological) have recorded a variety of non-physiological changes with increased probabilities for health hazards including several types of cancer, a lot of other studies find no connection between exposure to mobile telephony radiations and health risks. Inconsistencies observed between studies are partly expected since no identical conditions can ever be attained between different studies and different labs, but also they are explained by some authors to be due to biased samples. According to a recent article in which possible secret ties between industries and University researchers are discussed, (Hardell et al, 2007b). Since a large number of studies are funded by companies, a matter arises on how much independent these studies can be.

In the present review we shall emphasize on the studies that indicate different possible effects on living organisms, since we consider that we must take most seriously and focus the most on the possibility that is worse for living organisms and the natural environment. Additionally because of the large number of studies relating RF-microwave radiations in general, we shall concentrate on those that regard to radiations with frequencies and intensities close to those utilized by digital mobile telephony radiations (800-2450 MHz).

A. Biological Effects

Microwaves are found to produce thermally and non-thermally a large number of biological effects, in many cellular and animal studies, (Banik et al, 2003). In the case of radiations emitted by mobile telephony antennas at intensities that people are normally exposed, the effects are non-thermal as verified by different experimenters, (Diem et al, 2005; Panagopoulos et al, 2004; 2007a; 2007b; Leszczynski et al, 2002; Schirmacher et al, 2000; Velizarov et al, 1999)

Regarding non-thermal effects of RF radiations, it is a must to refer to the pioneer works of Bawin et. al. and Blackman et. al. back in the seventies and eighties although these works were relating lower frequency RF radiations. In those pioneer experiments, RF radiation with carrier frequencies 147 and 450 MHz, modulated by sinusoidal ELF signals 0-40 Hz, was found to decrease Ca^{2+} concentration in chicken brain cells. The effect was found to become maximum at modulation frequencies 6-20 Hz and at intensities 0.6-1 mW/cm^2, (Bawin et al 1975; 1978). Non-modulated RF signals were not found to be as bioactive as modulated ones by ELFs and additionally, these effects were found to be non-linearly depended on radiation intensity and frequency, exhibiting “windows” within which the phenomena appeared and then disappeared for values outside, (Blackman et al, 1980; 1989).

Repairable DNA damage and increased expression of heat shock protein Hsp 70 without changes in cell proliferation rates was detected in human lens epithelial cells after 2h exposure to 1.8GHz RF field, amplitude modulated at 217 Hz with 3 W/kg SAR. The DNA damage was determined by use of the comet assay, (Lixia et al, 2006).

Increased expression of genes encoding ribosomal proteins and consequently up-regulating the cellular metabolism in human cell types, was found after in vitro exposure to 900 and 1800MHz mobile phone radiation, (Remondini et al, 2006). In an other study, gene and protein expression were altered in human endothelial cell lines, after 900 MHz GSM mobile phone radiation exposure at an average SAR of 2.8 W/kg. Genes and proteins were differently affected by the exposure in each of the cell lines, suggesting that cell response to this type of radiation might be genome and proteome- dependent which in turn might explain to some extend the discrepancies in replication studies between different laboratories, (Nylund and Leszczynski, 2006).

Exposure of human endothelial cells in vitro, to GSM 900 MHz mobile phone radiation for 1h at non-thermal levels, average SAR 2 W/kg, caused transient increase in heat shock protein hsp27 phosphorylation and transient changes in protein expression levels, (Leszczynski et al, 2002).

Rapid (within minutes) induction of heat shock protein hsp70 synthesis, was found in the insect *Drosophila melanogaster*, after in vivo exposure to GSM 1900 MHz mobile phone radiation, (Weisbrot et al, 2003).

According to a theoretical report, repetitive stress caused by mobile phone radiation, leading to continuous expression of heat shock genes in exposed cells and tissues may result to cancer induction, (French et al, 2001).

Two hours of exposure by a cellular mobile phone, changed the structural and biochemical characteristics of acetylcholinesterase, an important central nervous system enzyme, resulting to a significant alteration of its activity. The enzyme was exposed within an aqueous solution at 5 cm distance from the mobile phone, (Barteri et al, 2004).

Exposure of myoglobin solution to 1.95 MHz microwave radiation for 3h at non-thermal levels was found to affect the folding of the protein and thereby changing its biochemical properties, (Mancinelli et al, 2004).

In vitro exposure for 1h of human skin fibroblasts to GSM radiation, induced alterations in cell morphology and increased the expression of mitogenic signal transduction genes, cell growth inhibitors and genes controlling apoptosis, (Pacini et al, 2002).

In an earlier study, 960 MHz GSM-like signal at SAR 0.021, 0.21 and 2.1 mW/cm^2 with exposure times 20, 30 and 40 min respectively, was found to decrease the proliferation rate of transformed human epithelial amnion cells. The maximum effect was reached at lower power level with a longer exposure time than at higher power level, (Kwee and Raskmark, 1998).

In another study, in vitro exposure of human peripheral blood lymphocytes to continuous 830 MHz radiation, with average SAR 1.6-8.8 W/kg, was found to produce losses and gains of chromosomes (aneuploidy), a somatic mutation leading to cancer. The effect was found to be activated via a non-thermal pathway, (Mashevich et al, 2003).

Long term exposure of rats to 900 MHz mobile phone radiation produced oxidative stress (increased oxidant products of free radicals) in retinal tissue. Melatonin and caffeic acid phenethyl ester (CAPE)- component of honeybee propolis administered daily to the animals prior to their EMR exposure, caused a significant reduction in the levels of the oxidant products, (Ozguner et al, 2006). In a previous study of the same group, melatonin was found to reverse oxidative tissue injury in rat kidneys, after 10 days exposure-30 min per day, to 900 MHz GSM radiation emitted by mobile phone, (Oktem et al, 2005).

Male mice were exposed to 1800 MHz GSM-like microwaves, 0.1 mW/cm^2 for two weeks on workdays, 2h per day. Then mice were anesthetized and blood samples were taken for hematology, serum chemistry and serum testosterone determinations. Additionaly, testicles, epididymes, adrenals, prostates and pituitary glands were removed for histology. Red blood cell count and serum testosterone level were found to be significantly higher in the exposed groups but no significant alterations were found in the other investigated variables, (Forgacs et al, 2005).

Mice prone to the development of lymphomas, exposed for two 30 min periods per day for up to 18 months, to 900 MHz pulsed microwave radiation with a 217 Hz pulse repetition frequency at SAR ranging from 0.007 to 4.3 W/kg, developed twice the number of tumors than the unexposed ones, (Repacholi et al, 1997).

Male Wistar 35-day-old rats were exposed to 2.45 GHz radiation for 2 h/day for a period of 35 days at a power density of 0.344 mW/cm^2, (SAR 0.11 W/kg). After 35 days the rats were sacrificed and whole brain tissue was isolated for protein kinase C (PKC) assay. The study revealed a decrease in PKC activity. Electron microscopy study showed an increase in the glial cell population in the exposed group. The results indicated that chronic exposures may affect brain growth and development, (Paulraj and Behari, 2006a). In another study of the same group, single strand DNA breaks were measured as tail length of comet. Fifty cells from each slide and two slides per animal were observed. The study showed that chronic exposure to microwave radiation at non-thermal levels (SAR 1 and 2 W/kg) causes statistically significant increase in DNA single strand breaks in rat brain cells, (Paulraj and Behari, 2006b).

In another study mice placed within an RF antenna park were repeatedly mated for five times while they were continuously exposed at very low levels of RF radiation (0.168-1.053 $\mu W/cm^2$). A progressive decrease in the number of newborns per maternal mouse was observed after each mating, which ended to irreversible infertility, (Magras and Xenos, 1997). In a more recent study of the same group, it was found that exposure of pregnant rats to GSM-like 940 MHz radiation at 5 $\mu W/cm^2$, resulted in aberrant expression of bone morphogenetic proteins (BMP)-(major endocrine and autocrine morphogens known to be involved in renal development), in the kidneys of newborn rats, (Pyrpasopoulou et al, 2004).

Increase in the number of micronuclei in rat bone marrow erythrocytes, a sign of genotoxicity, was observed after 30 days exposure for 2h daily, to 910 MHz microwave radiation, (Demsia et al, 2004).

In several other mammal studies, no effects were found, in regards to genotoxicity of second generation mobile telephony (GSM, DCS) and third generation, "universal mobile telecommunication system" (UTMS) radiations, (Sommer et al 2007; Oberto et al 2007; Juutilainen et al 2007;Tillmann et al 2007; Gatta et al 2003).

The mortality of chicken embryos was found to increase to 75% from 16% in the control group, after exposure to radiation from a GSM mobile phone, (Grigor'ev, 2003). This result is in agreement with the increased mortality of fertilised chicken eggs that was recorded after irradiation by low power 9.152 GHz pulsed and continuous-wave microwaves, (Xenos and Magras, 2003).

Several studies have reported that microwave exposures increase the permeability of the blood-brain barrier (BBB), an hydrophobic barrier made by endothelial cells to protect the mammalian brain from harmful compounds in the blood. A Swedish group has reported that 915 MHz microwaves at non-thermal intensities causes leakage of albumin into the brain through the BBB in rats, accumulating in the neurons and glial cells which surround the capillaries in the brain, (Salford et al, 1994). The same group reported that GSM mobile phone radiation from a test mobile phone with a programmable constant power output, opens the BBB for albumin, resulting to damage of brain cells in rats. The power density and SAR were within the ICNIRP limits, (Salford et al 2003). These were the first experiments that indicated cell damage caused by mobile phone radiation although this radiation was not a real mobile phone signal. However in an earlier study of the same group, continuous-wave and pulsed 915 MHz radiation at relatively high intensities, 1 W and 2 W respectively, was not found to damage brain or promote brain tumour development in rats, (Salford et al. 1993).

Exposure of an in vitro BBB model, consisted by rat brain cells growing in a culture with pig blood cells, exposed to 1800 MHz microwave radiation pulsed at 217 Hz repetition rate (DCS-like), at SAR 0.3-0.46 W/kg, increased the permeability to sucrose of the BBB twice compared to the control culture. No significant temperature rise was detected during the exposures, (Schirmacher et al, 2000). In a latter study of the same group, in vitro exposure of three other BBB models with distinctly higher barrier tightness than the previously used one, did not cause any effect on the permeability of the BBB of the models, (Franke et al, 2005).

In regards to DNA damage or cell death induction due to microwave exposure, in a series of early experiments, rats were exposed to pulsed and continuous-wave 2450 MHz radiation for two hours at an average power density of 2 mW/cm^2 and their brain cells were subsequently examined for DNA breaks by "comet" assay. The authors found a dose-dependent (0.6 and 1.2 W/kg whole body SAR) increase in DNA single-strand and double-strand breaks, four hours after the exposure to either the pulsed or the continuous-wave radiation, (Lai and Singh 1995; 1996). The same authors found that melatonin and PBN (N-tert-butyl-alpha-phenylnitrone) both known free radical scavengers, block the above effect of DNA damage by the microwave radiation, (Lai and Singh 1997). Although these experiments were the first to report DNA damage by microwaves, the radiation intensity ($2mW/cm^2$) was relatively high, exceeding the international exposure limits (ICNIRP 1998) and additionally the radiation frequency was the same as in microwave ovens. This is why the authors of this review cannot be sure on whether the reported effects were thermal or non-thermal.

In vitro exposure of mouse fibroblasts and human glioblastoma cells to 2450 MHz, (Malyapa et al, 1997a), 835.62 MHz and 847.74 MHz (Malyapa et al, 1997b), radiations at SAR 0.6 W/kg, was not reported to damage DNA as measured by comet assay.

A number of recent studies have reported DNA damage, or cell damage, or cell death, induced by mobile telephony or similar RF radiations at non-thermal intensity levels, (Aitken et al, 2005; Diem et al 2005; Panagopoulos et al 2007; Salford et al, 2003; Markova et al, 2005; Caraglia et al, 2005; Nikolova et al, 2005), while some other studies did not find any such connection, (Hook et al, 2004; Capri et al, 2004a; 2004b; Meltz 2003; Cranfield et al, 2003). Aitken et al 2005, reported damage to mitochondrial genome and the nuclear beta-globin locus in the spermatozoa of mice exposed to 900 MHz, 0.09 W/kg SAR, for 7 days, 12h per day. Diem et al 2005, reported single and double-strand DNA breakage in cultured human and rat cells exposed to 1800 MHz mobile phone-like radiation. Panagopoulos et al 2007a, found DNA fragmentation at a very high degree, caused in the reproductive cells of female *Drosophila* insects only by few min daily exposure to a real mobile phone signal for only few days. These were the first experiments that showed extensive DNA damage and cell death by real digital mobile phone GSM and DCS signals. Previous experiments of the same group had shown a large decrease in the reproductive capacity of the same insect, caused by real mobile phone similar exposures, (Panagopoulos et al, 2004).

B. Clinical Studies on Humans. Effects on EEG, EDA, Melatonin, etc

Mobile telephony radiation is found in several studies to affect electroencephalograms (EEG), electrodermal activity (EDA) and the synthesis rate of hormones like melatonin, in humans.

In a series of early experiments performed by a Finish group, GSM mobile phone exposure was found to alter the EEG oscillatory activity of healthy adult subjects, in the 6-8 and 8-10 Hz frequency bands during cognitive (visual memory) tasks, (Krause et al, 2000). In more recent experiments of the same group, exposure of 10-14 year old children to mobile phone GSM field while performing an auditory memory task, induced changes in their brain oscillatory EEG responses in the frequencies 4-8 Hz and 15 Hz, (Krause et al, 2006).

Exposure for 30 min to pulse modulated 900 MHz mobile phones-like EMF, increased waking regional cerebral blood flow (rCBF) and enhanced EEG power in the alpha frequency range (8-12 Hz) prior to sleep onset and during sleep. Exposure to the same field without pulse modulation did not enhance power in waking or sleep EEG, (Huber et al, 2002). In another set of experiments of the same group, 30 min exposure to the same 900 MHz GSM-like field during waking period preceding sleep, increased the spectral power of the EEG in non-rapid eye movement sleep. The maximum increase occurred in the 9.75-11.25 Hz and 12.5-13.25 Hz frequency ranges during the initial part of the sleep. Since exposure during waking, modified the EEG during subsequent sleep, the changes in the brain function induced by mobile telephony radiation are considered to outlast the exposure period, (Huber et al, 2000).

Mobile phone exposure prior to sleep was found to decrease rapid eye movement sleep latency and to increase EEG spectral power in the 11.5-12.5 Hz frequency, during the initial part of sleep following exposure, (Loughran et al, 2005).

Some other studies have failed to find any effects of mobile phone-microwave exposures on EEG during cognitive testing, or to replicate earlier findings, (Röschke and Mann, 1997; Wagner et al., 1998).

Mobile phone radiation was found to affect the evoked neuronal activity of the central nervous system (CNN) as represented by EDA, an index of the sympathetic nervous system. Mobile phone exposure was found to lengthen the latency of EDA (Skin Resistance Response), irrespectively of the head side next to mobile phone, (Esen and Esen, 2006). Therefore, mobile phone exposure may increase the response time of users with different negative consequences, like for example the increase in the risk of phone-related driving hazards, e.t.c.

A statistically significant increase of chromosomal damage was found in blood lymphocytes of people who used GSM 900 MHz mobile phones, compared to a control group of non-users, matched according to age, sex, health status, drinking and smoking habits, working habits, and professional careers. The increase was even greater for users who were smoker-alcoholic, (Gadhia et al, 2003)

In another type of clinical study, exposures of humans to GSM 900 MHz and DCS 1800 MHz mobile phones fields for 35 min, were not found to change significantly arterial blood pressure or heart rate during or after the exposure, (Tahvanainen et al, 2004).

Prolonged use of mobile phone, (more than 25 min per day), was found to induce a reduction in melatonin production among male users. The effect was enhanced by additional exposure to 60 Hz ELF magnetic field, (Burch et al, 2002).

Two studies about possible immediate- short term effects of GSM and UTMS (third generation of mobile networks)-like exposure on well being and cognitive performance in humans based on questionnaires, found contradictive results. The first (Zwamborn et al, 2003) reported no effects of GSM-like exposure, while the UTMS-like exposure was found to reduce well-being and cognitive performance. The second, (Regel et al, 2006) reported no effects at all from either type of radiation. The opinion of the authors of this review is that studies based on questionnaires cannot be as much objective as studies based on measurable indexes like EEG or EDA. Besides, it would be unlikely that subjects would report themselves immediate effects on their well-being.

C. Epidemiological Studies

According to the Swedish Prof. L. Hardell and his research group, the concluding results of up to date epidemiological studies among users for more than ten years use of mobile phones indicate consistently an increased risk for acoustic neuroma and glioma, especially for ipsilateral exposure, (Hardell et al, 2007a). Earlier work of the same research group had found a connection between digital (2nd generation) and analogue (1st generation) mobile phones use and malignant brain tumors, highest for more than ten years latency period, (Hardell et al, 2006).

Another review study of the Austrian Prof. M.Kundi conducted few years ago, states as the resume from several epidemiological and experimental studies, that long term exposure to mobile phone emissions (analogue and digital) constitutes a small to moderate increased risk for developing certain types of cancer, (Kundi, 2004).

Several other studies had not found any association between mobile phone use and cancer, (Inskip et al, 2001; Johansen et al, 2001; Muscat et al, 2002).

A major difficulty in epidemiological studies among mobile phone users is the variation of parameters governing the exposure from hand held mobile phones, i.e. the distance from the nearest base station which can considerably change the intensity of the radiation emitted by the phone, the actual duration of daily use, e.t.c. Nevertheless, the studies done on habitants living close to base stations are more consistent since the station emits a more constant radiation level on a daily basis and therefore a person residing nearby, receives a measurable radiation at least for several hours per day.

A recent Egyptian study (Abdel-Rassoul et al, 2007) found that inhabitants living nearby mobile telephony base stations may develop a number of neuropsychiatric problems like headaches, memory changes, dizziness, tremors, depression, sleep disturbances, reported also in previous studies as “microwave syndrome” (Navarro et al 2003), plus changes in the performance of neurobehavioral functions. Similar results were found by other studies in different countries like in France, (Santini et al 2003), Poland (Bortkiewicz et al 2004), Spain (Navarro et al 2003), Austria (Hutter et al 2006).

Other epidemiological studies have reported diminishes in the populations of birds around mobile telephony base stations at distances 100-600m from the masts in Belgium, (Everaert and Bauwens 2007) and within 200m from the masts in Spain (Balmori 2005). These studies are in agreement with earlier biological studies which had reported increased mortality of avian embryos, exposed to low levels (5-120 $\mu W/cm^2$) of RF antennae radiation, (Xenos and Magras, 2003).

The Design of Bioelectromagnetic Experiments and a Reason for Inconsistencies

As described in the previous paragraphs, there are frequently contradictory results in the bioelectromagnetic experiments performed by different labs. One factor that we have found to be very important and able to completely change the results of a biological experiment is the influence of the stray electromagnetic fields that exist inside any lab.

Within a usual room inside a house or laboratory there are 50-60 Hz fields due to the electric wirings and electrical appliances. Close to the walls, near to sockets or close to electrical appliances one can measure electric fields up to 50 V/m and magnetic fields up to 10 mG. Such fields are found to affect biomolecules, cells and whole organisms in different ways and therefore to affect the outcome of any biological experiment, (Goodman E. et al. 1995; Panagopoulos et al. 2002; Weaver and Astumian 1990). Prior to the design of any biological experiment, a careful scanning of stray fields inside the lab is necessary. The experiments should be performed at the place with the minimum stray fields and special care should be taken in having the control under identical conditions with the exposed groups except only for the factor studied. Temperature, light and humidity are additional important factors that should be identical between exposed and control groups.

Before the relatively recent evolution of knowledge in the field of Bioelectromagnetism, ambient electromagnetic fields within the labs were not taken into account in biological experiments. But living organisms are very sensitive to external electromagnetic fields, natural or artificial ones. Rooms or devices used as incubators, are constructed to keep a constant temperature, humidity, e.t.c. in their internal space, but usually are sources of EMFs from their own electrical circuits. A specialized physicist should always be member of any experimental team for taking good care of such factors.

Effects of Mobile Telephony Radiation on a Model Organism

Introduction

In order to study the ability of the electromagnetic signals emitted by cellular mobile telephony antennas to affect the biological function of living organisms, we used a biological model, the reproductive capacity of the insect *Drosophila melanogaster*, a well studied experimental animal with many advantages, including its short life cycle and the good timing of its metamorphic stages and developmental processes, (King 1970). Especially the good timing of this insect's early developmental stages (oogenesis, spermatogenesis, embryogenesis, larval and pupal stages), under certain environmental conditions (i.e. temperature, humidity, food e.t.c.), is a very important feature, on which our experimental protocols were based.

In order to study the effects of mobile telephony radiation on the reproductive capacity, we exposed the insects to real mobile phone signals, emitted by commercially available handsets.

The basic cellular processes are identical in insect and mammalian cells. In addition, insects (particularly *Drosophila*) are much more resistant, at least to ionizing electromagnetic radiation, than mammals, (Koval and Kazmar 1988, Koval et al 1979, 1977, Abrahamson et al 1973). Therefore, a proper experimental protocol relating *Drosophila* can be very useful in assessing the bioactivity of electromagnetic radiation in general, (including non-ionizing radiation and electromagnetic fields).

Our experiments, regarding few minutes daily exposure of this model organism for only few days, to cellular mobile phone signals, have shown a large decrease in the reproductive capacity, affecting both sexes (Panagopoulos et al 2004). Both systems of digital mobile telephony radiation used in Europe, GSM 900 MHz and DCS 1800 MHz were found to decrease the insect's reproductive capacity, but GSM 900 MHz was found to be even more bioactive than DCS 1800 MHz, mainly due to the higher intensity of GSM 900 MHz antennas compared to DCS 1800 MHz ones, (Panagopoulos et al 2007b; 2007a). The decrease in the reproductive capacity was found to be due to induced cell death (DNA fragmentation) in the gonads, caused by both types of mobile telephony signals, (Panagopoulos et al 2007a). Unpublished experiments of ours presented here for the first time, show that the bioactivity is strongly and non-linearly dependent on the intensity of the radiation, becoming maximum for intensities higher than 200 $\mu W/cm^2$ and within an "intensity window" around 10 $\mu W/cm^2$.

Materials and Methods

Experimental Animal

We used *Drosophila melanogaster* flies, wild-type strain, Oregon R, held in glass bottles with standard food, kept in incubator at 25 °C, with 12-h periods of light and darkness and 70% relative humidity, cultured according to standard methods, (Panagopoulos et al 2004).

The food consisted of 450ml water, 4g agar, 13g yeast, 32g rice flour, 16g sugar, 25g tomato pulp. The mixture was boiled for over 10min to ensure sterility, which was preserved by the addition of 2ml propionic acid and 2ml ethanol. This food quantity was enough for 25-30 glass vials which were sterilized before the food was added.

In each experiment, we collected newly emerged adult flies from the stock early in the afternoon, anesthetized them lightly with diethyl ether and separated males from females. We divided the collected flies in groups of ten in standard laboratory cylindrical glass vials, with 2.5cm diameter and 10cm height, with standard food, which formed a smooth plane surface, 1cm thick at the bottom of the vials. The vials were closed with cotton plugs.

Exposure System

Before each set of experiments we measured the mean power density of the radiation emitted by the mobile phone handset in the RF range at 900MHz and/or 1800MHz, with the field-meter, "RF Radiation Survey Meter, NARDA 8718", with its probe inside a glass vial similar to the ones we used for the insects in our experiments. In addition, we measured in the same way the mean electric and magnetic field intensities at the Extremely Low Frequency (ELF) range, with the field-meter, "Holaday HI-3604, ELF Survey Meter".

The experimenter's position in relation to the mobile phone during the measurements was the same as during the exposures. The mobile phone was held close to the experimenter's head with its antenna facing downward. The exposures and the field measurements, took place in a quiet but not sound-isolated room to simulate the actual conditions to which a user is subjected during a normal conversation on the mobile phone. The room conditions and the positions of all items around the experimental bench were always the same. Exposures and measurements of mobile phone emissions were always conducted at the same place where the mobile phone had full perception of both GSM and DCS signals. The handset was fully charged before each set of exposures or measurements.

In the most new digital cell phone handsets, the antenna is in the back and upper side of the device. This can be easily verified by measuring the emitted radiation holding the probe of the field meter in contact with different parts of the handset's surface.

The measured exposure values were in general within the established exposure limits, (ICNIRP 1998).

We used commercially available digital mobile phone handsets in all the sets of our experiments, in order to analyze effects of real mobile telephony exposure conditions. As far as we know, we were the first to use a commercially available mobile phone handset itself in biological experiments, (Panagopoulos et al 2000a). The obvious reason was that these devices are the most powerful RF transmitters in our immediate daily environment. Thus, instead of using simulations of digital mobile telephony signals with constant parameters (frequency, intensity etc), or even "test mobile phones" programmed to emit mobile telephony signals with controllable power or frequency, we used real GSM, DCS signals which are never constant, since there are continuous changes in their intensity and frequency. Electromagnetic fields with changing parameters are found to be more bioactive than fields with constant parameters, (Goodman E.M. et al 1995; Diem et al 2005), probably because it is more difficult for living organisms to get adapted to them. Experiments with constant GSM or DCS signals can be performed, but they do not simulate actual conditions. Later other experimenters also started to use mobile phone handsets as exposure devices apparently for the same reasons, (Weisbrot et al 2003; Barteri et al 2005).

We exposed the flies within the glass vials by placing the antenna of the mobile phone outside of the vials, in contact with or at different distances from the glass wall and parallel to the vial's axis. The total duration of exposure was 6min per day in one dose and we started the exposures on the first day of each experiment (day of eclosion). The exposures took place for a total of 2 to 6 days in each experiment depending on the kind of the experiment, as described below. The daily exposure duration of 6min, was chosen in order to have exposure conditions that can be compared with the established exposure criteria, (ICNIRP 1998). Besides, early experiments had shown that only few minutes of daily exposure were enough to produce a significant effect on the insect's reproductive capacity (Panagopoulos et al, 2000a).

The experimenter could speak on the mobile phone during connection (this we called, "modulated" or "speaking" emission), or could just stay silent, ("non-modulated" or "non-speaking" emission, or DTX mode). The intensity of the emitted radiation increases about ten times when the user speaks during connection, than when there is no speaking, (Panagopoulos et al, 2000a).

Exposure Procedures

We carried out six sets of experiments: In the first set, we exposed the insects to the mobile phone's GSM 900 MHz field while the mobile phone was operating in non-speaking mode, (non-modulated emission or DTX). In the second set of experiments, the mobile phone was operating in speaking mode, (modulated emission) during the exposures. In the third set of experiments we investigated the effect of the mobile phone signal on the reproductive capacity of each sex separately. In the fourth set of experiments we compared the bioactivity between GSM 900 MHz and DCS 1800 MHz types of mobile telephony signals. In the fifth set of experiments we exposed the insects to different distances (intensities), from the mobile phone antenna from 0 to 100 cm, for both types of radiation. Finally, in the sixth set of experiments we tested the ability of GSM and DCS fields to induce DNA fragmentation (cell death) in the ovarian cells of the female insects during oogenesis.

In every single experiment we separated the newly emerged collected adult flies to exposed (E) and sham-exposed (SE)/control (C) groups. Each one of the groups consisted always of ten female and ten male, newly emerged flies. The sham exposed groups had identical treatment as the exposed ones, except that the mobile phone during the 6-min "exposures", was turned off.

Every time before each exposure, the cotton plugs were pushed down in the glass vials in order to confine the flies to a small area of about 1cm height between the cotton and the food so as to provide roughly even exposure to all flies. After the exposure, the cotton plugs were pulled back to the top of the vials, and the vials were put back in the culture room.

In every group of insects in all the sets of experiments, we kept the ten males and the ten females for the first 48h of the experiment in separate glass tubes. At eclosion, adult female flies have already in their ovaries eggs at the first preyolk stages and oogenesis has already started. The eggs develop through 14 distinct stages, until they are ready to be fertilized and laid, and the whole process of oogenesis lasts about 48h. By the end of the second day of their adult life, the female flies have in their ovipositors the first fully developed egg chambers of stage 14th, ready to be fertilized and laid, (King 1970; Panagopoulos et al 2004). At the same time, the first mature spermatozoa, (about 6h after eclosion) and the necessary paragonial substances (about 12h after eclosion) in male flies have already been developed (King 1970; Stromnaes and Kvelland 1962; Connolly and Tully 1998). Keeping males separately from females for the first 48h of the experiment ensures that the flies are in complete sexual maturity and ready for immediate mating and laying of fertilized eggs.

After the first 48h of each experiment, the flies were anesthetized very lightly again and males and females of each group were put together (ten pairs) in another glass tube with fresh food, allowed to mate and lay eggs for 72h. During these three days, the daily egg production of *Drosophila* is at its maximum (from the 3rd to 5th day of its adult life), then stays at a plateau or declines slightly for the next 5 days and diminishes considerably after the 10th day of adult life (Bos and Boerema 1981; Shorrocks 1972; Ramirez et al 1983).

On the sixth day of each experiment in all six sets of experiments, the flies were removed from the glass vials and the vials were maintained in the culture room for six additional days, without further exposure.

After the last six days, most F_1 embryos (deriving from the laid eggs) are in the stage of pupation, where they can be clearly seen with bare eyes and easily counted on the walls of the glass tubes, as at the last stages before pupation, the larvae leave the food, crawling up the walls of the glass vials. There may be a few embryos still in the last stages as larvae, which are big enough and ready for pupation (on the surface or already away from the food), so that they can be easily counted. [If the remaining larvae are still many and the counting is imprecise, the experimenter can wait an additional day and recount the pupae]. There may be also already a few newly emerged F_1 adult flies, which can also be counted easily.

During the last six days, we inspected the surface of the food within the glass vials under the stereo-microscope for any non-developed laid eggs or dead larvae, something that we did not see in our experiments (empty egg-shells can be seen after hatching). The number of observed exceptions (non-developed eggs or dead larvae), both in exposed and control groups (less than 5%) was within the Standard Deviation of progeny number. [The insignificant percentage of F_1 egg and larvae mortality is due to the fact that the paternal-maternal flies were newly emerged during the first 2-5 days of their adult lives]. Therefore the number of pupae in our experiments corresponded to the number of laid eggs (oviposition). Furthermore, the counting of pupae can be done without any error at all, whereas the counting of laid eggs under a stereo-microscope is subject to considerable error.

The oviposition of *Drosophila* is influenced by many factors, like temperature, humidity, prior anesthesia, crowding, food, (King 1970). Special care must be taken to keep all these factors constant. Experience in handling the flies is necessary to prevent accidental deaths.

This number of F_1 pupae under the above described conditions, during the insect's three days of highest oviposition, is that we have defined as the Insect's Reproductive Capacity and this is the biological index we have used to examine the bioactivity of electromagnetic radiation-field.

The temperature during the exposures was monitored within the vials with a mercury thermometer with an accuracy of 0.05°C.

In the sixth set of experiments, after the additional last exposure in the morning of the sixth day from the beginning of each experiment, the flies were removed from the glass vials, and the ovaries of females were dissected into individual ovarioles and fixed for TUNEL assay. The vials were then maintained in the culture room for six additional days, without further exposure, in order to count the F_1 pupae as in all the sets of experiments.

TUNEL Assay

A widely used method for identifying cell death is TUNEL assay. By use of this method, fluorescein dUTP is bound through the action of terminal transferase, onto fragmented genomic DNA which then becomes labelled by characteristic fluorescence. The label incorporated at the damaged sites of DNA is visualized by fluorescence microscopy, (Gavrieli et al, 1992).

Each Drosophila ovary consists of 16 to 20 ovarioles. Each ovariole is an individual egg assembly line, with new egg chambers in the anterior moving toward the posterior as they develop, through the 14 successive stages as described, until the mature egg reaches the oviduct.

To determine the ability of GSM and DCS radiation to act as possible stress factors able to induce cell death during early and mid oogenesis, we used TUNEL assay, as follows: Ovaries were dissected in Ringer's solution and separated into individual ovarioles from which we took away egg chambers of stages 11-14. In egg chambers of stages 11-14 programmed cell death takes place normally in the nurse cells and follicle cells. Thereby we kept and treated ovarioles and individual egg chambers from germarium up to stage 10. Samples were fixed in PBS solution containing 4% formaldehyde plus 0.1% Triton X-100 (Sigma Chemical Co., Germany) for 30min and then rinsed three times and washed twice in PBS for 5 min each. Then samples were incubated with PBS containing 20 μg/ml proteinase K for 10 minutes and washed three times in PBS for 5 min each. In situ detection of fragmented genomic DNA was performed with Boehringer Mannheim kit containing fluorescein dUTP for 3h at 37°C in the dark. Samples were then washed six times in PBS for 1h and 30 min in the dark and finally mounted in antifading mounting medium (90% glycerol containing 1.4-diazabicyclo (2.2.2) octane (Sigma Chemical Co., Germany) to prevent from fading and viewed under a Nikon Eclipse TE 2000-S fluorescence microscope.

Results and Discussion

In the first two sets of experiments, we separated the insects into two groups: a) the Exposed group (E) and b) the Sham Exposed group (SE). The 6-min daily exposures took place for the first five days of each experiment.

In the first three sets of experiments, the exposures were performed by GSM 900 MHz mobile phone radiation-field. Before the exposures, we measured radiation and field intensities, as described above. In the RF range, the measured mean power density for 6min of modulated emission (M), with the antenna of the mobile phone outside of the glass vial in contact with the glass wall and parallel to the vial's axis was 0.436±0.060 mW/cm^2. The non-modulated (NM) corresponding measured mean value, was 0.041±0.006 mW/cm^2. In the ELF range, the measured values for modulated field, excluding the ambient electric and magnetic fields of 50Hz, were 6.05±1.62 V/m electric field intensity and 0.10±0.06 mG magnetic field intensity. The corresponding non-modulated values were 3.18±1.10 V/m and 0.030±0.003 mG. All given values are average from eight separate measurements of each kind ± Standard Deviation (SD). These values are typical for all commonly used GSM 900 MHz mobile phone handsets.

1. Effect of Non-Modulated GSM radiation-field on the Reproductive Capacity

We carried out four experiments (1.1-1.4) with non-modulated field, (non-speaking emission). The exposure parameters in this case simulate the situation when a user listens through the mobile phone during connection.

Results are listed in Table 1.

Table 1 shows the mean number of F_1 pupae (corresponding to the number of laid eggs) per maternal fly in the groups E(NM) exposed to Non-Modulated (NM), GSM 900 MHz mobile phone field and in the corresponding sham exposed (control) groups SE(NM) during the first three days of the insect's maximum oviposition.

The Non-Modulated GSM 900 MHz signalss, decreased the insect's reproductive capacity by up to 20% in relation to the unexposed groups with six min daily exposure for five days. No temperature increases were detected within the vials during the exposures.

Table 1. Effect of Non-Modulated GSM field on the Reproductive Capacity of *Drosophila melanogaster*

Experiment No	Groups	Mean Number of F_1 Pupae per Maternal Fly	Deviation from Control
1.1	E(NM)	9.7	-16.38%
	SE(NM)	11.6	
1.2	E(NM)	10	-15.96%
	SE(NM)	11.9	
1.3	E(NM)	9.8	-20.16%
	SE(NM)	12.4	
1.4	E(NM)	10.4	-19.38%
	SE(NM)	12.9	
Average ± SD	E(NM)	9.975 ± 0.31	-18.24%
	SE(NM)	12.2 ± 0.57	

Statistical analysis, (single factor ANOVA test) shows that the probability that mean oviposition differs between the exposed and the sham exposed groups, owing to random variations, is $P < 5\times10^{-4}$. Therefore, the decrease in the reproductive capacity is due to the effect of the GSM field.

2. Effect of Modulated GSM Radiation-field on the Reproductive Capacity

We carried out four experiments (2.1-2.4), with modulated emission (the experimenter was speaking close to the mobile phone's microphone, during the exposures). The exposure parameters in this case simulate the situation when a user speaks on the mobile phone during connection. Results are listed in Table 2.

Table 2 shows the mean number of F_1 pupae (corresponding to the number of laid eggs) per maternal fly in the groups E, exposed to "Modulated" GSM field and in the corresponding sham exposed groups, SE, during the first three days of the insect's maximum oviposition.

The Modulated GSM 900 MHz signals induced a large decrease in the insect's reproductive capacity up to 60% as compared to the unexposed groups. No temperature increases were detected during the exposures and thus these effects are considered as non-thermal.

Table 2. Effect of Modulated GSM field on the Reproductive Capacity of *Drosophila melanogaster*

Experiment No	Groups	Mean Number of F_1 Pupae per Maternal Fly	Deviation from Control
2.1	E(M)	6.7	-48.85%
	SE (M) (Control)	13.1	
2.2	E	5.1	-56.78%
	SE (M) (Control)	11.8	
2.3	E	5.6	-53.72%
	SE (M) (Control)	12.1	
2.4	E	6	-53.125%
	SE (M) (Control)	12.8	
Average ± SD	E (M)	5.85 ± 0.67	-53.01%
	SE (M) (Control)	12.45 ± 0.6	

The reproductive capacity was much more decreased by modulated emission, (50-60%), than by non-modulated emission, (15-20%). Thus the effect is strongly dependent on radiation-field intensity. At the same time, the intensity of the modulated signal, is about ten times more powerful than the non-modulated signal. Thereby, the effect is not linearly dependent on radiation intensity.

The results from the first two sets of experiments are represented, in Figure 1.

The statistical analysis shows that the probability that mean oviposition differs between the exposed and the sham exposed groups, owing to random variations, is very small, $P < 10^{-5}$. Thus the recorded effect is due to the GSM signal.

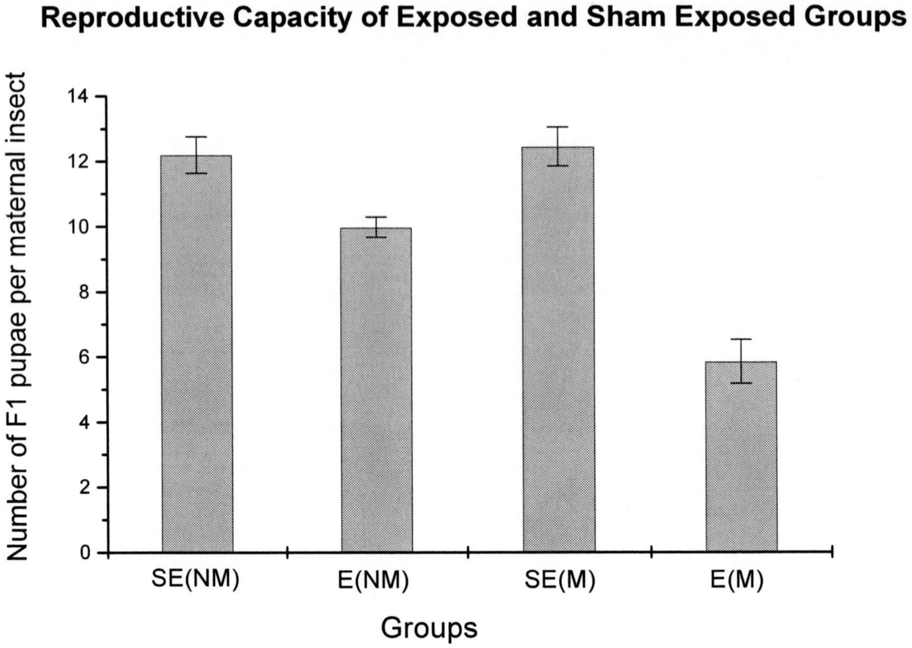

Figure 1. Reproductive Capacity of the groups exposed to non-modulated and modulated GSM 900 MHz field [E(NM), E(M)] and the corresponding sham exposed, [SE(NM), SE(M)], groups. [The error bars correspond to Standard Deviation].

3. Effects on the Reproductive Capacity of Each Sex

A third set of experiments (C) was carried out in order to record the effect of the GSM 900 MHz field on the reproductive capacity of each sex separately. The mobile phone was operating in speaking mode during the 6 min exposures, and the insects were separated into four groups (each one consisting again 10 male and 10 female insects): In the first group (E1), both male and female insects were exposed. In the second group (E2), only the females were exposed. In the third group (E3), we exposed only the males and the fourth group (SE) was sham exposed (control). Therefore in this third set of experiments, the 6-min daily exposures took place only during the first two days of each experiment while the males and females of each group were separated and the total number of exposures in each experiment was 2 instead of 5.

The results from this set of experiments are listed in Table 3 and represented graphically in Figure 2.

The results of this set of experiments show that the GSM field affects the reproductive capacity of both female and male insects. The female insects (E2) were more affected than males (E3) in these experiments. This is expected to be due to the fact that, by the time we started the exposures, spermatogenesis was already almost completed in male flies, while oogenesis had just started, (King 1970; Panagopoulos et al 2004).

Statistical analysis (single factor ANOVA test) shows that the probability that mean oviposition differs between the four groups because of random variations is $P < 10^{-7}$.

Table 3. Effect of "Modulated" GSM field on the Reproductive Capacity of each sex

Experiment No	Groups	Mean Number of F_1 Pupae Per Maternal Fly	Deviation from Control
3.1	SE(Control)	13.2	
	E1	8.5	-35.61%
	E2	9.4	-28.79%
	E3	11.7	-11.36%
3.2	SE (Control)	13.8	
	E1	7.6	-44.93%
	E2	8.9	-35.51%
	E3	12.1	-12.32%
3.3	SE (Control)	12.9	
	E1	7.8	-39.53%
	E2	9.3	-27.91%
	E3	11	-14.73%
3.4	SE (Control)	13.5	
	E1	6.9	-48.89%
	E2	7.8	-42.22%
	E3	12.2	-9.63%
Average ±SD	SE (Control)	13.35 ± 0.39	
	E1	7.7 ± 0.66	-42.32%
	E2	8.85 ± 0.73	-33.71%
	E3	11.75 ± 0.54	-11.985%

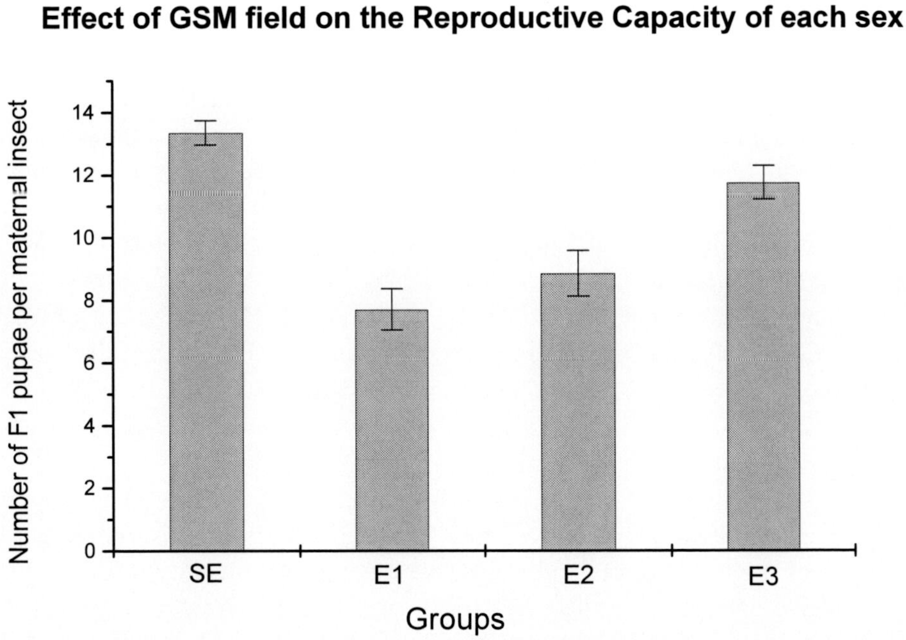

Figure 2. Effect of Modulated GSM field on the reproductive capacity of each sex of *Drosophila melanogaster*. Average mean number of F_1 pupae ±SD per maternal insect. SE: sham exposed groups, E1: groups that both sexes were exposed, E2: groups in which only the females were exposed, E3: groups in which only the males were exposed.

In the following fourth, fifth and sixth set of experiments, we used a dual band cellular mobile phone that could be connected to either GSM 900 or DCS 1800 networks simply by changing SIM ("Subscriber Identity Module") cards on the same handset. The highest Specific Absorption Rate (SAR), given by the manufacturer for human head, was 0.89 W/Kg. The exposure procedure was the same. The experimenter spoke on the mobile phone's microphone during the exposures. The GSM and DCS fields were thus "modulated" by the human voice, ("speaking emissions" or "GSM basic").

4. Comparison of Bioactivity between GSM 900 MHz and DCS 1800 MHz

In this set of experiments we separated the insects into four groups: a) the group Exposed to GSM 900MHz field with the mobile phone antenna in contact with the glass vial containing the flies (named as "900"), b) the group exposed to GSM 900MHz field with the antenna of the mobile phone at 1cm distance from the vial (named as "900A"), c) the group exposed to DCS 1800MHz field with the mobile phone antenna in contact with the glass vial (named as "1800"), and d) the Sham Exposed (Control) group (named as "SE"). The comparison between first and third group represents comparison with the usual exposure conditions between GSM 900 and DCS 1800 users, while comparison between second and third group represents comparison between possible effects of the RF frequencies of the two systems under equal radiation intensities. Therefore the second group (900A) was introduced for better comparison of effects between the two types of radiation.

Measured mean power densities in contact with the mobile phone antenna for six min of modulated emission, were 0.407 ± 0.061 mW/cm^2 for GSM 900 MHz and 0.283 ± 0.043 mW/cm^2 for DCS 1800 MHz. As was expected GSM 900 MHz intensity at the same distance from the antenna and with the same handset was higher than the corresponding DCS 1800 MHz. For the better comparison between the two systems of radiation we measured the GSM power density at different distances from the antenna and found that at 1cm distance, the GSM 900 MHz intensity was 0.286 ± 0.050 mW/cm^2, almost equal to DCS 1800 MHz at zero distance. Measured electric and magnetic field intensities in the ELF range for modulated field, excluding the ambient electric and magnetic fields of 50Hz, were 22.3 ± 2.2 V/m electric field intensity and 0.50 ± 0.08 mG magnetic field intensity for GSM at zero distance, 13.9 ± 1.6 V/m, 0.40 ± 0.07 mG correspondingly for GSM at 1 cm distance and 14.2 ± 1.7 V/m, 0.38 ± 0.07 mG correspondingly for DCS at zero distance. All these values are averaged over ten separate measurements of each kind ± standard deviation (SD).

Except for the power density - field measurements of the mobile phone emissions, we obtained the spectra of both types of radiation, plus the background spectrum in our lab, (Fig. 3). Each one of the two types of radiation gave a unique frequency spectrum. While GSM 900MHz gives a single peak around 900MHz, (Fig. 3b), DCS 1800MHz gives a main peak around 1800MHz and a smaller one around 900MHz, (Fig. 3c). The spectra were obtained by a Hewlett Packard 8595 E, (9 kHz-6.5 GHz), spectrum analyzer (USA).

We carried out ten replicate experiments. Results are listed in Table 4 and represented graphically, in Figure 4.

The results from this set of experiments show that the reproductive capacity in all the exposed groups is significantly decreased compared to the sham exposed groups. The decrease is maximum in the 900 groups, (48.25% compared to SE) and smaller in the 900A and the 1800 groups, (32.75% and 31.08% respectively), (Table 4). Although the decrease was even smaller in the 1800 groups than in 900A, differences between the 900A and 1800 groups were found to be within the standard deviation, (Table 4, Figure 4).

The statistical analysis shows that the probability that the reproductive capacity differs between groups, owing to random variations, is negligible, $P < 10^{-18}$.

Again, we did not detect any temperature increases, within the glass vials during the exposures.

The differences in the reproductive capacity between the groups were greater between 900 and 900A (owing to intensity differences between the two types of radiation) and much smaller between 900A and 1800, (owing to frequency differences between GSM and DCS), (Table 4).

This set of experiments shows that there is a difference in the bioactivity between GSM 900 MHz and DCS 1800 MHz and this difference is mainly due to the higher intensity of GSM 900 under the same exposure conditions, (differences between groups 900 and 900A) and not due to the different RF carrier frequencies, (differences between 900A and 1800 groups).

Intensity differences between the two types of cellular mobile telephony radiation depend also on the ability of communication between the antennas of the mobile phone and the corresponding base station. Even if GSM 900 usually has a higher intensity than DCS 1800, this situation can be reversed in certain places if GSM 900 has a much better signal perception between its antennas than DCS 1800, (Tisal 1998). Our results count for equal signal perception conditions between the two types of radiation.

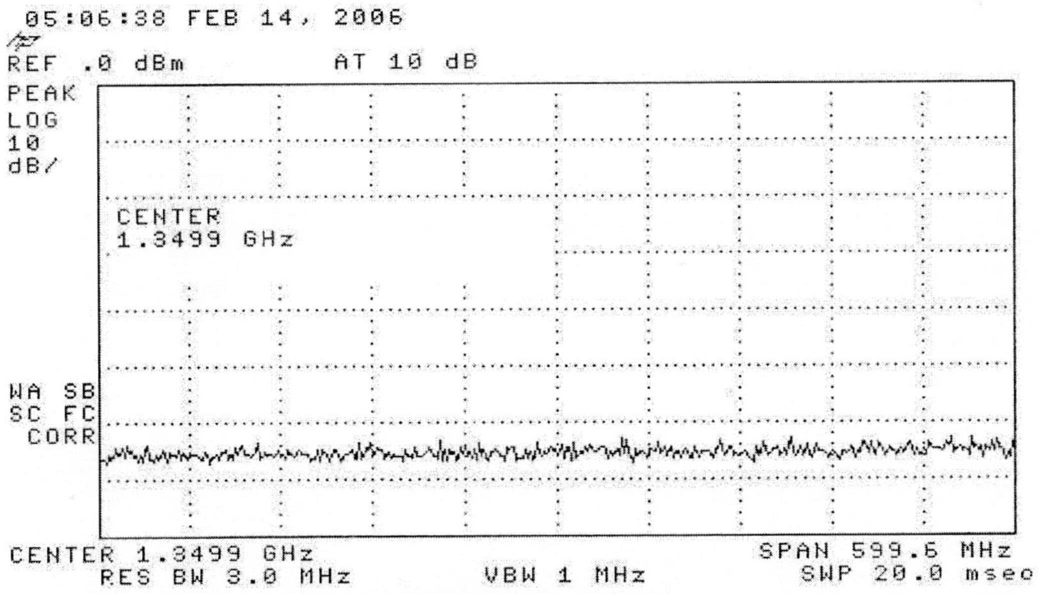

a. Background spectrum.

Figure 3. Continued on next page.

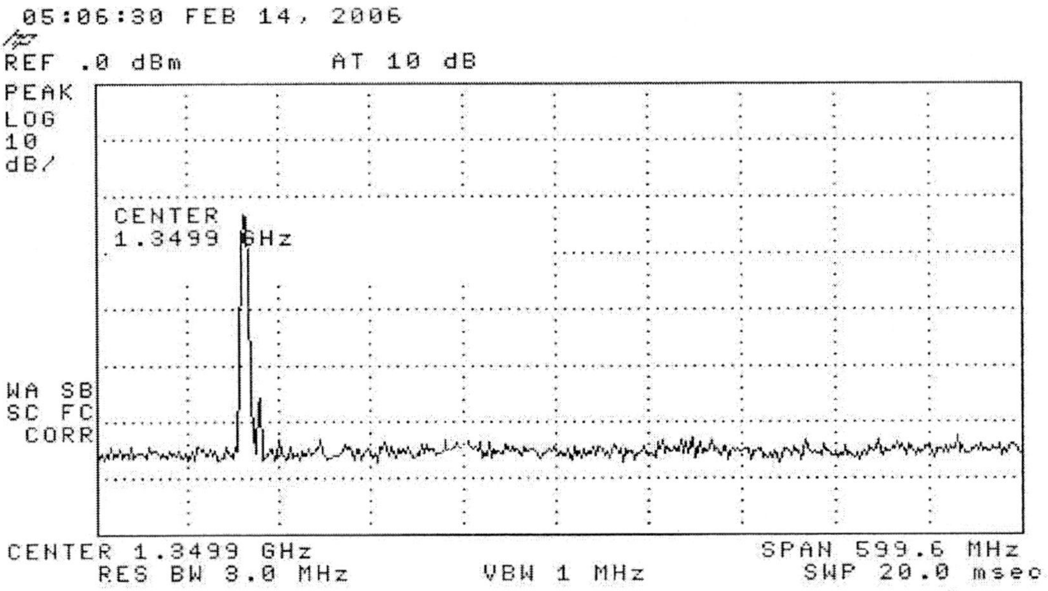

b. Spectrum of GSM 900 MHz.

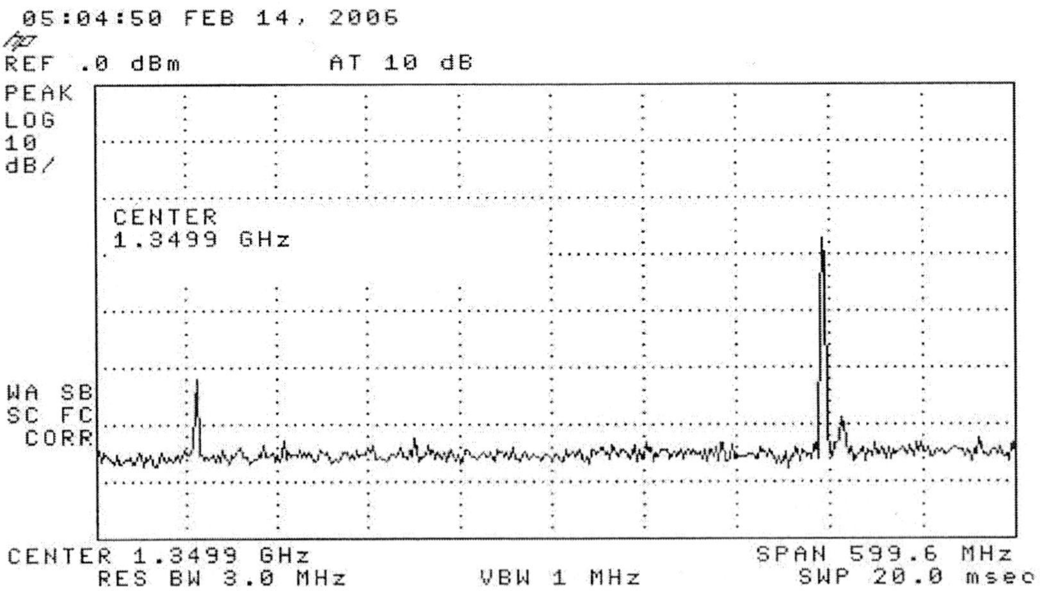

c. Spectrum of DCS 1800 MHz.

Figure 3. Background, GSM 900 MHz and DCS 1800 MHz spectra.

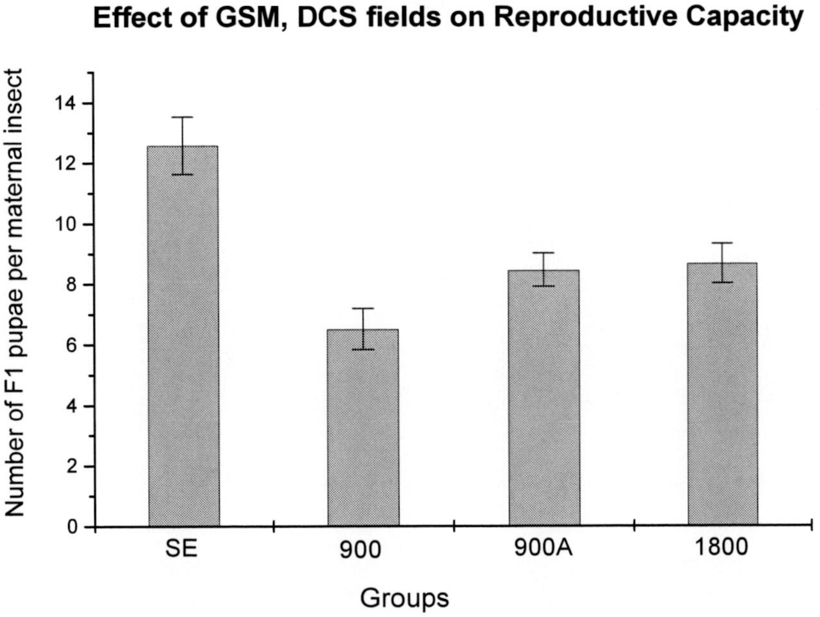

Figure 4. Reproductive Capacity (mean number of F1 pupae per maternal fly) of exposed (900, 900A, 1800) and sham exposed (SE) groups.

Table 4. Effect of Modulated GSM and DCS fields on the Reproductive Capacity of *Drosophila melanogaster*

Experiment No	Groups	Mean Number of F_1 Pupae per Maternal Fly	Deviation from Control
1	900	7.7	-42.54%
	900A	8.9	-33.58%
	1800	9.2	-31.34%
	SE (Control)	13.4	
2	900	5.8	-51.26%
	900A	8.1	-31.93%
	1800	7.9	-33.61%
	SE (Control)	11.9	
3	900	6.8	-46.03%
	900A	7.9	-37.30%
	1800	8.7	-30.95%
	SE (Control)	12.6	
4	900	7.4	-47.52%
	900A	9.7	-31.21%
	1800	9.9	-29.79%
	SE (Control)	14.1	
5	900	6.2	-52.31%
	900A	8.5	-34.62%
	1800	8.2	-36.92%
	SE (Control)	13	

Table 4. Continued

Experiment No	Groups	Mean Number of F_1 Pupae per Maternal Fly	Deviation from Control
6	900	6.1	-43.52%
	900A	8.2	-24.07%
	1800	7.8	-27.78%
	SE (Control)	10.8	
7	900	6.7	-47.66%
	900A	8.3	-35.16%
	1800	9	-29.69%
	SE (Control)	12.8	
8	900	6	-48.72%
	900A	7.9	-32.48%
	1800	8.4	-28.21%
	SE (Control)	11.7	
9	900	6.7	-49.24%
	900A	8.8	-33.33%
	1800	9.1	-31.06%
	SE (Control)	13.2	
10	900	5.7	-53.66%
	900A	8.3	-32.52%
	1800	8.5	-30.89%
	SE (Control)	12.3	
Average ± SD	900	6.51 ± 0.67	-48.25%
	900A	8.46 ± 0.55	-32.75%
	1800	8.67 ± 0.65	-31.08%
	SE (Control)	12.58 ± 0.95	

5. Radiation Bioactivity According to its Intensity (or According to the Distance from the Antenna)

The aim of this set of experiments was to investigate the dependence of GSM 900 MHz and DCS 1800 MHz bioactivity on their intensity, at different intensity levels that people are exposed to, from mobile phones and base station antennas. The radiation from base station antennas is almost identical to that of corresponding mobile phones but it is about 100 times stronger. Thus distances from mobile phones antennas correspond to about 100 times longer distances from base station antennas of the same type of radiation.

It is difficult to set up experiments regarding exposures from base station antennas since there is no way to have a sham exposed group of experimental animals under identical environmental conditions but without being exposed to the radiation at the same time. Thus we thought that the only way to simulate the reality of the exposure by a base station antenna is to expose the animals at different distances from a mobile phone within the lab.

Biological effects of mobile telephony signals at different intensities- distances from the antenna of a mobile phone handset, resembling effects from base station signals within residential areas, were not performed until now.

In each single experiment of this set, we separated the collected insects into thirteen groups: The first group (named “0”) was exposed to GSM 900 MHz or to DCS 1800 MHz field with the mobile phone antenna in contact with the glass vial containing the flies. The second (named "1"), was exposed to GSM 900 MHz or to DCS 1800 MHz field, at 1cm distance from the mobile phone antenna. The third group (named "10") was exposed to GSM 900 MHz or to DCS 1800 MHz field at 10 cm distance from the mobile phone antenna. The fourth group (named "20") was exposed to GSM 900 MHz or to DCS 1800 MHz field at 20 cm distance from the mobile phone antenna, etc, the twelveth group (named "100") was exposed to GSM 900 MHz or to DCS 1800 MHz field at 100 cm distance from the mobile phone antenna. Finally, the thirteenth group (named "SE") was the sham exposed. Each group consisted of ten male and ten female insects as previously.

Radiation and field measurements in contact and at different distances from the mobile phone antenna, for six min of modulated emission, for GSM 900 MHz and DCS 1800 MHz in the RF and ELF ranges excluding the background electric and magnetic fields of 50 Hz, are given in Table 5. All the values shown in Table 5, are averaged over ten separate measurements of each kind ± standard deviation (S.D.).

The measurements reveal that although ELF electric and magnetic fields fall at almost zero levels for distances longer than 50 cm from both GSM 900 and DCS 1800 mobile phone antennas, the RF components of the signals are still evident for distances up to 100 cm, (Table 5).

The Average mean values of reproductive capacity (number of F_1 pupae) from six identical experiments with each kind of radiation are shown in Table 6 and represented in Figures 5, 6. The statistical analysis (single factor Anova test) shows that the probability that the reproductive capacity differs between groups, owing to random variations, is negligible, $P < 10^{-8}$. Once again there was no temperature increases within the vials during the exposures.

The results show that the effect of mobile telephony radiation is maximum at zero distance (intensities higher than 200 μW/cm^2) and then becomes maximum at a distance of 20-30 cm from the antenna, depending on the intensity of radiation (GSM or DCS). This distance corresponds to an intensity around 10 μW/cm^2 for both types of radiation in regards to the RF components.

Table 5. Radiation and Field Intensities in the Microwave and ELF regions

Distance from Antenna (cm)	GSM Radiation Intensity at 900 MHz, (mW/cm^2)	GSM Electric Field Intensity at 217 Hz, (V/m)	GSM Magnetic Field Intensity at 217 Hz, (mG)	DCS Radiation Intensity at 1800 MHz, (mW/cm^2)	DCS Electric Field Intensity at 217 Hz, (V/m)	GSM Magnetic Field Intensity at 217 Hz, (mG)
0	0.380 ±0.058	19 ±2.5	0.9 ±0.15	0.250 ±0.048	13 ±2.1	0.6 ±0.08
1	0.260 ±0.047	12 ±1.7	0.7 ±0.13	0.068 ±0.015	6 ±0.8	0. 4 ±0.07
10	0.062 ±0.020	7 ±0.8	0.3 ±0.05	0.029 ±0.005	2.9 ±0.48	0. 2 ±0.05
20	0.032 ±0.008	2.8±0.4	0.2 ±0.04	0.012 ±0.002	0.7 ±0.12	0. 1±0.02
30	0.010 ±0.002	0.6 ±0.09	0.1 ±0.02	0.007 ±0.001	0.3 ±0.06	0.06 ±0.01
40	0.006 ±0.001	0.2 ±0.03	0.05 ±0.01	0.004 ±0.0007	0.1 ±0.04	0
50	0.003 ±0.0006	0.1 ±0.02	0	0.002 ±0.0003	0	0
60	0.002 ±0.0003	0	0	0.0016 ±0.0002	0	0
70	0.0017 ±0.0002	0	0	0.0014 ±0.0002	0	0
80	0.0012 ±0.0002	0	0	0.0008 ±0.0002	0	0
90	0.0010 ±0.0001	0	0	0.0005 ±0.0001	0	0
100	0.0004 ±0.0001	0	0	0.0002 ±0.0001	0	0

Table 6. Effect of Modulated GSM and DCS radiation-fields on the Reproductive Capacity at different Distances-Intensities from the antenna

Groups -Distance from mobile phone antenna, (cm)	Average Mean Number of F_1 Pupae per Maternal Fly, for GSM 900 MHz	Deviation from Sham Exposed Group	Average Mean Number of F_1 Pupae per Maternal Fly, for DCS 1800 MHz	Deviation from Sham Exposed Group
0	7.45 ± 0.72	-46.01 %	9.26 ± 0.68	-34.00 %
1	9.38 ± 0.61	-32.03 %	11.36 ± 0.54	-19.03 %
10	11.29 ± 0.80	-18.19 %	11.93 ± 0.71	-14.97 %
20	11.52 ± 0.79	-16.52 %	9.19 ± 0.62	-34.50 %
30	7.33 ± 0.58	-46.88 %	13.03 ± 0.83	-7.13 %
40	12.88 ± 0.98	-6.67 %	13.76 ± 0.85	-1.92 %
50	13.48 ± 0.81	-2.32 %	13.85 ± 0.74	-1.28 %
60	13.61 ± 0.84	-1.38 %	14.00 ± 0.91	-0.21 %
70	13.70 ± 0.91	-0.72 %	14.21 ± 0.89	+1.28 %
80	13.97 ± 0.77	+1.23 %	14.07 ± 0.79	+0.29 %
90	13.74 ± 0.96	-0.43 %	14.02 ± 1.03	-0.07 %
100	14.02 ± 1.01	+1.59 %	14.31 ± 1.08	+2.00 %

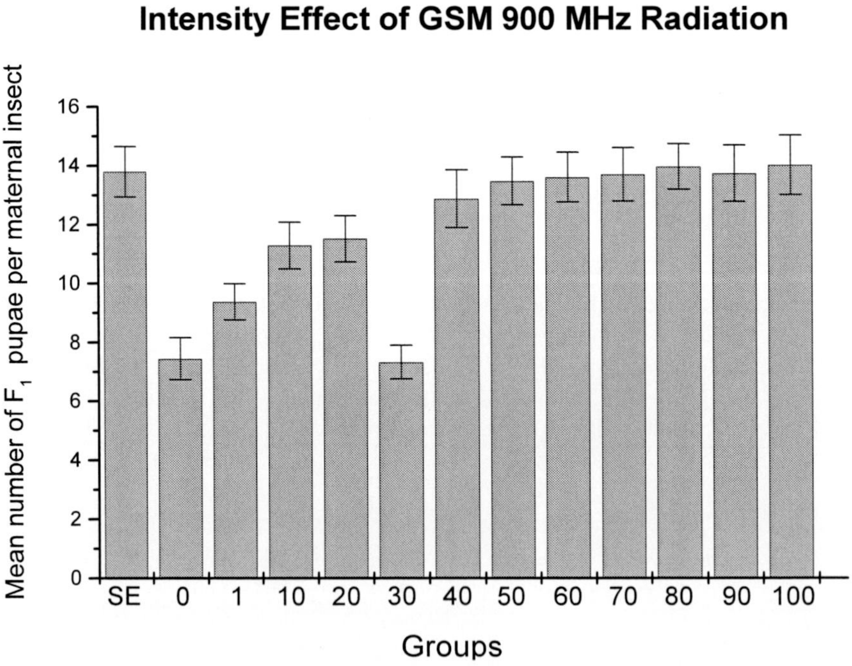

Figure 5. Reproductive Capacity in relation to the Distance from a GSM 900 MHz mobile phone antenna. The decrease in reproductive capacity is maximum at zero distance and at 30 cm distance from the antenna, corresponding to RF intensities 380μW/cm^2 and 10μW/cm^2 (Table 5).

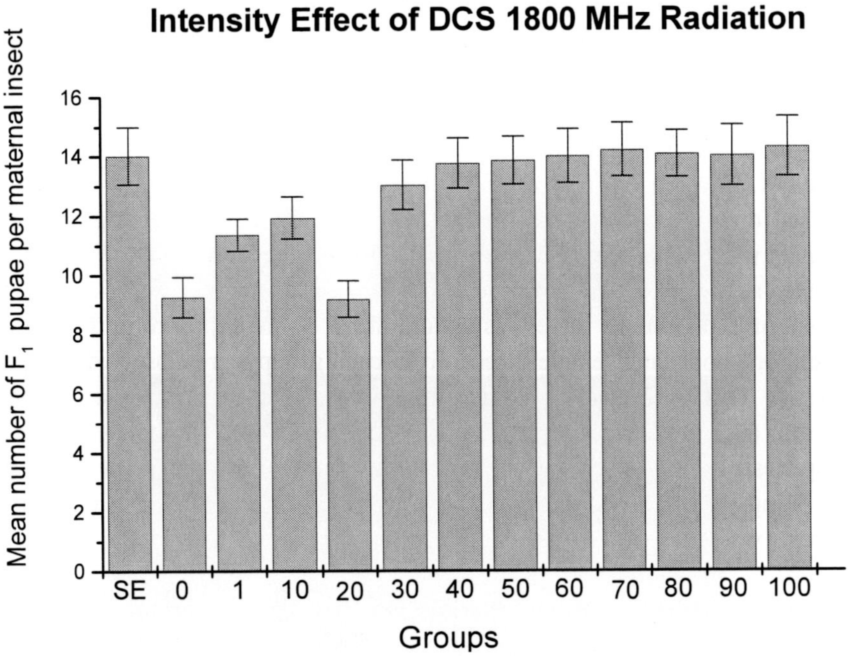

Figure 6. Reproductive Capacity in relation to the Distance from a DCS 1800 MHz mobile phone antenna. The decrease in reproductive capacity is maximum at zero distance and at 20 cm distance from the antenna, corresponding to RF intensities 250 $\mu W/cm^2$ and 12 $\mu W/cm^2$ (Table 5).

The effect on the reproductive capacity diminishes considerably for distances longer than 50 cm from the mobile phone antenna and disappears for distances longer than 80-90 cm, corresponding to radiation intensities smaller than 1 $\mu W/cm^2$. For distances longer than 50 cm where the ELF components fall within the background, the decrease in reproductive capacity is within the standard deviation. This might suggest that the ELF components of digital mobile telephony signals, play a key role in their bio-activity, alone or in conjunction with the RF carrier wave.

We have recorded the existence of an "intensity window" around 10 $\mu W/cm^2$ (in regards to the RF intensity) where the bio-effect becomes even more intense than at intensities higher than 200 $\mu W/cm^2$. This intensity window appears at a distance of 20-30 cm from a mobile phone antenna, which corresponds to a distance of about 20-30 meters from a base station antenna. Since mobile telephony base station antennas are usually located within residential areas, at distances 20-30 m from such antennas there are often houses and work places where people are exposed up to 24 hours per day.

Although intensity windows on the bio-effects of RF radiations have been recorded since many years, (Bawin et al 1975; 1978; Blackman et al, 1980), there is still no widely accepted explanation for their existence.

6. The Decrease in Reproductive Capacity is due to Cell Death in the Gonads

In each experiment of this final sixth set, we separated the collected insects into five groups. The first four groups were the same just as in the No 4 experiments: The first group (named

Table 7. Effect of GSM, DCS fields on Ovarian Cell Death

Groups	Dev. Stages	Ratio of TUNEL Positive to Total Number of Egg-chambers of each dev. stage	Sum Ratio of TUNEL Positive to Total Number of Egg-chambers of all stages	Percentage of TUNEL Positive Egg chambers	Deviation from Sham Exposed Groups
SE	Germarium 1-6 7-8 9-10	37/186 32/1148 78/364 7/282	154/1980	7.78%	0%
900	Germarium 1-6 7-8 9-10	165/189 675/1252 310/384 165/262	1315/2087	63.01%	+55.23%
900A	Germarium 1-6 7-8 9-10	116/184 484/1248 213/374 117/257	930/2063	45.08%	+37.30%
1800	Germarium 1-6 7-8 9-10	101/169 388/1202 196/358 91/239	776/1968	39.43%	+31.65%

Figure 7a, shows an ovariole from a sham exposed female insect, containing egg chambers from germarium to stage 8, all TUNEL negative. This was the typical picture in the vast majority of ovarioles and separate egg chambers from female insects of the sham exposed and control groups. In the SE groups, only 154 egg chambers (including germaria) out of a total of 1980 in 8 replicate experiments (7.78%), were TUNEL positive (Table 7), a result that is in full agreement with the rate of spontaneously degenerated egg chambers normally observed during *Drosophila* oogenesis, (Nezis et al., 2000; Baum et al., 2005).

Figure 7b shows an ovariole of exposed female insect (group 900A), with a TUNEL positive signal in the nurse cells at both checkpoints, germarium and stage 8, while egg chambers of intermediate stages are TUNEL negative. Corresponding pictures from 900 and 1800 (data not shown) had identical characteristics. The two checkpoints in all groups (exposed and SE/C) had the highest percentages of cell death compared to the other developmental stages 1-6 and 9-10, (Table 7). While in the SE groups the sum ratio of TUNEL positive to total number of egg chambers was slightly higher in stages 7-8 (78/364) than in the germarium (37/186), in all three exposed groups this ratio was higher in the germarium than in stages 7-8, (Table 7).

Figure 7c, shows ovarioles of exposed female insects (group 900A), with a TUNEL positive signal at all developmental stages from germarium to 7-8 and in all the cell types of the egg chamber, (nurse cells, follicle cells and the oocyte).

Although in most pictures the TUNEL positive signal was most evident in the nurse cells, in the majority of the egg chambers in all the exposed groups, a TUNEL positive signal was detected in all three kinds of egg chamber cells, (figures 1c).

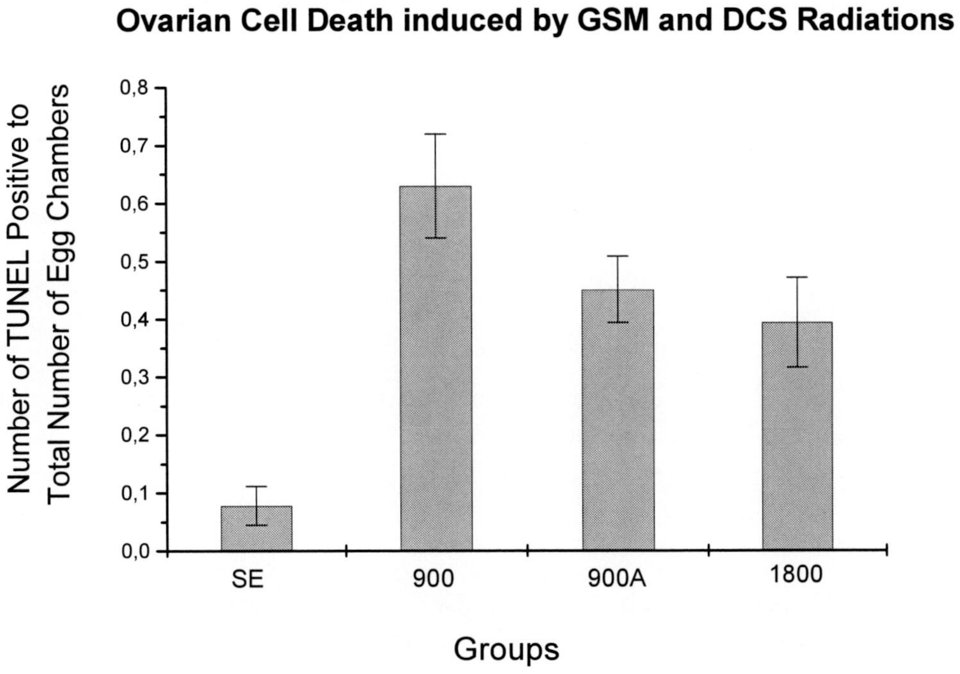

Figure 8. Mean ratio of Ovarian Cell Death (Number of TUNEL Positive to Total Number of Egg Chambers), in each experimental group ± SD, (0.078± 0.0335 in SE, 0.630± 0.0898 in 900, 0.451± 0.0574 in 900A and 0.394± 0.0777 in 1800).

In the SE groups the ratio of TUNEL positive egg chambers of stages 9-10 was very small (7/282). In contrast, the corresponding ratio in all three exposed groups was significantly higher, (165/262 in 900, 117/257 in 900A and 91/239 in 1800).

The summarised data of Table 7 are represented in Fig.8.

The statistical analysis, (single factor Analysis of Variance test), showed that the probability that groups differ between them because of random variations, is negligible, $P<10^{13}$.

Our experiments and the statistical analysis show that genomic DNA fragmentation of the egg chambers cells is induced by the mobile telephony radiation. Both types of radiation, GSM 900MHz and DCS 1800MHz induce cell death in a large number (up to 55% in relation to control), of ovarian egg chambers in the exposed insects with only 6 min exposure per day for a limited period of 6 days.

DNA fragmentation is induced in all cases predominantly at the two developmental stages named checkpoints, germarium and stages 7-8. Since the above check points were already known to be the most sensitive stages in response to other stress factors, (Chao and Nagoshi 1999; De Lorenzo et al., 1999; Nezis et al., 2000; Drummond-Barbosa and Spradling 2001; McCall 2004), such an observation could be expected. Our results show that these two checkpoints are the most sensitive stages also in response to electromagnetic stress. However the germarium checkpoint was found to be even more sensitive than stages 7-8 in response to this particular stress. Thereby the two check points are not equally responsive to distinct types of stress and may therefore also respond differentially to other types of stress stimuli. A possible explanation for the more sensitive germarium stage is that it may be more effective in evolutionary terms for the animal to block development of any defective egg chamber at the beginning rather than at later stages, in order to prevent the waste of precious nutrients.

In the sham exposed/control groups, induced DNA fragmentation was observed almost exclusively at the two developmental stages named check-points (37/186 in the germarium and 78/364 in stage 7-8) and only in few cases at the other provitellogenic and vitellogenic stages, 1-6 (32/1148) and 9-10 (7/282), correspondingly. In contrast, ovarian egg chambers of animals from all three exposed groups, were found to be TUNEL positive to a high degree at all developmental stages from germarium to stage 10, (Table 7).

In all cases (both in the sham exposed/control and also in the exposed groups), the TUNEL positive signal was more intense at the two check points, germarium and stages 7-8, than at the other developmental stages.

There was no detectable temperature increase within the vials during the exposures, therefore the effects are considered as non-thermal.

In this set of experiments, cell death was detected for the first time during all the developmental stages of early and mid oogenesis in *Drosophila*, from germarium to stage 10 and in all types of egg chamber cells, (nurse cells, follicle cells, oocyte). A possible explanation for these effects is that the electromagnetic stress induced in the ovarian cells by the GSM and DCS fields, is a new and probably more intense type of external stress, against which ovarian cells do not have adequate defence mechanisms like they do in the case of poor nutrition or chemical stress.

It is important to emphasize that the recorded effect in the oocyte which undergoes meiosis during the last stages of oogenesis, may result in heritable mutations upon DNA damage induction and repair, if not in cell death.

The results of this set of experiments reveal that the large decrease of reproductive capacity found in the previous sets of experiments is due to elimination of large numbers of egg chambers during early and mid oogenesis, either via stress induced apoptosis or necrosis of their constituent cells, caused by the mobile telephony radiation.

Our present results are in agreement with results of other experimenters reporting DNA damage in other cell types, assessed by different methods than ours, after *in vivo* or *in vitro* exposure to GSM radiation, (Diem et al., 2005; Markova et al., 2005; Salford et al., 2003; Lai and Singh 1995; 1996).

We do not know if the ovarian cell death found in our experiments to be induced by mobile telephony radiation is due to apoptosis, i.e. caused by the organism in response to the electromagnetic stress, or the result of necrosis caused directly by the electromagnetic radiation. This important issue remains to be uncovered.

A Plausible Mechanism for Mobile Telephony Radiation Bioeffects

As we have previously reported, (Panagopoulos et al. 2000b; 2002; Panagopoulos and Margaritis 2003b), any external oscillating electromagnetic field can induce a forced-vibration on the free ions that exist in large concentrations inside and outside all living cells in biological tissue playing a key role in all cellular functions initiating or accompanying all cellular biochemical processes.

The forced-vibrational movement of the free ions is described by the equation,

$$m_i \frac{d^2x}{dt^2} + \lambda \frac{dx}{dt} + m_i \omega_o^{\,2} x = E_o z q_e \sin\omega t \qquad [1]$$

in the case of an external harmonically oscillating electric field: $E = E_o \sin\omega t$ with circular frequency: $\omega = 2\pi\nu$, (ν, the frequency), where: z is the ion's valence, $q_e = 1.6\times10^{-19}$ Cb, the electron's charge, $F_2 = - m_i \omega_o^{\,2} x$, a restoration force proportional to the displacement distance x of the free ion, m_i the ion's mass and $\omega_o = 2\pi\nu_o$, with ν_o the ion's oscillation self-frequency if the ion were left free after its displacement x. In our case, this restoration force is found to be very small compared to the other forces and thus does not play any important role. $F_3 = -\lambda u$ is the damping force, where $u = \frac{dx}{dt}$, is the ion's velocity and λ, is the attenuation coefficient for the ion's movement, which for the cytoplasm or the extracellular medium is calculated to be $\lambda \cong 10^{-12}$ Kg/sec, while for ions moving inside channel proteins, is calculated to have a value: $\lambda \cong 6.4\times10^{-12}$ Kg/sec, (in the case of Na^+ ions, moving through open Na^+ channels), (Panagopoulos et al 2000b).

We have shown that the general solution of equation [1], is:

$$x = \frac{E_o z q_e}{\lambda\omega} \cos\omega t - \frac{E_o z q_e}{\lambda\omega} \qquad [2]$$

Since the second term of [2] is constant, the vibrational movement is described by the equation:

$$x = \frac{E_o z q_e}{\lambda\omega} \cos\omega t \qquad [3]$$

Eq. [3] shows that the forced - vibration is in phase with the external force.The amplitude of the free ions forced vibration is,

$$A = \frac{E_o z q_e}{\lambda\omega} \qquad [4]$$

Thus, the amplitude is proportional to the intensity and inversely proportional to the frequency of the external oscillating field.

Once this amplitude exceeds some critical value the coherent forces that the ions exert on the voltage sensors of voltage-gated membrane channels can trigger the irregular opening or closing of these channels, thus disrupting cell's electrochemical balance and function.

We have shown that in the most bioactive case of pulsed fields and for double valence cations (i.e. Ca^{+2}) interacting with the channel sensor, the condition for irregular gating of the channel becomes:

$$E_o \geq \nu \times 0.625 \times 10^{-4} \quad [5]$$

(ν in Hz, E_o in V/m). Whenever [5] is satisfied, the external field E can irregularly gate the ion channel.

Relation [5] declares that external ELF electric fields with intensities less than tenths of a mV/m should theoretically be able to disrupt cell function by irregular gating of ion channels (!)

According to this mechanism, lower frequency fields are the most bioactive ones and additionally pulsed fields are shown to be more bioactive than continuous, (uninterrupted), ones, (Panagopoulos et al., 2002).

Thereby, the ELF components of the mobile telephony signals are certainly within the criteria of this theory and thus able to produce the reported effects on living organisms.

Somebody may wonder, how could be possible that irregular gating of ionic channels on a cell membrane could lead to cell death.

Let us consider the irregular gating of ion channels on a cell's plasma membrane. If the electrochemical balance is destroyed by irregular increase of intracellular ion concentration, then water molecules may enter the cell driven by osmotic forces, proportional to the concentration increase. Such an effect could be able to cause the cell to swell out and the plasma membrane to get ruptured, resulting to cell necrosis.

It is known that perturbations of intracellular Ca^{+2} concentrations are responsible for apoptotic triggering, (Zhou et al., 1998; Sheikh and Huang, 2004; Santini et al. 2005). Therefore, another scenario of cell death, caused by irregular gating of ion channels, could be that due to altered intracellular Ca^{+2} concentrations, a false signal may be given to initiate apoptosis.

A common event leading to both apoptosis and necrosis is mitochondrial membrane permeabilization, (Armstrong 2006). This can also be done by direct action of an external EMF on mitochondrial membrane Ca^{+2} channels. Apoptosis is connected with increased mitochondrial concentration of Ca^{+2} ions, released from the endoplasmic reticulum, (Santini et al., 2005). A false uptake of Ca^{+2} ions by mitochondria can be due to irregular opening of mitochondrial Ca^{+2} channels, or due to increased cytosolic Ca^{+2} concentration, caused by irregular release either through the membrane of endoplasmic reticulum or through the plasma membrane. In all cases this could be done by irregular gating of electrosensitive Ca^{+2} channels which exist in all cell membranes.

We have just described few of the many hypothetical but very possible biochemical scenarios which could very explain by means of the above described biophysical theory, the effects of DNA damage recorded in our experiments as well as in other labs experiments, (Diem et al., 2005; Markova et al., 2005; Salford et al., 2003; Lai and Singh 1995; 1996).

Conclusions

As shown by increasing number of biological, clinical and epidemiological studies, the radiations emitted by mobile telephony, at levels that people are daily exposed, are highly bioactive producing a variety of effects on living organisms.

Our studies regarding the effects of mobile telephony radiations on a biological model, the reproductive capacity of the insect *Drosophila melanogaster*, have investigated different physical parameters of these radiations, like intensity, carrier frequency, pulse repetition frequency, distance from the antenna, e.t.c.

Our experiments have shown a large decrease in reproductive capacity caused by the GSM and DCS fiels-radiation. The recorded effect is due to extensive DNA fragmentation on reproductive cells of the experimental animal, induced by these fields-radiation.

Thus, digital mobile telephony radiations nowadays exert an intense biological action able to kill cells, damage DNA, or decrease dramatically the reproductive capacity of living organisms. Diminishes of bird and insect populations can be explained according to reproduction decreases. Phenomena like headaches, fatigue, sleep disturbances, memory loss e.t.c. reported as "microwave syndrome" can possibly be explained by cell death on a number of brain cells during daily exposures from mobile telephony antennas.

Our experiments show that radiation intensities higher than 1 $\mu W/cm^2$ are able to decrease reproduction of living organisms by killing reproductive cells. Our opinion is that the international exposure limits for these radiations should be set not higher than 1 $\mu W/cm^2$. Since short term exposures for few minutes per day are able to produce so intense effects on living organisms, the criteria should not be set according to average values but according to maximum values during the exposure periods.

Our experiments reveal that exposure at a distance of 20-30 cm from a mobile phone can be even more bioactive than exposure in contact with the antenna, due to the existence of an "intensity window" around 10 $\mu W/cm^2$. This intensity, in the case of a usual base station antenna corresponds to a distance of about 20-30 m from the antenna.

Although both types of radiation examined are found to be highly bioactive, GSM 900 MHz seems to be even more bioactive than DCS 1800 MHz, mainly due to higher intensity, but also even when it is emitted at almost the same intensity. Since differences in bioactivity between the two types of radiation under the same intensity are within standard deviation, it seems that RF carrier frequency plays a minimal role in the bioactivity of this radiation, in contrast to the ELF pulse repetition frequencies and the radiation and field intensities that seem to be of great importance in regards to bioactivity.

The ELF components of the mobile telephony signals, seem to play a key role on their bio-effects, since the recorded effects are considerably diminished at distances that these components fall within the background of stray 50 Hz electric and magnetic fields. This supports that lower frequency fields are more bioactive than higher frequency ones with the same rest characteristics, as it is predicted by our theory, (Panagopoulos et al 2000b; 2002), and supported by other experimental evidence, (Lin Liu and Adey 1982; Penafiel et al 1997).

A plausible explanation of the effects of mobile telephony radiations on living organisms is given by the biophysical mechanism that we have proposed, (Panagopoulos et al. 2000b; 2002; Panagopoulos and Margaritis 2003b). According to this mechanism, altered intracellular ionic concentrations due to irregular gating of ion channels on the cell membranes by an external electromagnetic field can initiate cell death through apoptosis or necrosis.

Similar effects on humans with those recorded in our experiments on insects, are considered to be possible because first, insects are found to be more resistant to radiations than mammals, (Koval and Kazmar 1988, Koval et al 1979, 1977, Abrahamson et al 1973) and second, our results are in agreement with reported effects on mammals, (Lai and Singh 1995; 1996; Aitken et al., 2005; Salford et al., 2003).

Scientific evidence implies the need of reconsideration of the current exposure criteria to account for non-thermal effects which constitute the large majority of the recorded biological and health effects. Since Mobile Telephony has become part of our daily life, a better design of base station antenna networks towards the least exposure of residential areas and a very cautious use of mobile phones, is necessary.

References

Abdel-Rassoul G, El-Fateh OA, Salem MA, Michael A, Farahat F, El-Batanouny M, Salem E. (2007): Neurobehavioral effects among inhabitants around mobile phone base stations. *Neurotoxicology*. **28**(2):434-40.

Abrahamson, S., Bender, M.A., Conger, A.D., Wolff, S. (1973), "Uniformity of radiation-induced mutation rates among different species", *Nature* **245**, 460-462.

Aitken RJ, Bennetts LE, Sawyer D, Wiklendt AM, King BV. (2005): Impact of radio frequency electromagnetic radiation on DNA integrity in the male germline. *Int J Androl*., **28**(3): 171-9.

Armstrong JS. (2006): Mitochondrial membrane permeabilization: the sine qua non for cell death. *Bioessays*. **28**(3):253-60. Review.

Balmori A, (2005): Possible effects of electromagnetic fields from phone masts on a population of White Stork (Ciconia ciconia), *Electromagnetic Biology and Medicine*, **24**(2); 109-119.

Banik S, Bandyopadhyay S, Ganguly S. (2003): Bioeffects of microwave-a brief review. *Bioresour Technol*. **87**(2):155-9. Review.

Barteri M, Pala A, and Rotella S, (2005): Structural and kinetic effects of mobile phone microwaves on acetylcholinesterase activity, *Biophysical Chemistry*, **113**(3), 245-253.

Baum JS, St George JP, McCall K. (2005): Programmed cell death in the germline. *Semin Cell Dev Biol*. **16**(2):245-59.

Bawin, S.M, Kaczmarek, L.K. and Adey, W.R., (1975): "Effects of modulated VMF fields, on the central nervous system", *Ann.N.Y. Acad.Sci*., **247**, 74-81.

Bawin S.M., Adey W.R., (1976): Sensitivity of calcium binding in cerebral tissue to weak environmental electric fields oscillating at low frequency, *Proc. Natl.Acad.Sci., U.S.A*., **73**, 1999-2003.

Bawin,S.M., Adey, W.R., Sabbot,I.M., (1978), "Ionic factors in release of $^{45}Ca^{2+}$ from chick cerebral tissue by electromagnetic fields". *Proc.Natl.Acad.Sci., U.S.A*., **75**, 6314-6318.

Belyaev IY, Hillert L, Protopopova M, Tamm C, Malmgren LO, Persson BR, Selivanova G, Harms-Ringdahl M., (2005): 915 MHz microwaves and 50 Hz magnetic field affect chromatin conformation and 53BP1 foci in human lymphocytes from hypersensitive and healthy persons, *Bioelectromagnetics* **26**(3), pp. 173-84.

Blackman,C.F., Benane,S.G., Elder,J.A., House, D.E., Lampe, J.A. and Faulk,J.M., (1980). "Induction of calcium - ion efflux from brain tissue by radiofrequency radiation: Effect of sample number and modulation frequency on the power - density window". *Bioelectromagnetics, (N.Y.)*, **1,** 35 - 43.

Blackman C.F., Kinney L.S., House D.E., Joines W.T., (1989), "Multiple power-density windows and their possible origin", *Bioelectromagnetics*, **10**(2), 115-128.

Bortkiewicz A, Zmyślony M, Szyjkowska A, Gadzicka E. (2004): Subjective symptoms reported by people living in the vicinity of cellular phone base stations: review. *Med Pr.* **55**(4):345-51. Review. Polish.

Bos M. and Boerema A. (1981): "Phenetic distances in the *Drosophila melanogaster* - subgroup species and oviposition-site preference for food components", *Genetica* **56,** 175-183.

Burch JB, Reif JS, Noonan CW, Ichinose T, Bachand AM, Koleber TL, Yost MG. (2002): Melatonin metabolite excretion among cellular telephone users. *Int J Radiat Biol.* **78**(11):1029-36.

Capri M, Scarcella E, Bianchi E, Fumelli C, Mesirca P, Agostini C, Remondini D, Schuderer J, Kuster N, Franceschi C, Bersani F. (2004a): 1800 MHz radiofrequency (mobile phones, different Global System for Mobile communication modulations) does not affect apoptosis and heat shock protein 70 level in peripheral blood mononuclear cells from young and old donors. *Int J Radiat Biol.* **80**(6):389-97.

Capri M, Scarcella E, Fumelli C, Bianchi E, Salvioli S, Mesirca P, Agostini C, Antolini A, Schiavoni A, Castellani G, Bersani F, Franceschi C., (2004b): In vitro exposure of human lymphocytes to 900 MHz CW and GSM modulated radiofrequency: studies of proliferation, apoptosis and mitochondrial membrane potential, *Radiat Res.*; **162**(2): 211-8.

Caraglia M, Marra M, Mancinelli F, D'Ambrosio G, Massa R, Giordano A, Budillon A, Abbruzzese A, Bismuto E. (2005): Electromagnetic fields at mobile phone frequency induce apoptosis and inactivation of the multi-chaperone complex in human epidermoid cancer cells. *J Cell Physiol.* **204**(2): 539-48.

Chao S. and Nagoshi R.N., (1999), Induction of apoptosis in the germline and follicle layer of *Drosophila* egg chambers, *Mech. Dev.* **88**, pp. 159–172.

Clark M.P., *Networks and Telecommunications,* 2nd ed., Wiley, 2001.

Cranfield CG, Wieser HG, Dobson J. (2003): Exposure of magnetic bacteria to simulated mobile phone-type RF radiation has no impact on mortality. *IEEE Trans Nanobioscience.* **2**(3):146-9.

Connolly J.B. and Tully T. (1998) *in* Drosophila; A Practical Approach, (Roberts D.B., ed.), Oxford University Press, New York.

De Lorenzo C., Strand D. and Mechler B.M., (1999), Requirement of *Drosophila l(2)gl* function for survival of the germline cells and organization of the follicle cells in a columnar epithelium during oogenesis, *Int. J. Dev. Biol.* **43**, pp. 207–217.

Demsia G, Vlastos D, Matthopoulos DP. (2004): Effect of 910-MHz electromagnetic field on rat bone marrow. *ScientificWorldJournal.* **4** Suppl 2:48-54.

Diem E, Schwarz C, Adlkofer F, Jahn O, Rudiger H., (2005): Non-thermal DNA breakage by mobile-phone radiation (1800 MHz) in human fibroblasts and in transformed GFSH-R17 rat granulosa cells in vitro. *Mutat Res.* **583**(2): 178-83.

Drummond-Barbosa D. and Spradling A.C., (2001), Stem cells and their progeny respond to nutritional changes during *Drosophila* oogenesis, *Dev. Biol.* **231**, pp. 265–278.

Dutta S.K., Subramaniam A., Ghosh B., Parshad, R., (1984), "Microwave radiation - induced calcium ion efflux from human neuroblastoma cells in culture". *Bioelectromagnetics, (N.Y.),* **5**, 71 - 78.

EMF world wide standards, International EMF project, www.who.int/docstore/peh-emf/EMFStandards/who-0102/Worldmap5.htm

Esen F, Esen H. (2006): Effect of electromagnetic fields emitted by cellular phones on the latency of evoked electrodermal activity. *Int J Neurosci.* **116**(3):321-9.

Everaert J, Bauwens D. (2007): A possible effect of electromagnetic radiation from mobile phone base stations on the number of breeding house sparrows (Passer domesticus). *Electromagn Biol Med.***26**(1):63-72.

Forgács Z, Kubinyi G, Sinay G, Bakos J, Hudák A, Surján A, Révész C, Thuróczy G. (2005): Effects of 1800 MHz GSM-like exposure on the gonadal function and hematological parameters of male mice, *Magy Onkol.* **49**(2):149-51.

Franke H, Ringelstein EB, Stögbauer F. (2005): Electromagnetic fields (GSM 1800) do not alter blood-brain barrier permeability to sucrose in models in vitro with high barrier tightness. *Bioelectromagnetics*. **26**(7):529-35.

French PW, Penny R, Laurence JA, McKenzie DR. (2001): Mobile phones, heat shock proteins and cancer. *Differentiation*. **67**(4-5):93-7.

Gadhia PK, Shah T, Mistry A, Pithawala M, Tamakuwala D, (2003): A preliminary study to assess possible chromosomal damage among users of digital mobile phones, *Electromagnetic Biology and Medicine*, **22**(2), 149-159.

Gatta L, Pinto R, Ubaldi V, Pace L, Galloni P, Lovisolo GA, Marino C, Pioli C. (2003): Effects of in vivo exposure to GSM-modulated 900 MHz radiation on mouse peripheral lymphocytes. *Radiat Res*. **160**(5):600-5.

Gavrieli Y, Sherman Y, Ben-Sasson SA. (1992): Identification of programmed cell death in situ via specific labeling of nuclear DNA fragmentation. *J Cell Biol*. **119**(3): 493-501.

Goodman E.M., Greenebaum B. and Marron M.T., (1995), "Effects of Electro- magnetic Fields on Mollecules and Cells", *International Rev. Cytol*. **158**, 279-338.

Grigor'ev IuG. (2003): Biological effects of mobile phone electromagnetic field on chick embryo (risk assessment using the mortality rate), *Radiats Biol Radioecol* **43**(5):541-3.

Hamnerius I, Uddmar Th, (2000): Microwave exposure from mobile phones and base stations in Sweden, *Proceedings, International Conference on Cell Tower Siting, Salzburg*, p.52-63, www.land-salzburg.gv.at/celltower

Hardell L, Carlberg M, Söderqvist F, Mild KH, Morgan LL. (2007): Long-term use of cellular phones and brain tumours: increased risk associated with use for > or =10 years. *Occup Environ Med.* **64**(9):626-32. Review.

Hardell L, Carlberg M and Hansson Mild K, (2006): Pooled analysis of two case-control studies on use of cellular and cordless telephones and the risk for malignant brain tumours diagnosed in 1997-2003, *Int Arch Occup Environ Health*. 2006 Sep;79(8):630-9.

Hardell L. and Hansson Mild K. (2006) Mobile phone use and risk of acoustic neuroma: results of the interphone case-control study in five North European countries. *Br J Cancer*. **94**(9):1348-9.

Hardell L, Walker MJ, Walhjalt B, Friedman LS, Richter ED. (2007b): Secret ties to industry and conflicting interests in cancer research. *Am J Ind Med.* **50**(3):227-33.

Hillebrand F. (Ed), GSM and UTMS, Wiley, 2002.

Hook GJ, Zhang P, Lagroye I, Li L, Higashikubo R, Moros EG, Straube WL, Pickard WF, Baty JD, Roti Roti JL. (2004): Measurement of DNA damage and apoptosis in Molt-4 cells after in vitro exposure to radiofrequency radiation. *Radiat Res*. **161**(2):193-200.

Hutter H-P, Moshammer H, Wallner P Kundi M, (2006): Subjective symptoms, sleeping problems, and cognitive performance in subjects living near mobile phone base stations. *Occupational and Environmental Medicine*; **63**:307-313.

Hyland G.J., (2000): Physics and Biology of Mobile Telephony, *Lancet*, **356**, 1833-36.

Huber R, Treyer V, Borbely AA, Schuderer J, Gottselig JM, Landolt HP, Werth E, Berthold T, Kuster N, Buck A, Achermann P. (2002): Electromagnetic fields, such as those from mobile phones, alter regional cerebral blood flow and sleep and waking EEG. *J Sleep Res*. **11**(4):289-95.

Huber R, Graf T, Cote KA, Wittmann L, Gallmann E, Matter D, Schuderer J, Kuster N, Borbély AA, Achermann P. (2000): Exposure to pulsed high-frequency electromagnetic field during waking affects human sleep EEG. *Neuroreport*. **11**(15):3321-5.

ICNIRP (1998), "Guidelines for limiting exposure to time-varying electric, magnetic and electromagnetic fields (up to 300GHz)", *Health Phys*. **74**, 494-522.

Inskip PD, Tarone RE, Hatch EE, Wilcosky TC, Shapiro WR, Selker RG, Fine HA, Black PM, Loeffler JS, Linet MS. (2001): Cellular-telephone use and brain tumors. *N Engl J Med*. **344**(2):79-86.

IRPA (1988), "Guidelines on limits of exposure to radiofrequency electromagnetic fields in the frequency range from 100kHz to 300GHz", *Health Phys.*, **54**, 115-123.

IRPA, (1990): "Interim Guidelines on Limits of Exposure to 50 / 60 Hz Electric and Magnetic Fields", *Health Physics*, Vol.58, No 1, 113-122.

Johansen C, Boice J Jr, McLaughlin J, Olsen J. (2001): Cellular telephones and cancer--a nationwide cohort study in Denmark. *J Natl Cancer Inst*. **93**(3):203-7.

Juutilainen J, Heikkinen P, Soikkeli H, Mäki-Paakkanen J. (2007): Micronucleus frequency in erythrocytes of mice after long-term exposure to radiofrequency radiation. *Int J Radiat Biol*. **83**(4):213-20.

King R.C. (1970), Ovarian Development in Drosophila Melanogaster. Academic Press.

Koval T.M., Hart R.W., Myser W.C. and Hink W.F. (1977), "A comparison of survival and repair of UV-induced DNA damage in cultured insect versus mammalian cells", *Genetics* **87**, 513-518.

Koval T.M. and Kazmar E.R. (1988), "DNA double-strand break repair in eukaryotic cell lines having radically different radiosensitivities", *Radiat. Res.* **113**, 268-277.

Koval T.M., Hart R.W., Myser W.C. and Hink W.F. (1979), "DNA single-strand break repair in cultured insect and mammalian cells after x-irradiation", *Int. J. Radiat. Biol. Relat. Stud. Phys. Chem. Med.* **35**, 183-188.

Krause CM, Sillanmäki L, Koivisto M, Häggqvist A, Saarela C, Revonsuo A, Laine M, Hämäläinen H. (2000): Effects of electromagnetic fields emitted by cellular phones on the electroencephalogram during a visual working memory task. *Int J Radiat Biol*. **76**(12):1659-67.

Krause CM, Björnberg CH, Pesonen M, Hulten A, Liesivuori T, Koivisto M, Revonsuo A, Laine M, Hämäläinen H. (2006): Mobile phone effects on children's event-related oscillatory EEG during an auditory memory task. *Int J Radiat Biol*. **82**(6):443-50.

Kundi M, (2004): Mobile phone use and cancer, *Occup Environ Med*, **61**, 560-570.

Kwee S. and Raskmark P., (1998), "Changes in cell proliferation due to environmental non-ionizing radiation: 2. Microwave radiation", *Bioelectrochemistry and Bioenergetics*, **44**, 251-255.

Lai H, Singh NP. (1995): Acute low-intensity microwave exposure increases DNA single-strand breaks in rat brain cells, *Bioelectromagnetics*, **16**(3): 207-10.

Lai H, Singh NP. (1996): Single- and double-strand DNA breaks in rat brain cells after acute exposure to radiofrequency electromagnetic radiation. *Int J Radiat Biol.* Apr;69(4): 513-21.

Lai H, Singh NP. (1997): Melatonin and a spin-trap compound block radiofrequency electromagnetic radiation-induced DNA strand breaks in rat brain cells. *Bioelectromagnetics.* **18**(6): 446-54.

Lai H, Singh NP. (2004): Magnetic-field-induced DNA strand breaks in brain cells of the rat. *Environ Health Perspect.* **112**(6): 687-94.

Leszczynski D, Joenväärä S, Reivinen J, Kuokka R. (2002): Non-thermal activation of the hsp27/p38MAPK stress pathway by mobile phone radiation in human endothelial cells: molecular mechanism for cancer- and blood-brain barrier-related effects. *Differentiation.* **70**(2-3):120-9.

Lin - Liu S. and Adey W. R. (1982), "Low frequency amplitude modulated microwave fields change calcium efflux rates from synaptosomes", *Bioelectromagnetics, (N.Y.),* **3**,309-322.

Lixia S, Yao K, Kaijun W, Deqiang L, Huajun H, Xiangwei G, Baohong W, Wei Z, Jianling L, Wei W. (2006): Effects of 1.8 GHz radiofrequency field on DNA damage and expression of heat shock protein 70 in human lens epithelial cells. *Mutat Res.* **602** (1-2):135-42.

Loughran SP, Wood AW, Barton JM, Croft RJ, Thompson B, Stough C. The effect of electromagnetic fields emitted by mobile phones on human sleep. *Neuroreport.* 2005 Nov 28;16(17):1973-6.

Magras, I.N.; Xenos, T.D. RF radiation-induced changes in the prenatal development of mice. *Bioelectromagnetics* 1997, 18, 455–461.

Malyapa RS, Ahern EW, Straube WL, Moros EG, Pickard WF, Roti Roti JL, (1997a): "Measurement of DNA damage after exposure to 2450 MHz electromagnetic radiation", *Radiation Research*, **148**(6), 608-617.

Malyapa RS, Ahern EW, Straube WL, Moros EG, Pickard WF, Roti Roti JL, (1997b): "Measurement of DNA damage after exposure to electromagnetic radiation in the cellular phone communication frequency band, (835.62 and 847.74 MHz)", *Radiation Research*, **148**(6), 618-627.

Mancinelli F, Caraglia M, Abbruzzese A, d'Ambrosio G, Massa R, Bismuto E. (2004): Non-thermal effects of electromagnetic fields at mobile phone frequency on the refolding of an intracellular protein: myoglobin. *J Cell Biochem.* **93**(1):188-96.

Markova E, Hillert L, Malmgren L, Persson BR, Belyaev IY. (2005): Microwaves from GSM mobile telephones affect 53BP1 and gamma-H2AX foci in human lymphocytes from hypersensitive and healthy persons. *Environ Health Perspect.* **113**(9): 1172-7.

Mashevich M, Folkman D, Kesar A, Barbul A, Korenstein R, Jerby E, Avivi L. (2003): Exposure of human peripheral blood lymphocytes to electromagnetic fields associated with cellular phones leads to chromosomal instability. *Bioelectromagnetics.* **24**(2):82-90.

McCall K., (2004): Eggs over easy: cell death in the Drosophila ovary, *Developmental Biology*, **274**(1), 2004, 3-14.

Meltz ML, (2003): Radiofrequency exposure and mammalian cell toxicity, genotoxicity, and transformation, *Bioelectromagnetics*, suppl. 6, 196-213.

Meyer FJC and Jakobus U, (2003): Numerical Techniques in RF Dosimetry for Mobile Phone Technology, In: Stavroulakis P. (Ed.), *Biological Effects of Electromagnetic Fields*, Springer, 171-237.

Muscat JE, Malkin MG, Shore RE, Thompson S, Neugut AI, Stellman SD, Bruce J. (2002): Handheld cellular telephones and risk of acoustic neuroma. *Neurology.* **58**(8):1304-6.

Navarro Enrique A., Segura J., Portolés M., Claudio Gómez- Perretta de Mateo, (2003): The Microwave Syndrome: A Preliminary Study in Spain, *Electromagnetic Biology and Medicine*, **22** (2-3), 161-169.

Nezis IP, Stravopodis DJ, Papassideri I, Robert-Nicoud M, Margaritis LH. (2000). Stage-specific apoptotic patterns during Drosophila oogenesis. *Eur J Cell Biol* **79**:610-620.

Nylund R, Leszczynski D. (2006): Mobile phone radiation causes changes in gene and protein expression in human endothelial cell lines and the response seems to be genome- and proteome-dependent. *Proteomics*. **6**(17): 4769-4780.

Oberto G, Rolfo K, Yu P, Carbonatto M, Peano S, Kuster N, Ebert S, Tofani S. (2007): Carcinogenicity study of 217 Hz pulsed 900 MHz electromagnetic fields in Pim1 transgenic mice. *Radiat Res.* **168**(3):316-26.

Oktem F, Ozguner F, Mollaoglu H, Koyu A, Uz E. (2005): Oxidative damage in the kidney induced by 900-MHz-emitted mobile phone: protection by melatonin. *Arch Med Res.* **36**(4):350-5.

Ozguner F, Bardak Y, Comlekci S. (2006): Protective effects of melatonin and caffeic acid phenethyl ester against retinal oxidative stress in long-term use of mobile phone: a comparative study. *Mol Cell Biochem*. **282**(1-2):83-8.

Paulraj R, Behari J. (2006a): Protein kinase C activity in developing rat brain cells exposed to 2.45 GHz radiation.*Electromagn Biol Med.* **25**(1):61-70.

Paulraj R, Behari J. (2006b): Single strand DNA breaks in rat brain cells exposed to microwave radiation. *Mutat Res*. Apr 11;596(1-2):76-80.

Pacini S, Ruggiero M, Sardi I, Aterini S, Gulisano F, Gulisano M.(2002): Exposure to global system for mobile communication (GSM) cellular phone radiofrequency alters gene expression, proliferation, and morphology of human skin fibroblasts. *Oncol Res.* **13**(1):19-24.

Panagopoulos DJ, Chavdoula ED, Nezis IP and Margaritis LH, (2007): Cell Death induced by GSM 900MHz and DCS 1800MHz Mobile Telephony Radiation, *Mutation Research*, 626, 69-78.

Panagopoulos DJ, Chavdoula ED, Karabarbounis A, and Margaritis LH, (2007): Comparison of Bioactivity between GSM 900 MHz and DCS 1800 MHz Mobile Telephony Radiation, *Electromagnetic Biology and Medicine* **26**(1).

Panagopoulos D.J., Karabarbounis A., and Margaritis L.H., (2004), Effect of GSM 900-MHz Mobile Phone Radiation on the Reproductive Capacity of *Drosophila melanogaster*, *Electromagnetic Biology and Medicine*, **23**(1), 29-43.

Panagopoulos D.J., Messini, N., Karabarbounis, A., Philippetis A.L., and Margaritis L.H., (2000a), "Radio Frequency Electromagnetic Radiation within "safety levels" Alters the Physiological Function of Insects", *In*: Kostarakis P, Stavroulakis P, (ed), *Millennium International Workshop on Biological Effects of Electromagnetic Fields,* Proceedings, Heraklion, Crete, Greece, 17-20 October, 2000, 169-175, ISBN: 960-86733-0-5.

Panagopoulos D.J, Margaritis L.H, (2003a): "Effects of electromagnetic fields on the Reproductive Capacity of *Drosophila melanogaster*, *In*: Stavroulakis P. (ed.), *Biological Effects of Electromagnetic Fields,* Springer, 545-578.

Panagopoulos DJ, Messini N, Karabarbounis A, Filippetis AL, and Margaritis LH, (2000b): "A mechanism for Action of Oscillating Electric Fields on Cells", *Biochemical and Biophysical Research Communications*, **272**(3), 634-640.

Panagopoulos D.J., Karabarbounis, A. and Margaritis L.H., (2002), "Mechanism for action of electromagnetic fields on cells", *Biochem. Biophys. Res. Commun.*, **298**(1), 95-102.

Panagopoulos D.J. and Margaritis L.H., (2003b), Theoretical Considerations for the Biological Effects of Electromagnetic Fields, *In*: Stavroulakis P. (Ed.) "Biological Effects of Electromagnetic Fields", Springer, 5-33.

Penafiel L.M., Litovitz T., Krause D., Desta A., Mullins J.M., (1997), Role of Modulation on the effects of microwaves on ornithine decarboxylase activity in L929 cells, *Bioelectromagnetics*, **18**, 132-141.

Port M, Abend M, Romer B, Van Beuningen D. (2003): Influence of high-frequency electromagnetic fields on different modes of cell death and gene expression. *Int J Radiat Biol.* **79**(9): 701-8.

Pyrpasopoulou A, Kotoula V, Cheva A, Hytiroglou P, Nikolakaki E, Magras IN, Xenos TD, Tsiboukis TD, Karkavelas G. (2004): Bone morphogenetic protein expression in newborn rat kidneys after prenatal exposure to radiofrequency radiation. *Bioelectromagnetics.* **25**(3):216-27.

Ramirez E., Monteagudo J.L., Garcia-Gracia M., Delgado J.M.R., (1983), "Electromagnetic Effects in Drosophila", *Bioelectromagnetics*, **4**, 315-326.

Regel SJ, Negovetic S, Röösli M, Berdiñas V, Schuderer J, Huss A, Lott U, Kuster N, Achermann P. (2006): UMTS base station-like exposure, well-being, and cognitive performance. *Environ Health Perspect.* **114**(8):1270-5.

Remondini D, Nylund R, Reivinen J, Poulletier de Gannes F, Veyret B, Lagroye I, Haro E, Trillo MA, Capri M, Franceschi C, Schlatterer K, Gminski R, Fitzner R, Tauber R, Schuderer J, Kuster N, Leszczynski D, Bersani F, Maercker C. (2006): Gene expression changes in human cells after exposure to mobile phone microwaves. *Proteomics.* **6**(17):4745-4754.

Repacholi MH, Basten A, Gebski V, Noonan D, Finnie J, Harris AW. (1997): Lymphomas in E mu-Pim1 transgenic mice exposed to pulsed 900 MHZ electromagnetic fields. *Radiat Res.* **147**(5):631-40.

Röschke J.and Mann K., (1997): No short-term effects of digital mobile radio telephone on the awake human electroencephalogram, *Bioelectromagnetics* **18**, pp. 172–176.

Salford LG, Brun A, Persson BR, Eberhardt JL,. (1993): Experimental studies of brain tumor development during exposure to continuous and pulsed 915 MHz radiofrequency radiation, *Bioelectrochemistry and Bioenergetics*, **27**(6):535-42.

Salford LG, Brun A, Sturesson K, Eberhardt JL, Persson BR. (1994): Permeability of the blood-brain barrier induced by 915 MHz electromagnetic radiation, continuous wave and modulated at 8, 16, 50, and 200 Hz. *Microsc Res Tech.* **27**(6):535-42.

Salford LG, Brun AE, Eberhardt JL, Marmgren L, and Persson BR, (2003): Nerve Cell Damage in Mammalian Brain after Exposure to Microwaves from GSM Mobile Phones, *Environmental Health Perspectives*, **111**(7), 881-883.

Santini MT, Ferrante A, Rainaldi G, Indovina P, Indovina PL. (2005): Extremely low frequency (ELF) magnetic fields and apoptosis: a review. *Int J Radiat Biol.* **81**(1):1-11. Review.

Santini R, Santini P, Danze JM, Le Ruz P, Seigne M. (2003): Symptoms experienced by people in vicinity of base stations: II/ Incidences of age, duration of exposure, location of subjects in relation to the antennas and other electromagnetic factors, *Pathol Biol (Paris).* **51**(7):412-5.

Sheikh MS, Huang Y. (2004): TRAIL death receptors, Bcl-2 protein family, and endoplasmic reticulum calcium pool. *Vitam Horm.***67**:169-88. Review.

Smith J.E. III, Cummings C.A. and Cronmiller C., (2002), Daughterless coordinates somatic cell proliferation, differentiation and germline cyst survival during follicle formation in *Drosophila, Development* **129**, pp. 3255–3267.

Sommer AM, Bitz AK, Streckert J, Hansen VW, Lerchl A. (2007): Lymphoma development in mice chronically exposed to UMTS-modulated radiofrequency electromagnetic fields. *Radiat Res*. **168**(1):72-80.

Schirmacher A, Winters S, Fischer S, Goeke J, Galla HJ, Kullnick U, Ringelstein EB, Stögbauer F. (2000): Electromagnetic fields (1.8 GHz) increase the permeability to sucrose of the blood-brain barrier in vitro. *Bioelectromagnetics*. **21**(5):338-45.

Shorrocks B., (1972), "Drosophila", Ginn, London.

Stromnaes O. and Kvelland I., (1962): "Sexual activity of *Drosophila melanogaster* males", *Hereditas*, **48**, 442-470.

Tahvanainen K, Nino J, Halonen P, Kuusela T, Laitinen T, Lansimies E, Hartikainen J, Hietanen M, Lindholm H. (2004). Cellular phone use does not acutely affect blood pressure or heart rate of humans. *Bioelectromagnetics*,**25**(2):73-83.

Tillmann T, Ernst H, Ebert S, Kuster N, Behnke W, Rittinghausen S, Dasenbrock C. (2007): Carcinogenicity study of GSM and DCS wireless communication signals in B6C3F1 mice. *Bioelectromagnetics.* **28**(3):173-87.

Tisal J., (1998), "GSM Cellular Radio Telephony", J.Wiley & Sons, West Sussex, England.

Velizarov S, Raskmark P, Kwee S, (1999): "The effects of radiofrequency fields on cell proliferation are non-thermal", *Bioelectrochemistry and Bioenergetics*, **48**, 177-180.

Wagner P., Roschke J., Mann K., Hiller W. and C. Frank, (1998): Human sleep under the influence of pulsed radiofrequency electromagnetic fields: a polysomnographic study using standardized conditions, *Bioelectromagnetics* **19**, pp. 199–202.

Weaver JC and Astumian RD, (1990): "The response of living cells to very weak electric fields: The thermal noise limit", *Science*, **247**, 459-462.

Weisbrot D, Lin H, Ye L, Blank M, Goodman R. (2003), Effects of mobile phone radiation on reproduction and development in Drosophila melanogaster. *J Cell Biochem.* **89**(1): 48-55.

Xenos T.D, Magras I.N, (2003): Low Power Density RF Radiation Effects on Experimental Animal Embryos and Foetuses, *In*: Stavroulakis P. (Ed.), *Biological Effects of Electromagnetic Fields,* Springer, 579-602.

Zhou YP, Teng D, Dralyuk F, Ostrega D, Roe MW, Philipson L, Polonsky KS. (1998): Apoptosis in insulin-secreting cells. Evidence for the role of intracellular Ca^{2+} stores and arachidonic acid metabolism. *J Clin Invest.* **101**(8):1623-32.

Zwamborn APM, Vossen SHJA, van Leersum BJAM, Ouwens MA, Mäkel WN. (2003). Effects of Global Communication System Radio-Frequency Fields on Well being and Cognitive Functions of Human Subjects with and without Subjective Complaints. FEL-03-C148. The Hague, the Netherlands: *TNO Physics and Electronics Laboratory.* Available: http: //home.tiscali.be/milieugezondheid/dossiers/gsm/TNO_rapport_Nederland_sept_2003.pdf.

In: Mobile Telephones... ISBN 978-1-60456-436-5
Editors: A.C. Harper and R.V. Buress, pp. 151-167

Chapter 4

SYMBIAN BASED TEST-BED IMPLEMENTATION OF WIRELESS COOPERATIVE SCENARIOS

***Leonardo Militano**, *Antonio Iera*† and *Antonella Molinaro*‡**
University Mediterranea of Reggio Calabria

Abstract

Evolution in the world of mobile telephones is moving toward a full functional and rich in features device. Examples of novel functionalities are the possibility to play and handle different multimedia contents, to have a wide number of communication capabilities through both remote (i.e. the Internet) and local network access. These examples testify to an evolutionary trend that will likely lead to mobile phones more and more comparable to portable computers. One of the most appealing features for a mobile phone appearing on the market is the number of built-in network interfaces (the most recent devices include interfaces to different networks, such as HSDPA, UMTS, EDGE, GPRS, WLAN, Bluetooth, and Infrared). This both pushes the providers to offer a wider portfolio of novel applications, which best exploit the augmented possibilities, and paves the way to Research and Development (R&D) activities in the field of wireless communications towards the deployment of real "Personal Communications" scenarios. *Cooperative Wireless Applications*, which we will focus on in the present chapter, is a significant example of a R&D area fuelled by the availability of heterogeneous wireless network interfaces, based on both cellular and short-range technologies. *Cooperation* covers many different aspects, which are being studied at the different protocol layers. We focus on a scenario where more devices can cooperate with each other on a short-range communication link basis. Usually a costly, high power demanding cellular network is the way users access a service (e.g. web content browsing, file download). A short range link has, instead, the following appealing features: high data rate, low power consumption, and low costs. By exploiting the synergy between a *long-range* (cellular) and a *short-range* network, different entities may share service costs through cooperation with peer devices. Currently, smartphones offer the possibility to implement the users own application to be run on the device. This offers us the opportunity of performing an analysis of telephone network interfaces

*E-mail address: leonardo.militano@unirc.it
†E-mail address: antonio.iera@unirc.it
‡E-mail address: antonella.molinaro@unirc.it

behaving in mutual co-operation and the possibility of evaluating the advantages actually attainable from the aforementioned innovative communication paradigm. In the chapter we first introduce the reader to the concept of cooperation in wireless scenarios and, subsequently, try to summarise the main implementation efforts made until now on mobile telephones. A performance evaluation campaign and a report on the main research results is given as a proof of theoretical advantages in cooperating under different coverage and traffic load conditions. Also a description of the ad-hoc implemented test-beds are given. Performance studies and relevant implemented evaluation tools are specifically oriented to Symbian Operating System mobile phones, although results are general and valid for any mobile terminal.

1. Introduction to Wireless Cooperative Networking

In recent years cooperation in wireless networking has gained a wide interest from both research and industry communities. With the advent of novel multimedia interactive services and high performance applications, the need for substantial improvements at both the system and the device level manifestly arises. Having in mind, for instance, TV-on-mobile and video streaming services, it clearly emerges that exacting system and terminal requirements are necessary to support the introduced services at an acceptably good Quality of Service. The promising features of cooperation, which are related to all layers of ISO/OSI network model, offer a viable solution to aforementioned issues. A comprehensive survey on cooperation in communication systems is given in [1] and in [2], which analyze the cooperation issue at the Physical, Medium Access Control, and Network layers. For the sake of clearness, we give here some very short examples of cooperation in wireless networks.

Multi-hop or relaying techniques are probably the most widely known cooperative techniques. These are based on the very simple concept that one or more nodes in a communication scenario act as repeater or amplifier of a given received signal. In doing so, not only the coverage area for communication among all the nodes will be extended, but also the power consumption will be reduced by the reduction of the average link distance between devices. Canonical relaying techniques are *Amplify and Forward, Compress and Forward*, and *Decode and Forward* [2]. In the *Amplify and Forward* technique the received signal is first amplified and then forwarded to the next node. As an extension of this, in the *Compress and Forward* technique the received signal is compressed before retransmission. The compression allows to forward the received signal also to a different communication standard. The *Decode and Forward* method decodes the received signal and re-encodes it prior to retransmission [7]. The advantage of this latter approach, which adds some complexity, is that received noise is not forwarded and neither amplified as in the Amplify and Forward approach. For additional reading about this interesting topic, several research contributions can be found in the literature that propose and study different solutions for relaying scenarios (i.e. [4], [8], [9]).

Cooperation also plays an important role to improve fading affected wireless channels. A common solution to fading affection is to create *diversity* in the transmission by generating different copies of the same signal. This solution generally requires more than one single antenna at the transmitter making it in many cases an unfeasible solution. Notwithstanding, to single-antenna devices cooperative communication techniques offer an opportunity to generate the wished diversity [4], [5] and [6]. Based on the concept that more single-antenna devices close to each other can share their antennas to form the so-called Virtual Antenna Arrays (VAAs), a transmitting device can achieve transmit diversity and enjoy the benefits of a virtual MIMO (Multiple Input Multiple Output) system.

Another application of cooperation in wireless networking is represented by *Cooperative network coding techniques*. These techniques are based on the concept that in a multi-node network all nodes retransmit all received packets by combining different information into a final forwarded packet. All nodes will receive a combination of information from many sources, this improving the overall information reception.

Besides the cited applications, cooperation may be effectively employed in future wireless communications as valuable mean to overcome intrinsic limitations of the systems, the terminals, and the conventional communication channels. Main issues are listed below:

- *energy/power consumption*: battery-powered devices, like smart-phones, PDAs, and laptops, are the weak points of communication paradigms with respect to energy/power consumption. The operating time of wireless devices suffers from constantly increasing energy needs; this adversely affecting new service support solutions;
- *high data rate capability*: wireless and mobile communication systems are currently characterized by strong limitations in terms of data rates. Real time video services and TV-on-mobile are just examples of appealing, high data rate demanding, novel services, whose deployment is clearly limited by conventional cellular systems;
- *Quality of Service (Qos)*: high-level Quality of Service requirements can be matched by addressing low-level target performance issues, such as data rate increase, transfer delay reduction, and energy saving;
- *service cost*: this is certainly one of the main limiting factors for a widespread subscription to new services. End-users often perceive costs related to a service as too high to invest on it;
- *cellular network capacity*: the recent explosion registered in the number of cellular subscribers requires a higher cell capacity in cellular networks and more efficient spectrum usage policies.

In order to address aforementioned issues in future wireless communication systems, the *inter-system cooperative communication* paradigm has been proposed as an alternative to conventional cellular solutions.

1.1. Inter-System Cooperative Scenarios

Fourth-Generation communications systems (and beyond-3G, as well) are emerging as an integrated "system of systems" and "network of networks" where different technologies work together to offer higher data rates in a "always" and "everywhere" connected network. The inter-system and multi-network cooperative communication paradigm perfectly fits to the envisaged scenario. This communication paradigm, extensively described in [3], envisages cooperation and synergy between heterogeneous networks involving both wide range cellular communication systems and short range communication systems.

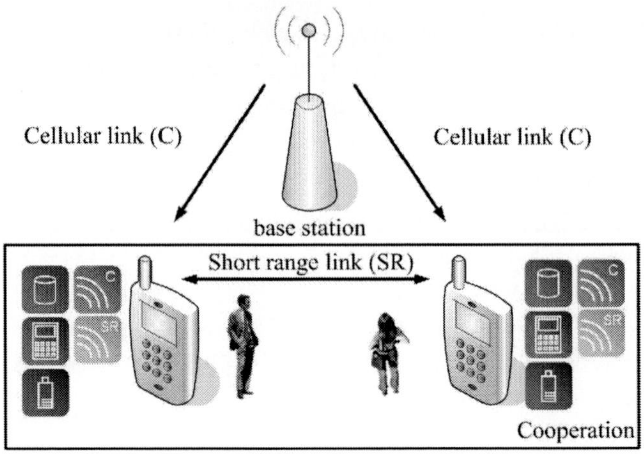

Figure 1. Cooperative wireless networking scenario [1].

In Fig. 1 the cooperative multi-network communication architecture, which is used as a reference scenario, is represented. Multimodal terminals, with both cellular and short-range communication capabilities, are required. Cellular technologies can be either General Packet Radio Service (GPRS), Enhanced Data rates for GSM Evolution (EDGE), Universal Mobile Telecommunication System (UMTS) or even Wireless Local Area Network (WLAN); Bluetooth and Wireless Local Area Network (WLAN) technologies are a viable solution to short-range connection needs. Current evolution in the world of mobile telephones is moving toward multimodal devices supporting all different radio interfaces offering a wide number of communication possibilities through both remote-network (the Internet) and local-network access. Currently, the most recent devices on the market usually include interfaces to many different networks such as HSDPA, UMTS, EDGE, GPRS, WLAN, Bluetooth and Infrared; thus being suitable for an implementation of a cooperative scenario.

For a realistic implementation scenario, we can imagine a group of users either sitting at the airport, or at their hotel, or waiting at the train station. The probability that more users are interested in spending their time by accessing the same web-content is high. We may think, for example, to big sport events showed on TV, news broadcast, weather forecast before departure to a foreign country, etc. In a conventional cellular communication architecture, all users would set up a single communication channel to access the data stream hosted in some web servers. Drawbacks we find out in this scenario are the following ones:

1. every single user is charged for the whole service with a time- or data-proportional billing;

2. in a low data-rate cellular network a service could suffer from long transfer time delay and low Quality of Service degree;

[1]Reproduced from "On the beneficial effects of cooperative wireless peer to peer networking", Militano, Fitzek, Iera and Molinaro, page 4, Figure 1, in "Wireless Communications: 2007 CNIT Thyrrenian Symposium, Springer Series on Signals and Communication Technology, S. Pupolin", with kind permission of Springer Science and Business Media

3. a battery-powered device could be unable to support the handling of the complete service due to lack of battery autonomy.

The inter-system cooperative communication paradigm not only offers a viable solution to address these issues, but also guarantees additional benefits to both network/service providers and to end-users. In the example of a group of users sitting in a waiting hall, we suggest the idea to agree on cooperating over short-range network links. By downloading only a portion of data each and by exchanging the downloaded portions in a peer-to-peer fashion, all of them will achieve the common objective of receiving the whole web content; most important, they will hit the target by sharing the correspond costs! The cellular link will play the key role of "service entry point", while the peer-to-peer short-range link implements the cooperative user's behavior.

Noteworthy, the data exchange over a short-range link will happen simultaneously to the cellular transmission, so that the devices will, in a sense, "accumulate" cellular links. In such a scenario, complementarities and capabilities of cellular and short-range radio networks will primarily benefit cellular subscribers. Indeed, thanks to the low consuming nature of short-range connections a reduced energy consumption is achieved. It is worth recalling that cellular links between base station and user devices need high power to cover some kilometer distances, while device-to-device short-range connections need low power to cover tens of meters only. This feature is the basis for all energy consumption related benefits offered by the novel architecture. Moreover, an increased downloading rate for end-users, and a consequent reduction of the experienced downloading delay, is achieved thanks to the fact that short-range connections can provide a much higher data rate than current cellular systems. Furthermore, the overall reduction of the cellular load on the down-link is obtained, because each user just downloads a portion of the data and exchanges other portions with the cooperating users. This avoids a replication of downloading of the same data more times on the cellular downlink and a consequent higher potential number of users that can access the network. Finally, also the service cost benefits from this behavior. In fact, a device that downloads, let's say, half of a file, will stay connected half of the time; this allowing a proportional billing reduction. As this is attractive for the end-user interest, service and network providers can design new business scenarios and cost models to incentive the support for cooperative networking and enlarge the subscribers basin. Noteworthy, for an utmost successful outcome of cooperation, the condition stated in (1) should be met, where R_{SR} is the data rate of the short-range link and R_C is the data rate for the cellular link.

$$R_{SR} \geq R_C \tag{1}$$

If condition (1) is not met both data rate and energetic gain will be adversely affected.

2. The Test-bed Implementation

We will describe in this section a test-bed implementation useful for a study and a performance evaluation of the introduced cooperative communication paradigm. To evaluate the effectiveness of terminals' cooperative behavior, analyze pros and cons of different combinations of networks, and investigate the right parameters for evaluating the results, a suitable test-bed application is required. Common commercial Nokia mobile phones (Nokia N70, Nokia N93, Nokia N95) revealed to be perfectly suitable to the deployment of the intended test-bed. These phones are based on the Symbian Operating System and belong to the Series 60 platform.

2.1. Symbian OS

The developed test-bed is based on the implementations of an ad-hoc conceived application. For this reason, we will now briefly give some basic knowledge of Symbian OS, stressing the attention on those features which are of interest for our activity. Symbian is a multitasking Operating System, that is gaining a big success in many devices developed by key manufacturers (Nokia, Siemens, Samsung, Panasonic and Sendo). It is expected that in a few years Symbian will be the most common Operating System among smartphones[2]. Symbian is an *object-oriented* operating system specifically designed for devices which are limited in resources (battery, CPU, memory) and which guarantees high stability and reliability. Efficient management of simultaneous events behaving asynchronously and the good interaction level with other services on the device are features that give the opportunity to effectively support a cooperative behavior. The following features are important for the success of this operating system [10]:

- *multitasking*: parallel application management;
- *performance*: device-specific power management to maximize battery lifetime;
- *standards*: inter-operability with other platforms, guaranteed by the use of standard based technologies;
- *modular architecture*: a modular architecture offering the opportunity for a clear and structural application development;
- *memory management*: optimized memory management.

The relevance of listed features in a resource-constrained device like a mobile phone is manifest. Multitasking is very efficient and far more used than multi-threading, which is less efficient at the application level. Also memory management is very performing, this being a very important feature on a mobile phone with limited memory resources.

The Series 60 Nokia platform is the world's leading smartphone software platform, offering a feature-rich software basis for phones with advanced data capabilities. The Series 60 architecture is build on the Symbian Operating System, completing it with some features, for example the GUI library. C++ is the programming language mostly used in Symbian OS, it is therefore convenient to program new applications in Symbian C++, so to have direct access to System API's. The Series 60 C++ SDK (Software Development Kit) offers the developer API's using standard technologies and open standards, to design and program his own application. This guarantees an interoperability with other Symbian based mobile phones. Developing in Symbian C++ the programmer has full and easy access to many functionalities like Bluetooth, Infrared and networking functions.

[2]A *smartphone* is a mobile phone with additional computer-like features and capabilities

Series 60 SDK and additional programming tools are freely available at Nokia's website *(http://www.forum.nokia.com/)* and provide the programmer with tools, sample code, documentation and an emulator to support programming, testing and debugging for new Symbian C++ applications.

2.2. Cooperation: the Implemented Application

As a first step we put our attention on a situation in which two sample users are interested in downloading the same data, for example a video file, hosted by a remote server. The two users will download different portions of the same video and, by setting up a cooperative short-range link connection, they will exchange the correspondent portions of data. More in details, the achievement of the main objective passes through the study and design of an overall architecture that requires:

- the integration of a cellular connection set-up mechanism into the application (GPRS, EDGE, UMTS and WLAN);
- the implementation of the HTTP transaction management on cellular links;
- the exploitation of HTTP/1.1 Range Request [12] feature for a dynamic, server independent and peer-failure tolerant file-sharing;
- the accomplishment of the cooperative terminal behavior;
- the development of an ad-hoc conceived graphical user interface both for the cooperative application and for the user-friendly visualization of evaluation metrics;
- the integration of a short-range connection set up into the application (Bluetooth, infrastructured WLAN, and ad-hoc WLAN).

To set up either a short-range or a cellular long-range link one needs to implement a TCP/IP transaction based on a socket communication. After opening a specific socket he/she can connect to a remote device and start transferring data. For the short-range networks a service discovery procedure is actually needed. While for the Bluetooth network the standard Service Discovery Protocol (SDP) is already available, when a WLAN is used the assumption that the IP address of the other peer is known a priori and used to connect to it is necessary. Within the application, a HTTP transaction on a GPRS/EDGE/UMTS/WLAN link is implemented for data downloading, whereas a Bluetooth/WLAN/ad-hoc transaction implements the peer-to-peer cooperative behavior. The test-bed foresees the presence of a remote server, where mobile terminals can download an mp4 file from. A *dynamic server-unaware* data delivery approach is implemented. According to it, the server will deliver a complete file to the non cooperating clients, while partial files will be delivered to cooperating clients. The 1.1 version of HTTP protocol is exploited to match the requirement of a cooperative *dynamic server-unaware* scenario. HTTP/1.1 *Range Request* [12] is one of the main features introduced to improve previous protocol versions. The Range Request header is intended to support dynamic data delivery which *does not require* any modification to servers or to content. In the implemented test-bed, the cooperating peers will perform an HTTP HEAD call to the server to obtain meta-information about the file to access. This call will be made over the cellular link, but it will not cause the transfer of the entity-body itself. The data size information, retrieved from the information deriving from the HTTP HEAD call, will be used to set the Range-Request field value of the subsequent HTTP GET call. The main advantage of implementing HTTP/1.1 *Range Request* feature is that we are not forced to upload additional files to the server for the cooperative service. Moreover, in case of a short-range link failure, a peer will not loose previously made efforts, but simply perform an additional GET call to the server to fetch the missing data over the cellular link. For these reasons, we refer to this approach as a *dynamic server-unaware* approach. The main limitation to a wide-spread use of this algorithm, is that old HTTP/1.0 servers present on the Web do not support the Range Request header. Moreover, as stated in [13], HTPP/1.1 servers, when a client makes a Range Request specifying one or more contiguous ranges of bytes, can either return one/more ranges in the response or ignore the Range Request header.

The application is provided with five different views. The first view shows ongoing network activities. During the download process data rates of the involved networks are displayed, split into six values: Bluetooth In, Bluetooth Out, WLAN In, WLAN Out, Cellular In and Total In. The latter one is the sum of the incoming cellular and short range communication data rates. Once the transaction is completed, mean data rate and overall transfer time are displayed in this first view. The second and the third views display information about Bluetooth and WLAN operations respectively, whereas the fourth view is dedicated to the operations on the cellular link. The last of the five views lists the downloaded files, which can be displayed following the transaction completion.

In Fig. 2 we report two sample screen-shots of the application graphical user interface. The instantaneous data rate values on involved radio interfaces and the total incoming data rate at device (which we refer to as *Virtual Data Rate*) are plotted. Specifically, instantaneous data rate values in the Figure refer to a cooperative Edge/Bluetooth and a Edge/WLAN scenario. In the Figure *Bs* stands for Bluetooth sending, *Br* stands for Bluetooth receiving, *Ws* stands for WLAN sending, *Wr* stands for WLAN receiving, *Hr* stands for Http receiving (cellular link) and *Tr* stands for Total receiving (Virtual Data Rate). From this glint of the application one can notice how the instantaneous virtual data rate, in both cases, is about twice the value of the cellular rate.

3. Performances and Results

Clearly, the two-device test-bed we described, is a simplified representation of a real scenario, in which reasonably more than two terminals will cooperate. Notwithstanding, it is an effective test-bed for experimental measurements aiming at assessing the benefits/drawbacks introduced by the cooperative behavior in terms of energy consumption, data rate and transfer time.

Energy consumption is evaluated through a series circuit where the mobile phone is connected to a multimeter and a power supply device. The multimeter is set to read the average current value used by the phone to download the desired data flow over the needed downloading period. According to Equation 2, the energy value is obtained by multiplying the voltage by the current and by the time.

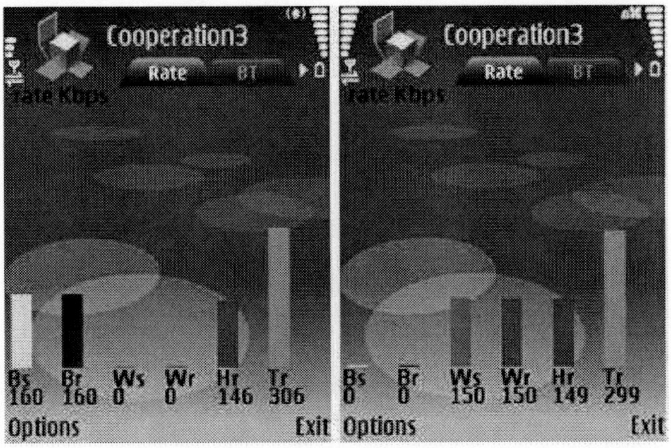

Figure 2. Application screenshots reporting instantaneous data rate values in cooperative Edge/Bluetooth and Edge/WLAN scenarios, respectively.

$$\vec{Energy}(J) = \vec{Voltage}(V) \times \vec{Current}(I) \times \vec{Time}(s) \tag{2}$$

For data rate measurements, timers have been implemented in the application to measure the elapsed time and, consequently, the data rate values. These values refer to both *send* and *receive* operations in the cooperative scenario and the *receive only* operation in the non-cooperative scenario on the correspondent communication links. In the stand-alone case, a timer is started upon reception of the first data chunk over the cellular link. This is afterwards stopped upon reception of the last data chunk. In the cooperative situation an additional similar behaving timer is implemented for the short-range link. For evaluating the virtual data rate, the highest value of the two timers is taken as overall transaction time in the cooperative scenario.

3.1. Cellular Link Performances

Before considering the benefits related to cooperative scenarios, we want to give an overview on the cellular radio interface performances which is the actually term of comparison when defining the benefits of cooperation. To this aim, our measurement campaigns first consider mobile phones downloading a video file from a remote server by using conventional cellular communication systems in a stand-alone fashion. We report here the results in terms of data rate and energy per Kbit consumption values. Our analysis focuses on GPRS, EDGE, UMTS and WLAN (802.11g) networks in standard conditions when using Nokia N93 mobile phones. The reader is referred to Table 1 to have an overview of the results which are of easy interpretation.

3.2. Bluetooth as Cooperative Short-Range Link

As already stated, Bluetooth is one of the viable solutions to the issue of short-range link availability in the reference cooperative scenario. In the implementation of the Bluetooth link set-up we assume that the two mobile phones before starting the data downloading

Table 1. Cellular interface performance overview

Cellular interface	Data rate	Energy consumption
GPRS	42 Kbps	0.0135 J/Kbit
EDGE	150 Kbps	0.0103 J/Kbit
UMTS	345 Kbps	0.0041 J/Kbit
WLAN (802.11g)	1734 Kbps	0.0007 J/Kbit

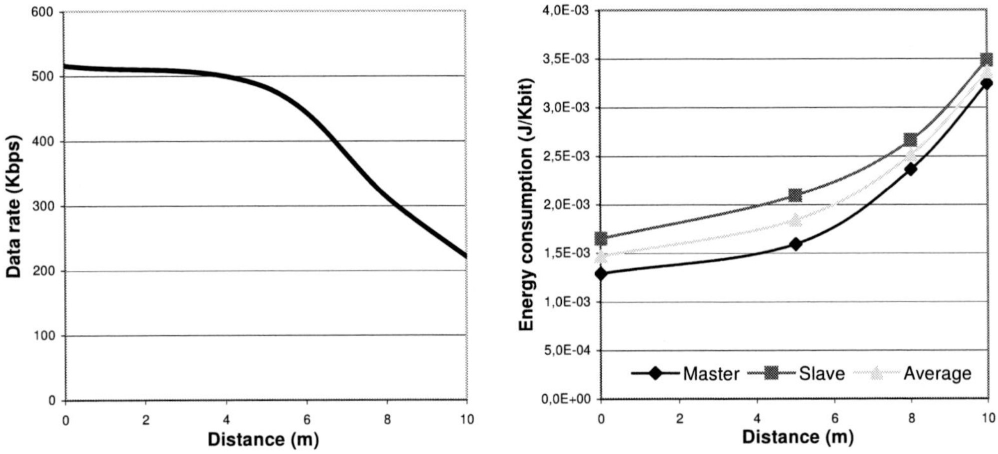

Figure 3. Bluetooth data rate and energy per Kbit consumption versus the distance between devices in a symmetric communication channel without cooperation.

session decide whether to cooperate or not. In case they are going to cooperate, the data downloading and sharing modalities are strongly related to the master or slave role played by each device in the Bluetooth link.

Before showing the results of the cooperative behavior, an analysis of the Bluetooth interface performance under different conditions is needed. From our tests it clearly appears how data rate and energy consumption are strongly influenced by parameters like *distance between devices*, played *master or slave role*, and *interference* from a possible active co-located 802.11g network interface. The influence of the first two parameters on the Bluetooth performance is clearly plotted in Fig. 3 (a symmetric communication channel without cooperation is considered). We notice in the Figure how data rate decreases with the distance, while the energy per Kbit value increases with it. Moreover, it is worth underlining that, whatever the distance, the energy consumption is always higher for the slave terminal with respect to the master terminal.

In Fig. 4 Bluetooth data rate and energy consumption are reported to show how their values are not only influenced by the parameters considered in Fig. 3, but also by the presence of a co-located active 802.11g network interface. Bluetooth and 802.11g networks are both working in the 2.4 GHz-ISM-Band and mutual interference effects between them

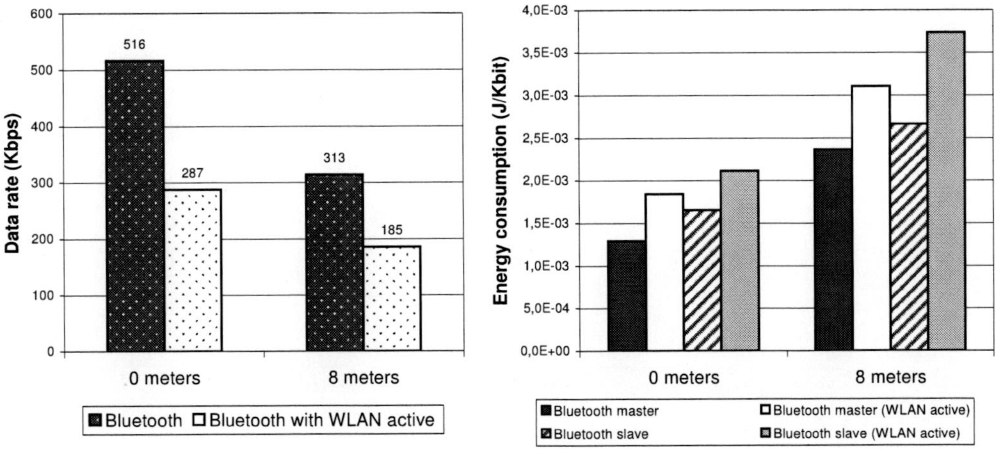

Figure 4. Bluetooth data rate and energy consumption versus the distance between devices, the role played by the device (master or slave) in the link, and the contingent presence of an active co-located 802.11g network interface.

will result in a loss of performance for both systems. Extensive work has been made by the research community on this subject and viable coexistence solutions have been proposed. A deeper analysis on this matter is given in Section 3.2.3. when dealing with the WLAN-Bluetooth cooperative scenario results.

3.2.1. The GPRS-Bluetooth Scenario

The benefits of a cooperative scenario involving GPRS download channels and Bluetooth cooperation channel have been addressed in [3]. The results refer to a 0 meters distance between the devices scenario when no active 802.11g network interface is present.

Comparing the energy consumption, data rate and transfer time values of the cooperative and the non-cooperative scenarios, manifest gains in the former case have been obtained. As for the energy consumption, we registered a 44.3% gain in the average value. The reason for this gain, is that the short-range Bluetooth connection consumes very much less energy than the long range cellular GPRS connection. Each device, downloading half of the data over GPRS and exchanging the received data with another device over Bluetooth connections, will experience the cited energy consumption gain. As for the data rate and transfer time results, they showed respectively a doubled and a halved value. This is an expected result as the two devices accumulate the cellular links. Detailed results on this scenario are reported in Table 2.

3.2.2. Bluetooth Short-Range Link in Cooperation with EDGE and UMTS Cellular Networks

For an exhaustive analysis of Bluetooth based solution for cooperative scenarios, measurement campaigns have been performed by adopting cellular links of EDGE and UMTS networks. These aimed at comparing the stand-alone downloading process and the cooperative

Table 2. GPRS-Bluetooth scenario results [3]

	No cooperation	Cooperation	Gain
$Energy_{avg}(J)$	282.6	157.3	44.3%
$Energy_{max}(J)$	302.1	167,1	45.7%
$Energy_{min}(J)$	270.9	146	46.1%
$Rate_{avg}(Kbps)$	41.5	82.8	99.4%
$Rate_{max}(Kbps)$	42.6	84.8	99.1%
$Rate_{min}(Kbps)$	39.8	79.1	99%
$Time_{avg}(sec)$	491	246	49.9%
$Time_{max}(sec)$	511	257	49.6%
$Time_{min}(sec)$	478	241	49.7%

case in terms of energy consumption gain. The tight relation to the distance between the devices and the dependence on the role played by the device (master or slave) in the communication link still emerge. When UMTS is used, the range of values for the energy consumption gain is wider than in the case of EDGE.

The analysis of the data rate performance is quite straightforward. For all cooperative scenarios but one, the data rate is almost doubled compared to the non-cooperative case. The transfer time is consequently almost halved. We have an exception to this assertion when relation (1) (see Section 1.1.) is not satisfied. This happens in the UMTS-Bluetooth scenario when the devices are at about 10 meters distance. The Bluetooth data rate strongly decreases, not meeting anymore relation (1) and causing the Bluetooth link to be now the bottleneck of the whole communication architecture. In this case, the data rate gain decreases; a consequent reduced positive impact on the overall transfer time is observed.

3.2.3. The WLAN-Bluetooth Scenario

As already stated, relation (1) should be met for an utmost successful outcome of cooperation. When considering WLAN (802.11g) as the main service entry channel and Bluetooth as the short range link, the relation is not satisfied. This adversely affects data rate and energetic gains. In standard conditions, Bluetooth data rate cannot compete with WLAN, as shown by results in Table 1 and Fig. 3. Notwithstanding, we want to analyze the effect that an increase in the number of transmitting WLAN terminals, with the consequent reduction of the overall WLAN capacity, could have. This phenomenon, in fact, may bring the rate available to each WLAN user down to values comparable with the typical Bluetooth transmission rates. In this conditions the users will likely gain advantages from cooperation and their willingness to cooperate increases. A Bluetooth based cooperation among users in the WLAN will, at the same time, reduce the data transmission rate on the WLAN links of each cooperating user; this increasing the WLAN capacity in terms of number of users which contemporary access the WLAN.

[3]Reproduced from "On the beneficial effects of cooperative wireless peer to peer networking", Militano, Fitzek, Iera and Molinaro, page 11, Table 1, in "Wireless Communications: 2007 CNIT Thyrrenian Sym-

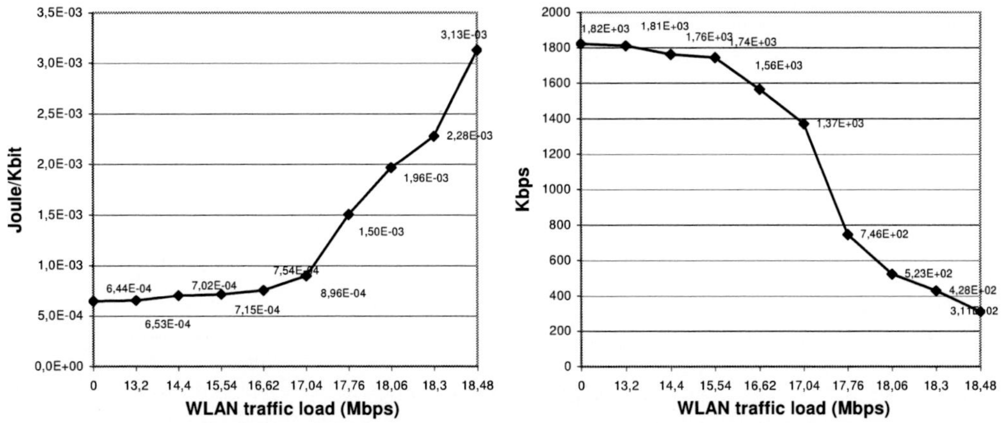

Figure 5. Energy consumption in J/Kbit and data rate values of WLAN network versus the overall traffic load in the network.

The phenomenon envisaged above is substantiated by the experimental results shown in Fig. 5. The energy consumption per Kbit of data and the data rate (Kbps) are plotted in the case of two cellular phones downloading the required data from the WLAN Access Point by exploiting their IEEE 802.11g interfaces. In this phase the terminals do not co-operate with one another. Curves are sketched by varying the WLAN background traffic load (generated through the open-source MGEN tool). It clearly appears that by increasing the overall traffic load, both energy consumption increases and the data rate of each cellular phone is reduced. For overall traffic load values higher than 18,5 Mbps, observed cellular phones are not able anymore to access the network through their WLAN card. Merging this figure with Fig. 3 we notice that now there are some ranges of parameters values in which maybe it is worth letting the two terminals cooperate through their Bluetooth interfaces while downloading data from the WLAN Access Point. For a distance between terminals of about 5 meters or less and an overall WLAN traffic load higher than about 18 Mbps we observe advantageous conditions to the cooperating users in terms of both achievable data rate and energy consumption.

A further step in our analysis foresees the investigation on a key factor to be kept under control to actually allow terminals to attain advantages from a cooperative behavior within the cited applicability ranges. By this we mean the Bluetooth and IEEE 802.11g mutual interference in the 2.4 GHz-ISM-Band, which we already introduced in section 3.2.. The suspect is that the coexistence of the two interfering systems will further reduce its applicability range.

Our worries are confirmed by the outputs of our test campaigns, as summarized in Figure 6. It is sketched by considering the case of two cellular phones downloading a portion of the requested file from the WLAN AP and at the same time exchanging the relevant downloaded portions with one another. This experimental campaign is conducted under different WLAN traffic loading levels (by fixing at 1 meter the distance between

---

posium, Springer Series on Signals and Communication Technology, S. Pupolin", with kind permission of Springer Science and Business Media

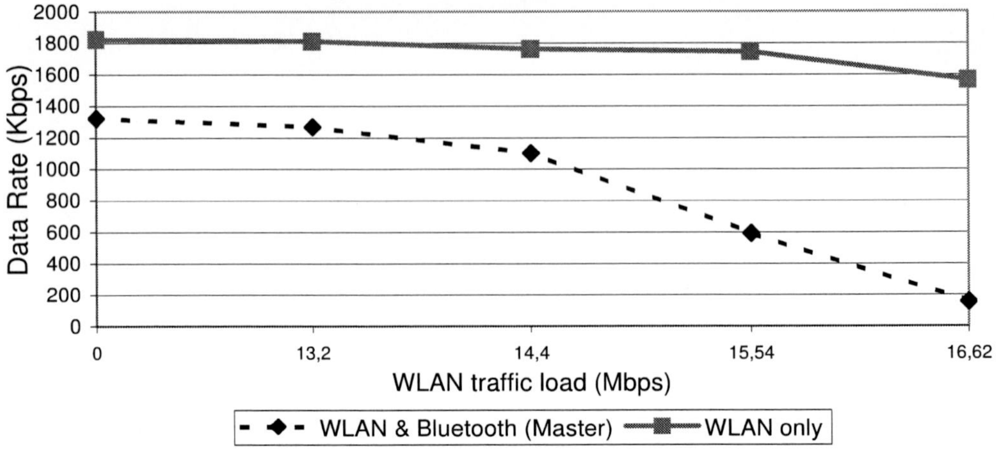

Figure 6. WLAN data rate in a stand-alone and in a WLAN-Bluetooth cooperative scenario.

the two cooperating cellular phones) to study the impact of Bluetooth interference on the WLAN achievable rate. As expected, compared to the case of WLAN transmission only, the downloading rate each user experiences over the WLAN link is reduced; this making Bluetooth cooperation no more beneficial. From the comparison in Figure 6, it clearly appears that:

- a reduction of the rate over the 802.11g link is always measured, and it is more manifest by increasing the loading level of the 802.11g network;
- if there is cooperation activity over the Bluetooth link, then an overall loading level of the WLAN reaching the value of 16.62 Mbps implies that the two cooperating mobile phones cannot access anymore the WLAN (the reader is invited to compare this value with the border value achievable under non-cooperative WLAN behavior).

Based on the obtained results, at this stage of the analysis, we can state that even when operating within the theoretical applicability ranges for WLAN-Bluetooth cooperation, the mutual interference of the considered radios seems to exclude any real advantage. Nevertheless, extensive research has been conducted dealing with 802.11 and Bluetooth coexistence and mutual interference; as a result, many possible solutions have been proposed. These can be classified into two main classes: either based on collaborative (with information exchange between the two networks) or based on non collaborative (with no information exchange between the two networks) schemes. An example of a non collaborative scheme is Adaptive Frequency Hopping (AFH) in which the device regularly scans the spectrum to identify and avoid channels with interference. This technique improves the performance of physical links when in the presence of interference; it also reduces the interference caused by physical links on other devices in the 2.4 GHz frequency band. Adaptive Frequency Hopping (AFH) is introduced into the Bluetooth standard since version 1.2. Other examples of non collaborative techniques within the Bluetooth standard are Adaptive Packet Selection and Adaptive Packet Scheduling. According to these schemes, a decision is taken on the packet type to use and the best moment to transmit it, based on the knowledge of the environment interference.

Collaborative schemes instead, are based on information exchange between co-located radio transmitters to locally control transmissions and interference. In our reference cooperative scenario we are actually in such a situation where the two radio transmitters are within the same device, which is unfortunately also the worst case for interference. Among these collaborative schemes we can mention: a time division multiplexing approach at the MAC (Medium Access Control) layer where the two radios alternate wireless medium access to avoid interference (namely, AWMA); Packet Traffic Arbitration (PTA) which provides per-packet transmissions authorization; channel skipping solutions which use a static range of frequencies for 802.11, which are skipped by Bluetooth as possible hopping frequencies [18]; and hybrid solutions which use a combination of the aforementioned algorithms and require a 2 or 3-wire connection between the 802.11 and Bluetooth radios. The reader is referred to [14],[15], [16] and [17] for more details about this matter.

The introduced solutions could be a viable add-on to the implemented cooperative WLAN-Bluetooth scenario. This gives us the opportunity to affirm that still a deeper analysis needs to be performed before excluding the WLAN-Bluetooth cooperative scenario from the set of feasible network combinations.

3.3. WLAN and WLAN Ad-hoc as Short-Range Links in Cooperation with EDGE and UMTS Cellular Links

In the view of a thorough analysis of a WLAN employment as short-range link in cooperation, it is worth recalling the key differences between a WLAN working in an "infrastructured" configuration and a WLAN working in ad-hoc configuration. A WLAN network has an infrastructure-based nature, which makes cooperation dependent on a pre-existing network and on the accessibility to the available Access Point (AP). Cooperating devices should exchange information (downloaded from cellular links) through the Access Point. The latter could be a limitation in some particular cases. As we reported in section 3.2.3., the accessibility to the AP was limited by a traffic load higher than 18, 5 Mbps. For higher values mobile phones couldn't access the network. This is probably caused by lack of resources and implementation of admission control policies.

Contrariwise, the decentralized feature of the ad-hoc WLAN network is an appealing characteristic for the cooperative communication paradigm. It is, therefore, a more interesting approach than infrastructured WLAN. Cooperating devices can directly exchange information without passing through the AP. Notwithstanding, the ad-hoc approach has the drawback of a higher energy per Kbit consumption.

The two approaches, when exploited as cooperative short-range links, offer an expected data rate gain of about 100% in both scenarios with EDGE and UMTS cellular links; consequently the transfer time is halved. Moreover, the results of an ad-hoc conducted assessment campaign have shown that the 802.11g ad-hoc approach is less efficient from the energetic point of view, due to signalling overhead.

4. Conclusion

In this chapter a novel communication paradigm has been introduced, offering an answer to limitative features of conventional cellular links in the view of supporting novel quality-, data rate-, and energy-demanding services. Based on modern mobile phones, which offer a wide range of network interfaces to access either the web or local networks, we implemented a test-bed for a cooperative wireless communication scenario. For a better analysis of the novel communication paradigm, a performance overview of both cellular and short-range networks has been given. All results focused on data rate, energy consumption, and transfer time as main benefits introduced by the cooperative approach. We proved that the novel cooperative communication paradigm guarantees very interesting benefits for the end-user basing our conclusions on experimental measurements campaigns performed on the deployed test-bed.

References

[1] Frank H.P. Fitzek and Marcos D. Katz *Cooperation in Wireless Networks: Principles and Applications – Real Egoistic Behavior is to Cooperate!*, Springer 2006, ISBN 1-4020-4710-X.

[2] Mischa Dohler, Djamal-Eddine Meddour, Sidi-Mohammed Senouci and Ahmed Saadani, Cooperation in 4G Hype or Ripe?, *IEEE Technology and Society Magazine*, in press, 2008.

[3] L. Militano, F.H.P. Fitzek, A. Iera and A. Molinaro, On the beneficial effects of co-operative wireless peer to peer networking, *Tyrrhenian International Workshop on Digital Communications (TIWDC'07)*, Ischia, Italy, September 2007.

[4] Laneman, J.N. Tse, D.N.C. Wornell and G.W. Wornell, Cooperative Diversity in wireless Networks: Efficient Protocols and Outage Behaviour , *IEEE Transaction on Information Theory*, December 2004.

[5] Laneman and G.W. Wornell, *Distributed space-time-coded protocols for exploiting cooperative diversity in wireless networks*, IEEE Transaction on Information Theory, 2003.

[6] A. Nosratinia, T.E. Hunter and A. Hedayat Cooperative Communications in Wireless Networks, *IEEE Communications Magazine*, October 2005.

[7] Sendonaris, A.Erkip and E. Aazhang *Increasing uplink capacity via user cooperation diversity*, IEEE ISIT, August 1998.

[8] R. Pabst, B. Walke, D. Schultz et al., Relay-based deployment concepts for wireless and mobile broadband radio, *IEEE Communications Magazine*, May 2004.

[9] M. Dohler and A.H. Aghvami, *Distributed Cooperative Communication Networks with application to cellular, ad hoc and sensor networks*, Tutorial at IEEE VTC Spring 2006, Melbourne, Australia, May 2006.

[10] Leigh Edwards, Richard Barker and the Staff of EMCC Software Ltd., *Developing Series 60 Applications: A Guide for Symbian OS C++ Developers*, Addison-Wesley, 2004.

[11] *S60 Platform: Basics Version 1.0 October 16th, 2006*, www.forum.nokia.com.

[12] R. Fielding, J. Gettys, J. Mogul, H. Frystyk and T. Berners-Lee, *Hypertext Transfer Protocol - HTTP/1.1 RFC 2068* January 1997.

[13] B. Krishnamurphy, J.C. Mogul and D.M. Kristol, Key differences between HTTP/1.0 and HTTP/1.1 WWW '99, *Proceeding of the eight international conference on World Wide Web*, Toronto, Canada, May 1999.

[14] A. Doufexi, A. Arumugam, S. Armour and A. Nix, An investigation of the impact of Bluetooth interference on the performance of 802.11g wireless local area networks *IEEE Semiannual Vehicular Technology Conference* VTC Spring, Vol. 1, pp. 680-684, September 2003.

[15] Axel Sikora and Voicu F. Groza, Coexistence of IEEE 802.15.4 with other Systems in the 2.4 GHz-ISM-Band, *Proceedings of the IEEE IMTC*, 2005.

[16] N. Golmie, N. Chevrollier, and O. Rebala Bluetooth and WLAN Coexistence: Challenges and Solutions, *IEEE Wireless Communications.* v10 i6. 22-29.

[17] C.F. Chiasserini and R.R. Rao, Coexistence Mechanisms for Interference Mitigation between IEEE 802.11 WLANs and Bluetooth, *INFOCOMM 2002*, pp. 590-598.

[18] Nick Hunn, *Bluetooth and 802.11 coexistence*, CTO - Ezurio Ltd, EZURIO white paper, 2006.

In: Mobile Telephones...
Editors: A.C. Harper and R.V. Buress, pp. 169-189
ISBN: 978-1-60456-436-5

Chapter 5

ELECTROENCEPHALOGRAPHIC ALTERATIONS IN HUMAN BEINGS DUE TO MOBILE TELEPHONES

***J.L. Bardasano*[1], *R. Goya*[2] *and J. Alvarez-Ude*[2]**
[1]Department of Medical Specialties,
[2]Department of Physics; University of Alcalá de Henares, Madrid, Spain

Abstract

The method of analysis applied here, Fast Fourier transform (FFT), allows us to discern between the brain integrated biological signal and noise, invalid impure signal.

Alterations induced by mobile telephones in the EEGs of healthy subjects are the outcome of electromagnetic oscillation.

The non-linear analysis, Higuchi algorithm, of electromagnetic alterations, due to the use of mobile telephones, yields a statistically relevant decrease of the fractal dimension (D)

There is a correlation between the biological brain signal and fractal dimension decrease.

Auditory cortex activation does not affect integration of oscillation from a mobile telephone.

Oscillations from active mobile telephones alter EEG recordings, at least in the fronto-parieto-temporal areas, at the frequency band of beta brain waves.

Artefacts introduced by the system (subject and recording devices) alter EEG recordings in all areas of the cranial calotte, at the frequency band of beta brain waves.

Introduction

There is a growing awareness in our society about how dangerous for the human body electromagnetic fields in radiofrequency wavelengths can be. [1]

Electromagnetic fields from mobile phones, Global System Mobile Communications, and their base stations, can affect living beings, man in particular, in different ways. There is increasing evidence, from different areas of knowledge, about this environmental electromagnetic pollution. [2-6]

The main line of research to assess the damage from exposure to electromagnetic fields is basically linear, which is adequate to measure thermal effects. On the other hand, non-thermal effects do necessarily depend on the condition of the organism at the moment of exposure. They do not only vary between different individuals but also on the condition of the same individual at different times. Therefore, the nature of object under study is non-linear. And, yet, there is a resistance, on the part of official institutions, to approach the problem from a non-linear perspective. Roughly speaking, there is greater relevance and publicity given to favourable data towards the market than to independent research.

Main Scientific Studies on Living Organisms Exposed to the Mobile Telephone

Bawin and col. [7] assessed that a 147 MHz modulated radiation and a 1 mW/cm magnetic field or radiant intensity could seriously alter the EEG of a cat. Adey [8], observed that chicken brains subjected to 147 MHz radiation and between 6 and 20Hz modulations increased the flow of calcium up to 20%. By studying the incidence of nervous system cancer in two-year mobile phone exposed rats from the 19th day of gestation, Adey and co-workers found out answers pointing to carcinogenic effects of the signals. [9,10]

Repacholi and co-workers use lymphoma-prone genetically modified mice and expose them to intense radiofrequencies of modulated signals (typical of mobile telephones) for 18 month, 1 h a day, periods. In the end a small, but statistically significant, increase in the frequency of lymphomas was found. [11]

Utteridge and team [12] tried to replicate the former project with transgenic mice, but they did not observe statistically relevant increases. [13,14] found out that electromagnetic fields can alter the brain rhythms in rats.

In order to assess how the human brain is affected by mobile telephones, [15-18] studied EEG recording as a basic working tool.

Neurons can respond to the electromagnetic fields by being microwave-stimulated. [19,20] Physiological neuronal alterations of relevance, which can be linked to radio frequencies from the mobile phones, have also been found. [21-25]

Some studies show that exposure to 900 Hz electromagnetic fields has an effect on electroencephalograms. [26-31] These electromagnetic field effects from mobiles on the human EEG were studied during a drill for aural memory. Energy alters the four basic bands (8 Hz–13 Hz, 14 Hz–30 Hz, 0 Hz–4 Hz and 5 Hz–7Hz) Results show evidence of a strong cause and effect relationship of radiation on EEG energy and on the amplitude peaks of each of the four rhythms. [32]

Griogoriev [33] has recently suggested the existence of negative effects related to the use of mobiles phones. The Polish researcher, Wlodzimierz Klonowski, has begun to apply the fractal theory to EEG recordings; thus, moving away from linear-method signal analysis and suggests non-linear dynamic ones such as Higuchi algorithm to study EEG recordings of the brain signal. Approximate entropy assigns a non-negative number to a time series and the highest values point to a greater irregularity of the analysed data. The sample entropy, which measures a time series regularity and the complexity of Lempel-Ziv, which is an estimate of the complexity of the signal. In general terms, the data suggest that EEG recordings for Alzheimer disease are the most regular and less complex particularly at the electrodes on the anterior side of the cranial calotte.

Higuchi [34] develops a non-lineal time series analysis to compute the fractional dimension. He applies it to natural phenomena and compares it with other BK and FFT methods. By setting the non-linear parameters: dimensional correlation, Lyapunov's and Hurst's exponent and approximate entropy significantly lower values than in the basal state can be found, through reflexology and musical effects. Therefore, these stimuli lead to a decrease of the EEG signal disorder.

Proximal Field

Electromagnetic waves are an immediate after-effect of the Ampère-Maxwell laws and of the Faraday-Henry's classical electromagnetism. Hence, alternating electricity creates an alternating magnetic field, which in turn recreates an alternating electric field, and starts over again. At each point of the space we find a field made up of two fields with the same frequency, an electric and a magnetic one, which is called *electromagnetic field.* The behaviour of electromagnetic waves is not uniform in the space, due to the fact that there are two areas which behave differently. These areas are called *proximal field* and *distal field.* In a proximal field the electrical (*E*) and magnetic (*H*) fields strongly depend on the distributions of charge and currents that generate them, thus they are studied as independent stationary fields. In a distal field, the electromagnetic field sets its behaviour. The border of these areas is not clear-cut and it depends on the geometry of the currents which yield these fields. In radiofrequencies, the distal field takes place at distances above three wave lengths.

In a proximal field *E* and *H* are at odds, not in phase. There could be an intense *E* field and a weak *H* field and the other way round; there is not a constant correlation, so separate measurements for them are required.

As regard to mobile telephones the emission frequencies range between 880 and 960 MHz; that is to say, the wave lengths are between 31 and 34 cm. In this case the proximal field expands to 3x31 = 93 cm. Mobile telephones are used in proximal fields. Being in a proximal field is relevant as magnetic permeability of living matter equals that of the air for all purposes. So, magnetic fields go into living matter without altering it, spreading throughout the organism.

Artefacts Obtained from an EEG Recording

We must consider those artefacts generated by the measuring system and the biologic artefacts, which have to be eliminated when it comes down to study the EEG.

Chaos and Fractals in Human Physiology

In the last twenty five years, Goldberg and team [35] have found out that that the heart, the nervous system and other biological systems can behave extremely at random when they are young and healthy. Contrary to what used to be thought, ageing and illness sometimes go hand in hand with growing regular behaviours. Irregularity and unpredictability are characteristic and marked features of good health. Decrease of variability and stressed periodicity are associated with a state of illness. It is common to think that stress alters the well paced rhythms and leads to erratic responses and several sequels, of which illness and old age can be part. Different pathologies (epilepsy, Parkinson disease, bipolar dysfunction) show a loss of variability and an increase in periodicity. Systems which are controlled by the nervous system are chaotic which operate under a wide range of conditions and so they are adaptable and flexible, which allows the systems to adapt to the requirements of changing and unpredictable electromagnetic environments.

Application of Fractal Theory to the EEG Signal by Means of Higuchi Algorithm

EEG recordings translate numerically in time series. [36] The complexity of the signal can be analysed in time domain, frequency or in space phase where the data must be studied in a multi-dimensionally. Fractal geometry is expressed in algorithms; in other words, by means of rules and procedural instructions which need the help of a computer in order to turn them into forms and structures.

Complexity measurements of electroencephalography with non-linear quantifiers [37]:

- Approximate entropy
- Spectral entropy
- Sample entropy
- Multi-scalar entropy
- Lempel-Ziv complexity
- Lyapunov exponent
- Maragos Algorithm
- Higuchi Algorithm

To process the quantifiers, mentioned above, as far as chaos theory is concerned, the reconstruction of an attractor in the multidimensional space phase of the EEG signal is certainly complex. It takes time, demands high power computation and the results are hard for most clinics to interpret. On the other side, Higuchi algorithm is simpler, can be calculated straight from the original signal, does not require preliminary reconstruction of the space phase, it is much faster, more intuitive in its application and easier to understand, and so more promising in its medical application. [38]

Higuchi Algorithm: Development and Application

Klonowski states that biomedical signals are generated by complex self-regulated systems. Higuchi proposes an algorithm to calculate fractal dimension in a time domain for this kind of series. A fractal curve can be subdivided in similar *k* curves, ($k = k_1, k_2 ..., k_{max.}$) in such a way that the length of the curve can be expressed as k^{-D}, D being the fractal dimension. Let us apply these analyses of signals, recorded in *a basal situation, during exposure to external physical agents* (mobile on and listening to a reading from a home phone number).

Procedure 1

It consists of reducing numerous experimental data to a simpler but more significant representation. Here we reduce about 80 000 experimental data, obtained for each electrode to approximately 550 significant data according to the following procedure:

$$\varphi(1), \varphi(2), \varphi(3), ..., \varphi(N) \qquad N = 80\,000$$

$$\varphi[1] = \frac{\varphi(1) + \varphi(2) + ... + \varphi(250)}{\Delta t(1\text{–}250)}$$

$$\varphi[2] = \frac{\varphi(100) + \varphi(101) + ... + \varphi(350)}{\Delta t_{(100\,-\,350)}}$$

...

...

...

$$\varphi[\cong 534] = \frac{\varphi(79750) + \varphi(79751) + ... + \varphi(80000)}{\Delta t(78750\text{–}80000)}$$

With these approximately 550 data we apply procedure 2, as indicated.

Procedure 2

Basal Situation

The recorded signal in a channel is a time series, in a computerised system of data procurement:

$$\varphi(1), \varphi(2), \varphi(3), ..., \varphi(N)$$

where ($\varphi(i)$ is the amplitude of the signal i at moment times i = 1,2,3 ..., N; N being the total number of dots.

From these data, a new k time series is constructed this way:

$$\varphi_m^k = \varphi(m,k);\ \varphi(m), \varphi(m+k), \varphi(m+2k),\ ...,\ \varphi\left(m + \text{int}\left[\frac{(N-m)}{k}\right]k\right)$$

Where,

$$\text{int}\left[\frac{(N-m)}{k}\right]$$

called Gauss formulation, stands for the whole of the quotient, where k and m are already integrated; m stands for the initial time series and k indicates the time interval; and finding k groupings of new time series, $k = k_1, k_2, k_3\ ..., k_{max}$.

Here we show different representations for m and k values:

Basal Situation

The fractal curve is divided in k similar curves, (k_{min}, k_1, k_2, k_{max}) [41] and applying the same criteria as before we obtain (Figure 1):

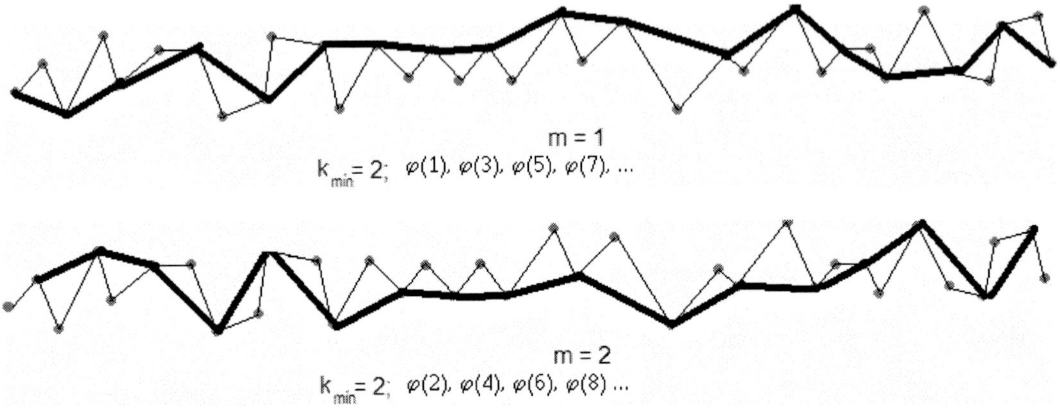

Figure 1.

By calculating the lengths of the curves for each m value, with the same $k_{min} = 2$ value and obtaining the average value of the length of the curve $L_m(k)$ (Figure 2):

$$L_1(k_{mín}) = \frac{1}{k}\sum_{1}^{k} L_m(k)$$

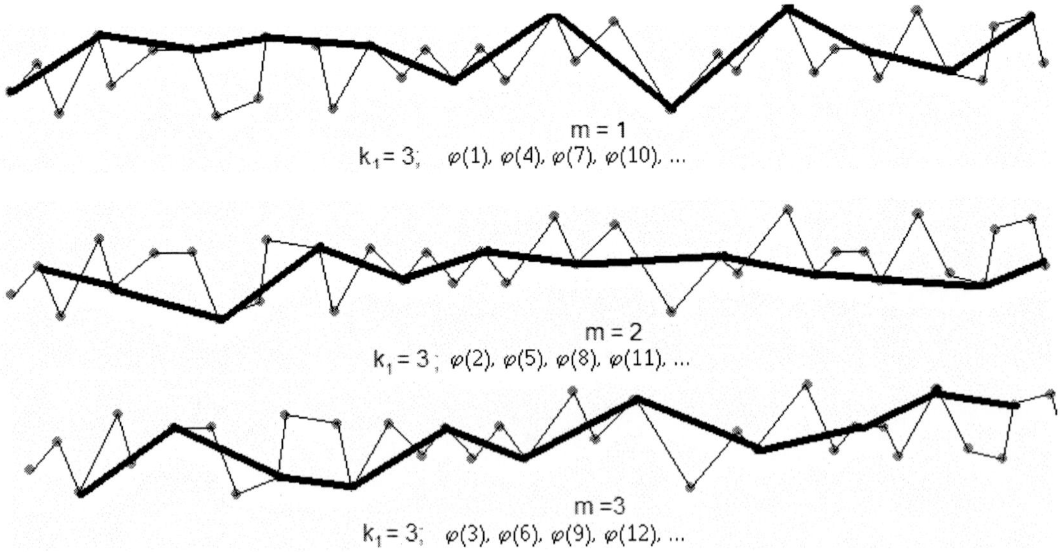

Figure 2.

By calculating the lengths of the three curves, for each value of m, with the same value, $k_1 = 3$, we obtain the mean value (Figure 3):

$$L_2(k_1) = \frac{1}{k}\sum_{1}^{k} L_m(k)$$

If we calculate the length of the curve, for each m value, with the same k_{max} value, we obtain:

$$L_5(k_{máx}) = \frac{1}{k}\sum_{1}^{k} L_m(k)$$

With the influence of external physical agents: We carry out a similar study of the values of the EEG recording for the same subject with the mobile on and listening to a reading from a home phone number.

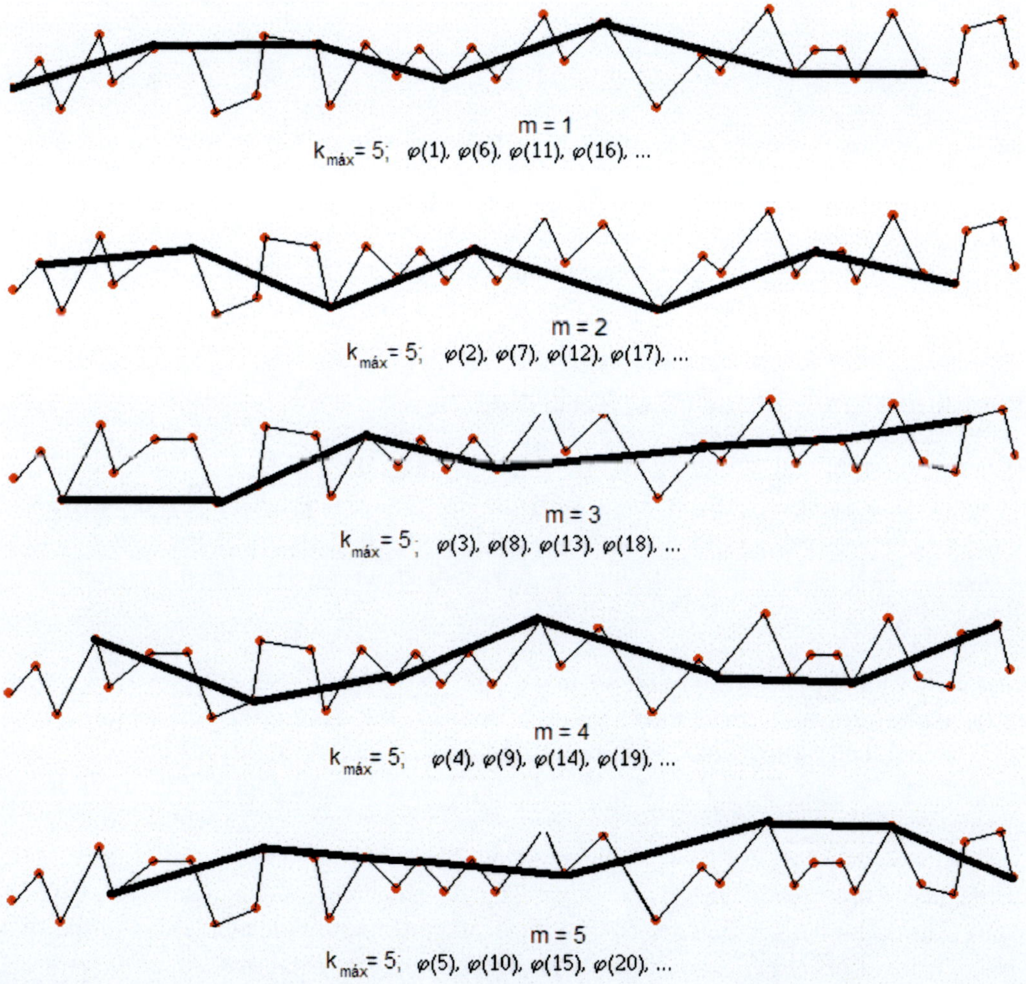

Figure 3.

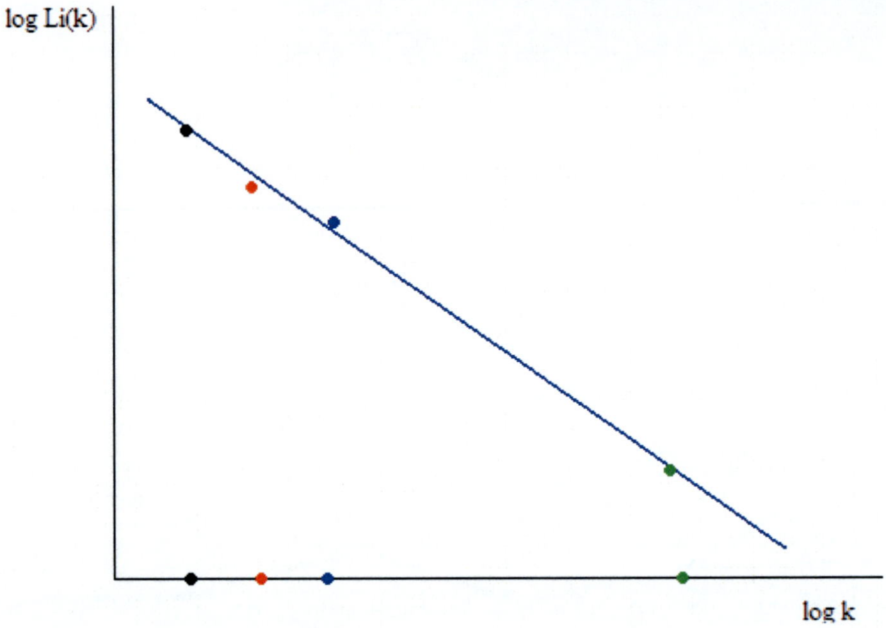

Figure 4. Ordinate axis, average log length of curve defined by Higuchi and abscissa axis time interval log. Subject in basal situation.

The graphic representation of the average length logarithm, of the curve defined by Higuchi, face to face the time interval logarithm leads to straight lines with a lesser steep and, therefore, with a lesser fractal dimension than in the basal approach (Figure 4).

Quantitative Study of the EEG Recording

Power lines and domestic electricity, and other communication systems such as mobiles telephones generate electrical and magnetic fields which have been added to those existing in nature since the beginning of time. The degree of production of electrical and magnetic fields has reached such a proportion that we can now talk about environmental electromagnetic pollution.

It could be relevant to know a given method to study the waves or oscillations which we receive with the mobile on and so to be able to study its influence on the neuronal electric system. An ambitious project would be to study the changes in electric potentials in healthy individuals, in subjects with either low or no adapting ability to electromagnetic environments in which we find ourselves and also in subjects with neurological pathologies.

The EEG recording signal is complex, the addition of activities of millions of brain neurons and extracerebral electric activity. The development of computers allows us to process EEG recordings in new ways. The application of mathematical analysis to the study of biological signals always leaves the doubt of producing artificial mathematical results. In order to develop our project , without moving astray from changes in electrical potentials in healthy subjects in a basal situation and with the mobile on and listening, we have a digital electroencephalograph, model SAM 32 FOFC 1, which contains software for mathematical analysis which must be validated.

The mathematical operating procedure does not differentiate between cerebral and extracerebral signals. Therefore, any study whose specific data are to be subjected to a process of mathematical analysis, requires a previous assessment of the process.

From a Physical and Medical perspective, the comparison of the EEG recordings, frequency maps and amplitude maps, in healthy subjects in a basal situation and with the mobile on and listening, will allow us to know:

1. whether a EEG recording signal, in a basal situation, is basically cerebral.
2. whether the EEG recording signal, with the mobile on and listening, is basically cerebral.
3. whether the EEG recordings carried out are contaminated by extracerebral signals and cannot be erased.

We carry out EEG recordings on a phantom, inert being, in basal conditions and with mobile on. When we compare results we will obtain physicomedical information of the process.

In order to avoid electromagnetic environmental interference, measurements are taken as the subject is inside the Faraday cage and the transmission goes through optic fibre to the analysis panel.

Methods

Population Sample

24 healthy subjects, 9 women and 15 men, between 20 and 73 years old, the average age being 36, without ingesta of medicines and neurologically healthy.

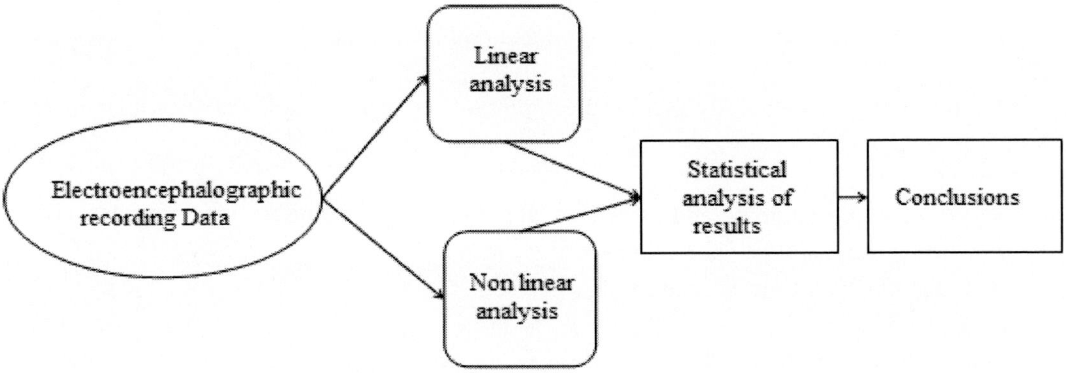

Figure 5. Sketch of the process to follow in the analysis of the work.

Phases of the Method

1. A basal EEG recording is done to each subject, with eyes closed, for 5 minutes.

We obtain an EEG recording with artefacts and, the frequency and amplitude maps, after applying the fast Fourier transform to the values for electrical potentials.

The EEG recordings are then cleared of recognisable artefacts, (noise).

After we compare the two conditions, we find that:

- There was a decrease of delta waves decrease in 100% of cases.
- There was an increase in the percentage of theta waves in 20% of cases.
- There was an increase in the percentage of alpha waves in 100% of cases.
- There was a decrease in the percentage of beta waves in 75% of cases.
- There was a decrease of average interval of amplitude in 100% of cases.

Summary of results: the extracerebral potentials (artefacts and noise), in basal conditions, always increase the average interval of amplitude and the percentage of delta waves.

2. An EEG recording is carried out on each subject with the mobile on, in the neighbourhood of the right ear, listening to a reading from a home phone call with eyes closed for 5 minutes. The EEG recording of the subject with the mobile on and listening has been cleansed of artefacts like blinking and that of the experimenter getting out of the cage. And yet it is not free of the sound of chewing or other artefacts which are associated with active thinking on the part of the subject and which might be confused with the functioning of the mobile.

Results:

a. the presence of the mobile phone activated modifies the percentage of frequencies. Delta waves increase significantly, theta increase slightly and both alpha and beta decrease.
b. Mobile telephones increase the average amplitude interval.

In all the cases under study there is an important increase of delta waves which we know is only possible

- In sleeping subjects and/or with a diagnosed brain tumour. This increase can not occur, from a physiological point of view.
- Due to errors from the recording system.

As we cannot eliminate all known artefacts, the results are not realistically accurate.

Mobile telephones modify the EEG recordings and, thus the potential values depending on the time. They alter wave amplitude relevantly and the percentage of waves with coinciding frequencies with beta waves frequencies.

3. We perform an EEG recording on a phantom, or inert being, for 5 minutes:

- In basal conditions
- With the mobile on and reading a text from a home telephone.

Results:

The presence of the mobile on and reading a text from a home telephone increases the interval of amplitudes of the brain waves significantly, which are not true brain waves.

All that leads us to conclude that the quantitative mathematical analysis of the recordings and the colour maps of frequencies and amplitudes with their corresponding tables, from the program, System PLUS by Micromed, do not respond to the question of whether the brain waves are modified or not due to mobile phones.

4. We have created programs which, with the EEG recording data, allow us to specify

- Figure 6 (a) algorithm, allows us to obtain power spectra.
- Figure 6 (b) algorithm, allows us to obtain a straight falling phase line and calculates the fractal dimension.
- Figure 6 (c) algorithm, allows us to obtain the fractal variability with time.
- Figure 6 (d) algorithm, allows us to obtain a histogram of the fractal dimension.

5. In an inert system or phantom for any areas of the cranial calotte, we find, in basal situation, frequency waves coincidental with cerebral delta waves, but of greater amplitude and very low fractal dimension. No cerebral signal. With the mobile on and reading a text from a home phone, waves with maximums with regular frequencies and amplitudes much superior to basal values, never coinciding with the features of cerebral waves, and very high fractal dimension. No cerebral signal or non-integrated cerebral signal.

Therefore, the fractal dimension is going to oscillate between these two extreme values in the system we are going to study.

If we can show that the fractal dimension decreases when we have the mobile on with regards to the basal state we will be able to infer that it is an integrated cerebral signal and, so, pollution free, since the mobile only yields a very high fractal dimension.

In order to show the concept of integrated cerebral signal we present one of the cases studied.

We obtain the curves of power spectrum from the electrodes on the right side of the cranial calotte, as indicated in Figure 8. We also observe the evolution of the graphs from a recording of highest contamination at the electrode Fp_2F_8 to no contamination at electrode T_6O_2.

The graphs A' and B' of power spectra, contain a series of peaks, with regular frequencies and very high amplitudes, as befits a non integrated signal.

Graphs C' and D', which are similar, with a maximum value at the ordinate axis which matches the area of frequencies of theta and alpha waves, are integrated signals.

When we apply the Higuchi algorithm we observe that for the most contaminated electrodes the fractal dimension increases and, according to the table below, the difference is negative between the fractal dimensions, Dbasal, Dmobile.

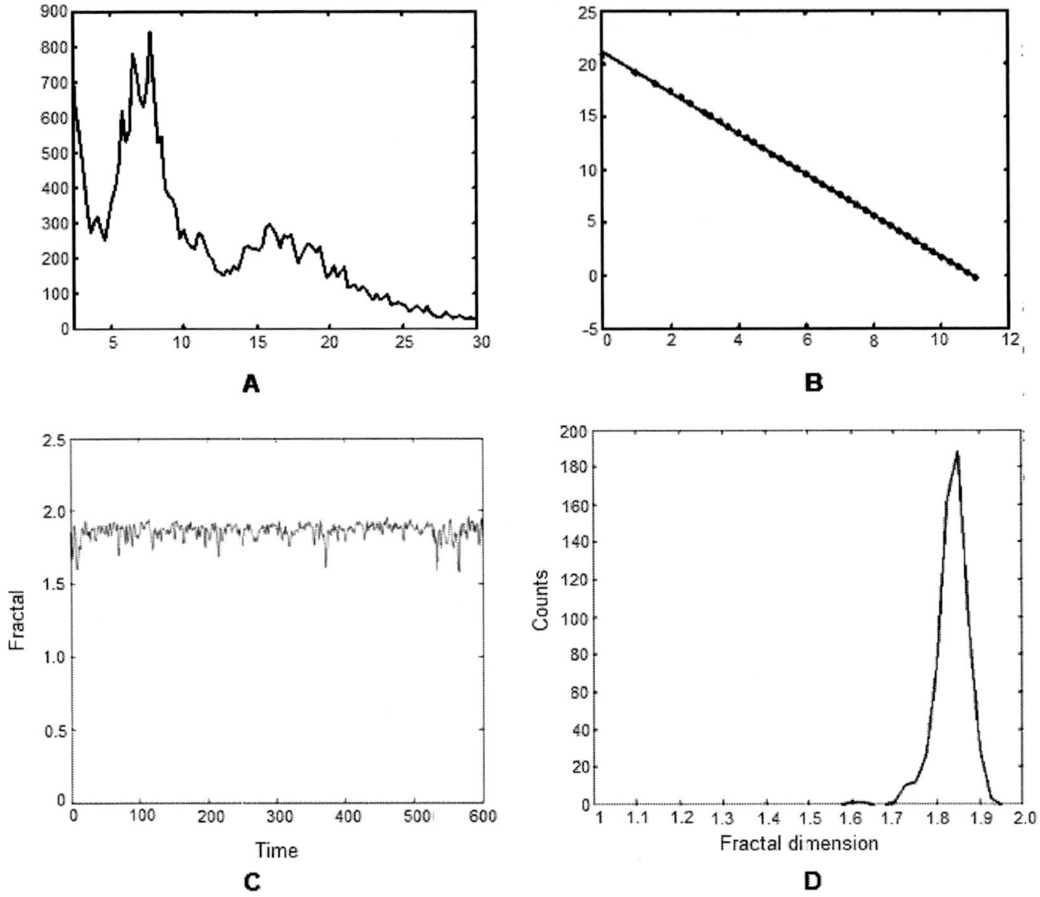

Figure 6. A: Power spectral density. B: Straight line of falling phase and fractal dimension value. C: Fractal variability with time. D: Value of fractal dimension.

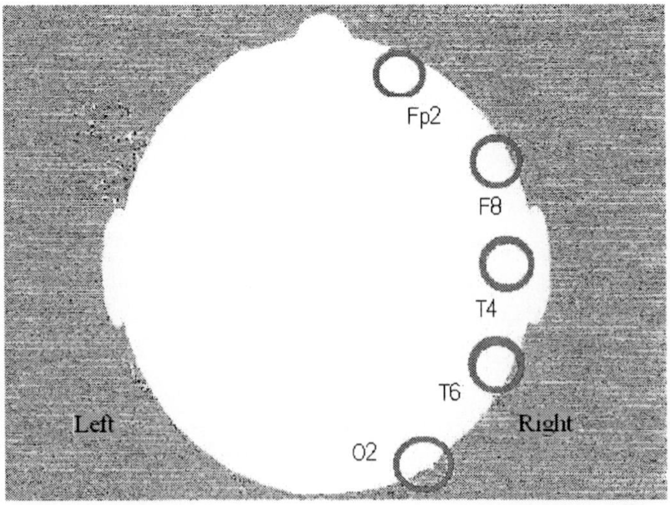

Figure 7. Electrodes in cranial calotte closest to the mobile telephone.

	Fractal dimension		
Electrodes	**Basal**	**Mobile on**	**Basal Mobile**
Fp2F8	1.8908.	1.9422	–0.0514
F8T4	1.808	1.8721	–0.0541
T4T6	1.7595	1.7475	0.012
T6O2	1.791	1.7354	0.0556

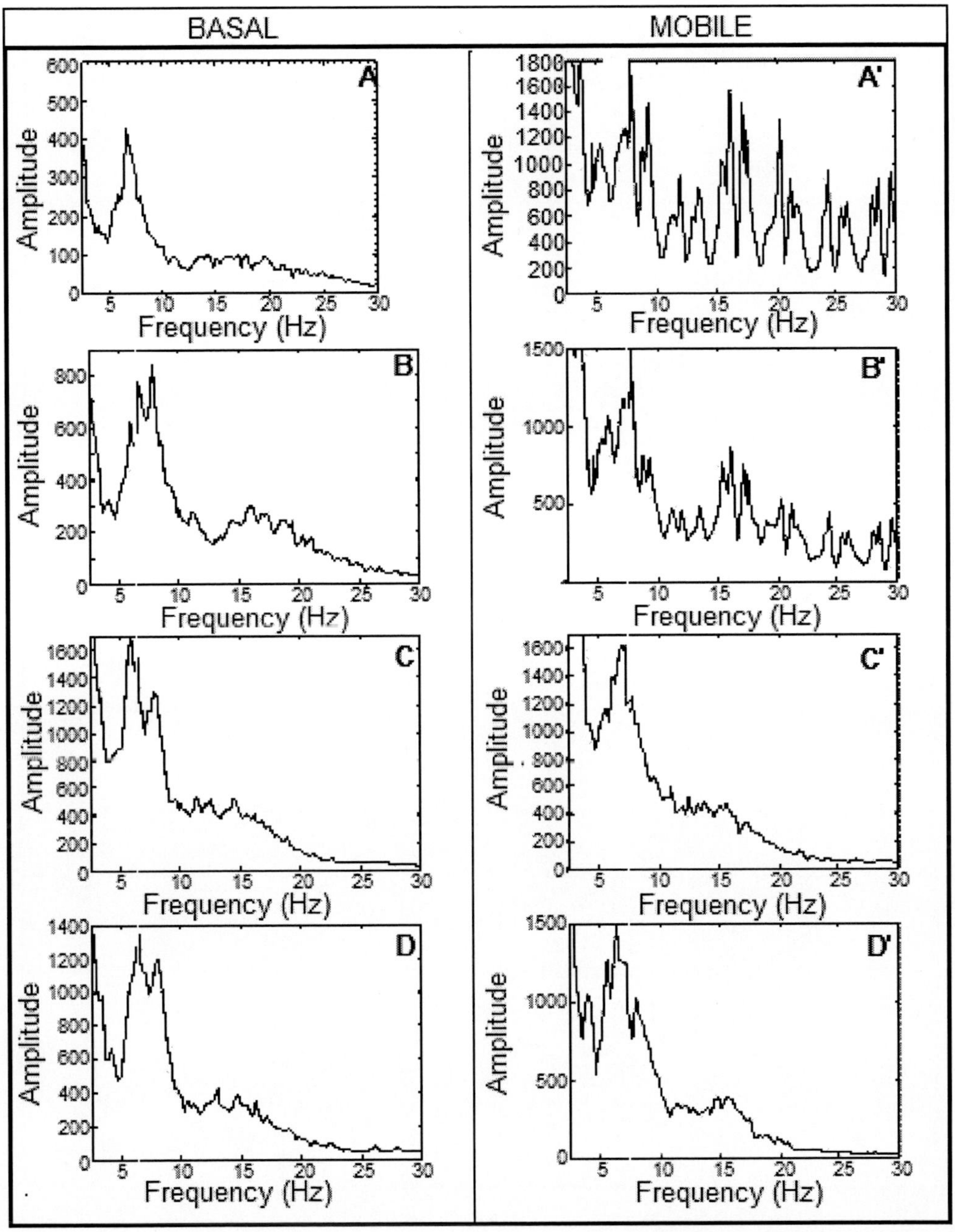

Figure 8. Graphs: Power spectra; A' and B' non-integrated signal; C' and D' integrated signal.

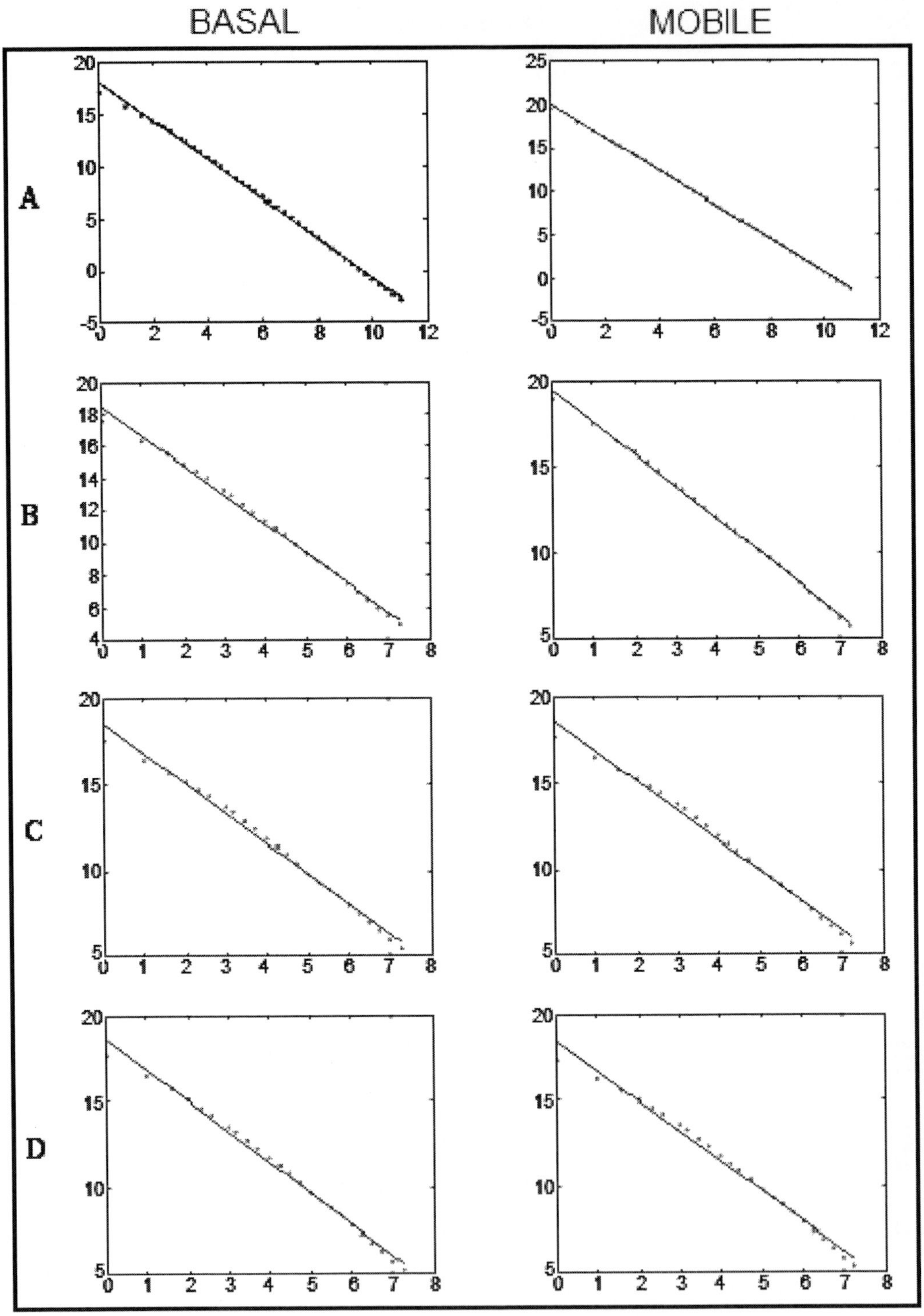

Figure 9. Lines whose slope is the fractal dimension. A, B, C and D in basal situation, electrodes $F_{p2}F_8$ F_8T_4, T_4T_6 and T_6O_2 and A'B',C' and D' mobile on, for the same electrodes.

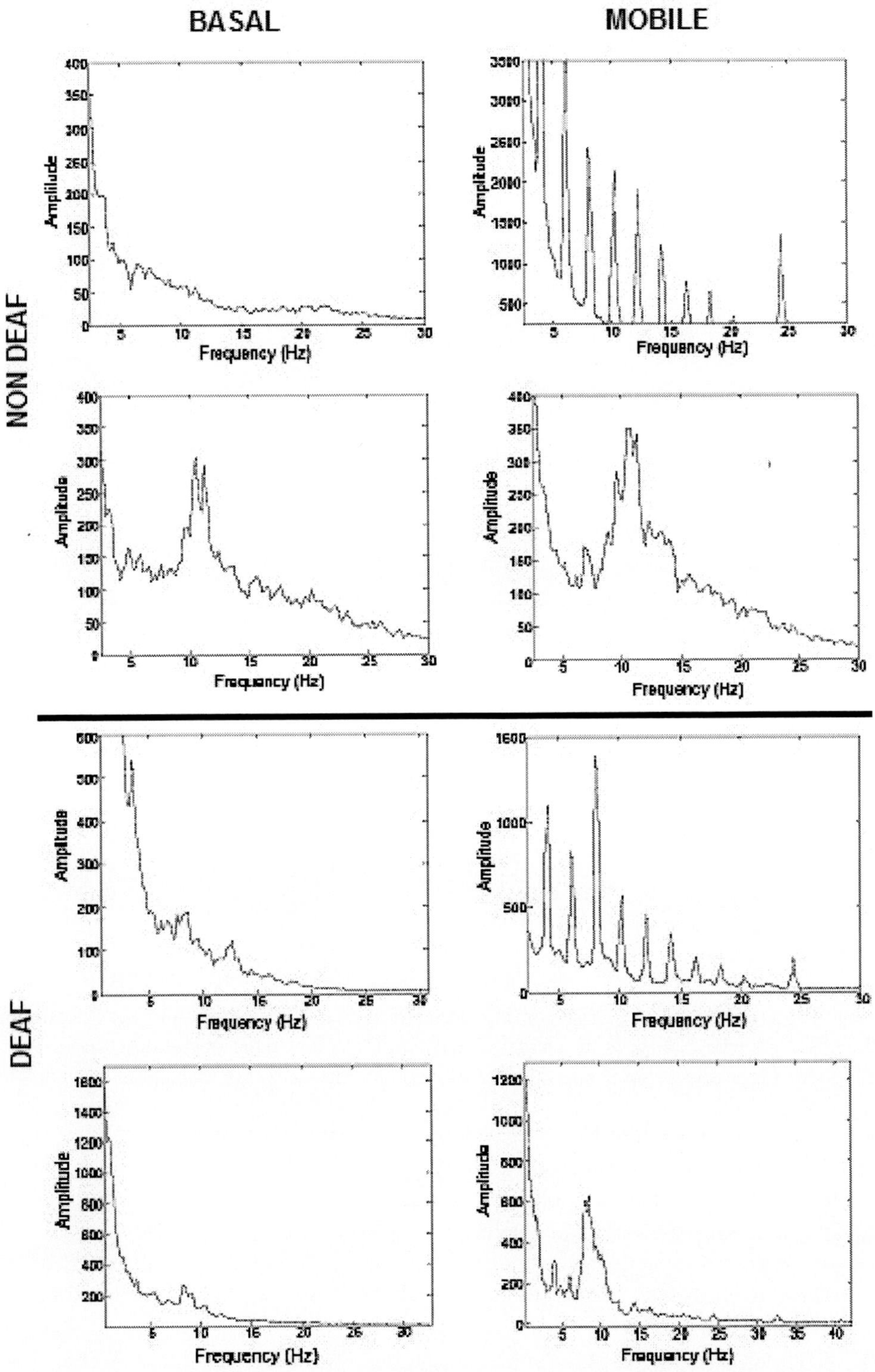

Figure 10. Power spectral density for the electrodes F $_{p1}F_3$ and T_3T_5 in non deaf subject and a deaf subject.

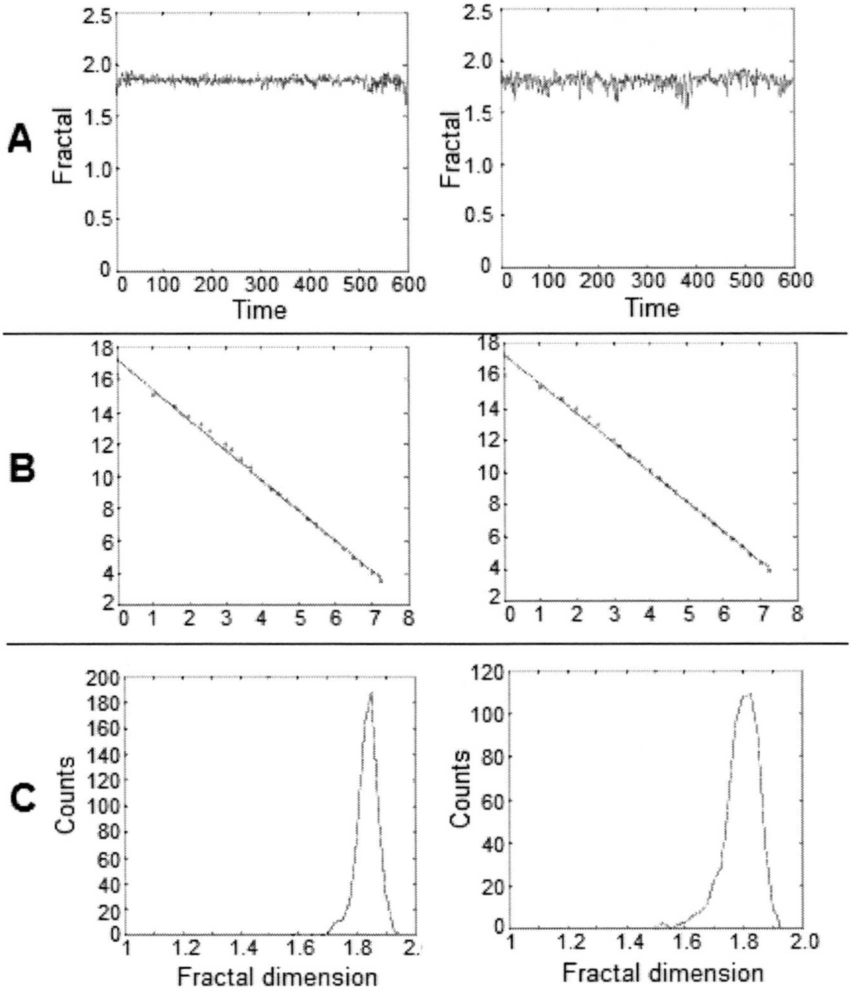

Figure 11. These graphs of the EEG recording at electrode FzCz. A. Variation of the fractal dimension with time. B. Line whose slope is the fractal dimension. C. Value of the fractal dimension. The figures on the left side represent the basal situation and those on the right side the mobile on.

The difference is positive at the electrodes, where we observe the integrated signal, and the fractal dimension decreases.

For all the subjects studied, the different electrodes of the EEG recording, in any area of the cranial area, show two ways of behaving:

A. The power spectra of the subject, in a basal situation and with the mobile on and listening, are similar.
B. different frequencies and great amplitudes. No cerebral signal or non-integrated cerebral signal present.

The fractal dimension of the subject with the mobile on and listening is greater than the fractal dimension of the same subject under basal conditions.

Influence of Sound on EEG

Under the experimental conditions applied to the cerebral waves of a subject, who is placed in a Faraday cage, if the mobile is on and listening to a reading from a home phone, there is an interaction between:

a. the sound waves which move via aural cortex
b. the oscillations from the mobile phone on with its physical features.

In order to know the effect the sound waves have on the cerebral waves a first trial is carried out on two healthy subjects and two conduction deaf subjects who undergo two EEG recordings.

1. under basal conditions
2. with the mobile on and reading a text from a home phone number.

We analyse the results from the application of algorithm 1 using power spectrum curves.

By comparing the power spectrum of the electrodes for the same cerebral areas, $F_{p1}F_3$ and T_3T_5, under the conditions already mentioned, we can observe that:

Both the deaf and non-deaf subjects at electrode $F_{p1}F_3$, do not integrate the signal in the brain but they do integrate it at electrode T_3T_5. The response is similar for both subjects.Similar responses have been found at other electrodes of cranial calotte for the same subjects.We have not found significant differences between the EEG recordings of non-deaf and deaf subjects.

These preliminary results show that activation of the aural cortex does non have any bearing on the integration of the signal.

Classification and percentage of the behaviour of the fractal dimension of the subjects for the three electrodes under study

	Electrodes				
	T_6O_2	C_3P_3	FzPz	Total	Percentage
Diminished fractal dimension	7	13	18	38	52.75%
Augmented fractal dimension	12	9	6	27	37.50%
Very augmented fractal dimension	5	2	0	7	9.72%

These data indicate that in most cases we obtain a decrease of the fractal dimension which is associated to the mobile on and being listened to, which is a consequence of the integration of the signal in the cerebral actividy.

When we apply Higuchi algorithm and obtain fractal dimensions, we observe

A Data from the subject, whose fractal dimension (D) is larger than 1.9 are discarded, since that is the highest normal value according to our phantom pattern with the mobile on. The number of cases (7 in all) is grouped under the *very augmented* fractal dimension epigraph. As we analyse the EEG recording of these patients, we notice a great network of artefacts which make the EEG recording correlate with a normal outline impossible. That is so as the amplitude is larger than that of the physiological cerebral field and as beta activity is very high, which would correspond with a great muscular activity and noise due to oscillations of the mobile on.

B After this first analysis, for the remaining cases (65 in all) we find two phenomena:

- $D_{mobile} < D_{basal}$. In these cases, part of the energy that gets to the brain from the oscillation of the mobile is integrated and modifies the brain waves under basal conditions. We can then assume this is an effect of the mobile phone.
- $D_{mobile} > D_{basal}$. This can be due to two reasons. For most of the recordings we find an occultation of signal integration, due to the existence of artefacts corresponding to extracerebral noise. In the other cases (25%) there seems to take place a real integration of the signal, but, for unknown reasons, the fractal dimension increases. It can not be explained as due to an artefact as in the previous case. These data will also be discarded.

These data indicate that in most cases we obtain a decrease of the fractal dimension which is associated to the mobile being on and being listened to, which is a consequence of the integration of the signal in the cerebral activity.

Here we summarise the statistics for both variables to check whether the differences between thae two samples are statistically significant.

	Statistical summary	
	Basal	**Mobile (listening)**
Frequency	38	38
Average	1.79739	1.74321
Median	1.7971	1.75085
Variance	0.0033292	0.00494648
Standard deviation	0.0576992	0,0703312
Minimum	1.6654	1.5987
Maximum	1.8952	1.8753
Range	0.2298	0.2766
Standardised asymmetry	-0.796364	-0.559097
Standardised Kurtosis	-0.238946	-0.424463

For the data of the basal fractal dimension as well as for the mobile on and listening fractal dimension, the standardised asymmetry and kurtosis are within the -2/+2 range: they arise from normal distributions.

The results we have obtained suggest that the cerebral activity, at T_6O_2 and C_3P_3 electrodes, is modified by the oscillation from the mobile telephone.

Results

Activation of the aural cortex does not have an influence on signal integration the EEG recording.

The mobile telephone has a certain influence on the brain waves, as we can see by comparing, with the help of Higuchi algorithm, the non-linear quantifiers of the EEG signal complexities which take us to the fractal dimension, in the basal situation and with the mobile on and listening.

The results obtained, which are statistically significant, suggest that the cerebral activity over all the cranial calotte becomes modified due to having the mobile telephone on. [39-41]

Conclusion

The method of analysis, fast Fourier transform (FFT), allows us to discriminate between the integrated biological cerebral signal and the contaminated non-valid signal.

The changes observed, in the EEG, of healthy subjects affected by the mobile telephone are the effect of electromagnetic oscillations.

The non linear analysis, Higuchi algorithm, of electroencephalographic changes produced by mobile telephones, allows us to see a decrease, statistically significant, of the fractal dimension (D).

There is a correspondence between the integrated biological cerebral signal and the decrease of the fractal dimension.

The activation of the aural cortex does not have a bearing on the integration of the oscillation from the mobile phone.

The oscillation produced by mobile telephones on, modify the EEG recordings, at least in fronto-parieto-temporal areas, at the frequency range of cerebral beta waves.

The artefacts introduced by the system (individual and recording equipment) modify the EEG recordings, in all areas of the cranial calotte, at the frequency range of cerebral beta waves.

References

[1] Hyland G. Physics and Biology of mobile telephony. *Lancet* 2000; 356: 1833-1836.

[2] Navarro EA, Segura J, Portolés M, Gómez Perretta C. The microwave syndrome: a preliminary study in Spain. *Electromagnetic in Biology and Medicine.* 2003; 22: 161-171.

[3] Santini R, Santini P, Danze JM, LeRuz P, Seigne M. Survey study of people living in the vecinity of cellular phone base stations. *Electromagn. Biol. Med.* 2003;22 (1): 41.

[4] Balmori A. Posibles efectos de las ondas electromagnéticas utilizadas en la telefonía inalámbrica sobre los seres vivo. *Ardeola.* 2004; 51(2): 477-490.

[5] Balmori, A., Posible effects of electromagnetic fields from phone masts on a population of white storks. *Electromagnetic Biology and Medicine*. 2005; 24: 109-119.

[6] Hutter HP, Moshammer H, Wallner P, Kundi M. Subjective symptoms, sleeping problems, and cognitive performance in subjects living near mobile phone base stations. *Occup. Environ. Med.* 2006; 63: 307-313.

[7] Bawin SM, Gavalas-Medici RJ, Adey WR. Reinforcement of transient brain rhythms by amplitude modulated VHF fields. In Springfield IL, Charles C Thomas, editors. *Biological and clinical effects of low frecuency magnetic and electric fields.* 1974: 172-186.

[8] Adey WR. Interaction mechanisms of low-level electromagnetic fields in living systems. Collective properties of cell membranes. In B. Norden, C. Ramel, editors. *Symposium,* Royal Swedish Academy of Sciences;. Stockholm. 1992. p. 47-77.

[9] Adey WR, et al. Spontaneous and nitrosourea-induced primary tumors of the central nervous system in fischer 344 rats chronically exposed to 836 MHz modulated microwaves. *Radiation Research* 1999; 152: 293-302.

[10] Adey WR, et al. Spontaneous and nitrosourea-induced primary tumors of the central nervous system in Fischer 344 rats exposed to frequency-modulated microwave fields. Cancer. *Radiation Research.* 2000; 60: 1857-1863.

[11] Repacholi MH, et al. Lymphomas in Eμ–Piṃıtransgenic mice to pulsed 900 MHz electromagnetic fields. *Radiation Research.* 1997; 147: 631-640.

[12] Utteridge TD, et al. Long-term exposure to Eμ–Piṃıtransgenic mice to 898.4 MHz microwavesdoes not increase lymphoma incidence. *Radiation Research.* 2002; 158: 357-364.

[13] Bawin SM, Satmary WM, Jones RA, Adey WR, Zimmerman G. Extremely-low-frecuency magnetic fields disrupt rhythmic slow activity in rat hippocampal slices. *Bioelectromagmetics.* 1996; 17: 388-395.

[14] Vorobyof VV, Galchenko AA, Kukushkin NI, Akoev IG. Effects of weak microwave fields amplitude modulated at ELF of symmetric brains areas in rats. *Bioelectromagnetics;* 1997;18: 293-298.

[15] Krause CM, Haarala C, Sillanmaki L, Koivisto M, et al. Effects of electromagnetic field emitted by cellular phones on the EEG during an auditory memory task: A double blind replication study. *Bioelectromagnetics.* 2004; 25: 33-40.

[16] Lin JC. *Human electroencephalograms (EEG) and mobile-phone radiation.* 2004; 308: 52-54.

[17] Curcio G, Ferrara M, Moroni F, D'Inzeo G, Bertini M, De Gennaro L. Is the brain influenced by a phone call?: an EEG study of resting wakefulness. *Neuroscience Research.* 2005; 53:265-270.

[18] Loughran SP, Word AW, Barton JM, Croft RJ, et al. The effect of electromagnetic fields emitted by mobile phones. *Neuroreport.* 2005; 16:1973-1976.

[19] Kholodov Y, Lebedeva NN. *Reactions of the central nervous system of man on electromagnetic fields.* M: Nauka. 1979. (in Russian).

[20] Beasond RC, Semm P. Responses of neurons to an amplitude modulated microwave stimulus. *Neuroscience Letters.* 2002; 33:175-178.

[21] Hossmann KA, Hermann DM. Effects of Electromagnetic Radiation of Mobile Phones on the Central Nervous System. *Bioelectromagnetics.* 2003; 24:49-62.

[22] Westerman R. Hocking B. Diseases of modern living; neurological changes associated with mobile phones and radiofrequency radiation in humans. *Neuroscience Letters.* 2004; 361:13-16.

[23] Huber R, Graf T, Cote KA, Wittman L, et al. Exposure to pulsed high-frequency electromagnetic field during waking affects human sleep EEG. *Neuroreport.* 2000; 11:3321-3325.

[24] Szyjkowska A, Bortkiewicz A, Szymczak W, Makowiec-Dabrowska T. subjective symptoms related to mobile phone use. *Pol. Merkuriusz. Lek.* 2005; 19 (112): 529-532.

[25] Papageorgiou CC, Nanou ED, Tsiafakis VG, Kapareliotis E et al. Acute mobile phone effects on pre-attentive operation. *Neurosci. Lett.* 2006; 397(1-2):99-103.

[26] Mann K, Röschke J. Effects of pulsed high-frequency electromagnetic fields on human sleep. *Neuropsychobiology.* 1996; 33: 41-47.

[27] Eulitz C, Ullsperger P, Freude G, Elber T. Mobile phones modulate response patterns of human brain activity. *Neuro Report.* 1998; 9: 3229-3232.

[28] Freude G, Ullsperger P, Eggert S, Ruppe I. Microwaves emitted by cellular telephones affect human slow brain potentials. *European Journal of applied physiology.* 2000; 81:18-27.

[29] Krause CM, Sillanmäki L, Koivisto M, Häggqvist A, et al. Effects of electromagnetic field emitted by cellular phones on the EEG during a memory task. *Neuroreport.* 2000; 11:761-764.

[30] Koivisto M, Revonsuo A, Krause C, Haarala C, Sillanmäki L, Laine L, Hämäläinen H. Effects of 902 MHz electromagnetic field emitted by cellular telephones on response times in humans". *Cognitive Neurosci. Neuropsychol.* 2000; 11 (2): 413-415.

[31] Kramarenko AV, Tan U. Effects of high-frequency electromagnetic fields on human EEG. A brain mapping study. *Int. J. Neuroscienc.* 2003; 113 (7): 1007-1019.

[32] Nanou E, Tsiafakis V, Kapereliotis E, Papageorgiou C, Rabavilas A, Capsalis C. Influence of the interaction of a 900 MHz signal with gender on EEG energy: experimental study on the influence of 900 MHz radiation on EEG. *The Environmentalist.* 2005; 25(2-4): 173-179.

[33] Grigoriev AI, Repacholi MH, Onischenco GG, Mild KH, Ilion LA, Izmerov NF. In *International conference mobile communication and health: medical, biological and social problems.* Moscow. 2000 September, 20-23.

[34] Higuchi T. Approach to an irregular time series on the basis of the fractal theory. *Physics D.* 1988; 31: 277-283.

[35] Goldberger AL, Rigney DR, West BJ. Fractals in Physiology and Medicine. *Yale Journal of Bology and Medicine.* 1987; 60: 421-435.

[36] Klonowski W, Olejarczy E, Stepien R. Nonlinear quantifiers of EEG-signal complexity. *Proceedings of 2000 International Symposium on Nonlinear Theory and its applications* (*NOLTA'2000*), Dresden, 2000; 261-264.

[37] Accardo A, Affinito M, Carrozzi M, Bouquet F. Use of the fractal dimension for the analysis of electroencephalographic time series. *Biological Cybernetics.* 1997; 77: 339-350.

[38] Klonowski W, Ciszewski J, Jernajczyk W, Niedzielska K. Application of chaos theory and fractal analysis for EEG-signal processing in patients with seasonal affective disorder. In *Proceedings of 1999 international symposium on nonlinear theory and its applications* (NOLTA'99). Waikoloa. Hawaii, 1999; 339-342.

[39] Bardasano JL, Álvarez-Ude J, Gutiérrez I, Goya R. New device against non-thermal effects from mobile telephones. *The Environmentalist.* 2005; 25:257-263.

[40] Goya R. *Electroencephalographic changes in humans due to the use of mobile telephone. Doctoral Thesis.* University of Alcalá de Henares, Madrid (in spanish). 2007.

[41] Bardasano JL, Álvarez-Ude J, Gutiérrez I, Raposo M, Goya R. EEG bioeffects on cochlear deaf from cellular phones. *The Environmentalist,* 2007; 27:519-523.

In: Mobile Telephones...
Editors: A.C. Harper and R.V. Buress, pp. 191-210
ISBN: 978-1-60456-436-5

Chapter 6

PSYCHOLOGY OF MOBILE PHONE USE AND MISUSE

J.G. Phillips[1], L. Saling[1] and A. Blaszczynski[2]
[1]Monash University
[2]University of Sydney

Abstract

The extensive community adoption of mobile phones coupled with the development of its own specific slang is seen by some as an emerging new phenomena. Nevertheless existing psychological approaches to the impulse control disorders and the addictions can definitely serve as useful starting points for understanding phone use. We discuss whether heavier mobile phone use could be viewed as an addiction. Building upon previous research on internet use, studies demonstrate that specific psychological traits can explain mobile phone use and misuse. For instance, both mobile phone use and internet use are linked to lower levels of the agreeableness trait. Nevertheless, in contrast to the introverted stereotype of the heavier internet user, mobile phone use is more strongly linked to extraversion. As the mobile phone is a communication device, it is not surprising that its use can be predicted by psychological traits such as extraversion, and as extraversion is also associated with sensation seeking, this trait can also explain the self stimulatory behaviours and risk taking behaviour associated with mobile phone (e.g. use while driving). In particular, the mobile phone elicits problems for heavier users, with a section of the community reporting difficulty managing their time, financial problems, and aches and pains. The development of problems can be understood in terms of distraction and traits like procrastination. Tendencies towards procrastination can explain messaging behaviour, distraction from important tasks, and hence loss of resources (i.e. time or money) that influences quality of work. It is likely that some elements of problematic use will be resolved by innovations such as Bluetooth (hands free while driving), gesture or handwriting recognition (aches and pains), and video (antisocial behaviour). But other innovations such as location aware advertising and gambling are likely to create new problems such as distraction and financial loss for a proportion of the community.

Psychology of Mobile Phone Use and Misuse

In 2006 the number of mobile phones in Australia exceeded the number of people in Australia. The Australian Mobile Telecommunications Association (AMTA) estimate that they provide service to enough mobile phones, that if the number was expressed as a proportion of the Australian population, it would be 96% (http://www.amta.org.au/). It is perhaps no surprise given such levels of market acceptance and penetration that manufacturers and service providers sometimes describe their product as "addictive" as they seek a share of a market reaching saturation (http://www.crackberry.com/). Nevertheless technical capability sometimes outstrips consumer acceptance (Mahatanankoon, Wen, & Lim, 2005). The present chapter explores those problematic aspects of mobile phone use that are sometimes described as addictive, in an effort to understand better the consumers of mobile phone technology. The present chapter will perform an analysis upon individual items from questionnaires from several surveys to identify features that predispose some people to heavier mobile phone use (Griffiths, 2003; Wood, Griffiths, Chappell, & Davies, 2004).

Along with the wholesale acceptance of mobile phones has come a number of new forms of behaviour and laws. The nature of SMS has engendered new forms of slang (http://www.smsslang.com/). In addition, the portability of the mobile phone has led to laws against mobile phone use while driving in some jurisdictions. And chat capability has led to a perceived need to issue advice on safe chatting (http://www.virginmobile.com.au/companyinfo/corporateresponsibility/safechattingtips.html) and bullying (http://www.virginmobile.com.au/companyinfo/corporateresponsibility/bullyingtipsheet.html) at the corporate, industry body (http://www.amta.org.au/default.asp?Page=1363), and government regulatory levels (http://www.acma.gov.au/WEB/STANDARD/pc=PC_310631).

Although, the development and adoption of these new technological capabilities could be argued to have created a profoundly different cohort of individuals (Leary, 1994; Prensky, 2001a,b), this position may not be so useful from a forensic viewpoint (Warden, Phillips, & Ogloff, 2004). Arguing that these behaviours are completely novel removes them from the realm of the generally accepted, and thus causes difficulties for expert testimony (Freckelton, 2002a,b). In the present chapter, we seek to address mobile phone use from within the framework of the psychology of the impulse control disorders and the addictions, in order to link mobile phone behaviours to that which is generally accepted scientifically.

Disorders of Impulse Control and Addiction

A number of psychiatric conditions are associated with problems controlling impulses. To attain a formal diagnosis for all of these conditions, the symptoms must be sufficiently severe to cause social or occupation dysfunction. Of these conditions causing problems controlling impulses, DSM-IV-TR classifies Impulse Control Disorders (not otherwise specified) as persistent, clinically significant impulsive behaviours that are not better explained by other disorders in DSM-IV-TR. The Impulse Control Disorders are characterised by behaviours that are repetitive and difficult to resist. An individual with an Impulse Control Disorder has difficulty resisting impulses to perform an action that is harmful to self or others. A number of conditions are listed amongst the impulse control disorders. The most relevant to this discussion is pathological gambling, but there are other conditions such as intermittent explosive disorder and kleptomania. As will be explained, the motivations for these Impulse Control Disorders differ from that for Obsessive-Compulsive Disorder. In the Impulse Control Disorders the behaviour is performed for pleasure, while for Obsessive-Compulsive Disorder the behaviours serve to reduce anxiety.

Although commonly associated with the use of drugs, the term addiction is used to describe a recurring tendency by an individual to engage in some specific activity, that is difficult to resist, despite harmful consequences to the individual's health, mental state or social life. When a drug is abused, the addiction is classified as a substance use disorder. When a behaviour such as gambling is at issue, it may be classified as an impulse control disorder. When discussing substance dependence, addiction is associated with symptoms such as tolerance and withdrawal, and problems controlling the use of the substance. There is a difference between a habit and an addiction. Habits can be considered to shape our personalities and enrich our lives, but when they generate problems such as failure to meet obligations, or occur when it is dangerous, then they could be considered to be addictions.

Mobile Phone Addiction

Claims that a behaviour such as mobile phone use can be addictive can elicit censure from pharmacologists (Holden, 2001). But this response is somewhat naïve (Holden, 2001; Redish, Jensen, & Johnson, in press). Drugs of abuse function by mimicking or in other ways stimulating existing neurotransmitter systems. For instance morphine mimics the neurotransmitter endorphin, nicotine acts on cholinergic receptors, and amphetamine mimics the neurotransmitter epinephrine (Craig & Stitzel, 1990). Thus it should not be surprising that damage to specific brain systems can produce problems of impaired control resembling those seen in the impulse control disorders spectrum (Comings, 1995; Stein, Hollander, & Cohen, 1994) or the addictions (Redish, Jensen, & Johnson, in press). For instance people with Obsessive Compulsive Disorder and Tourette's Syndrome both evince difficulty resisting an urge or an impulse to act (Phillips & Stelmach, 1996). Similarly Blum, Cull, Braverman, and Comings' (1996) suggest that a weakness in dopamine receptors can lead to reward deficiency syndrome. In the former cases a biological weakness can lead to problems controlling urges, while the latter cases the weakness appears to cause problems controlling the use of substances. Even though some of the behavioural addictions are sometimes considered to be of questionable validity (Shaffer, Hall, & Vander Bilt, 2000; Warden, Phillips, & Ogloff, 2004), it is reasonable to believe that there can be structural problems within the nervous system; for instance neural substrates associated with the limbic circuits of the basal ganglia (Blum et al, 1996) and dopaminergic dysregulation that contribute to problems of impulse control. Nevertheless, although there may be commonalties in the circuitry that may be stimulated or disrupted, the actual nature of the stimulation or disruption may be different from substance-related addictions. The issue is more whether behavioural addictions should be given equivalent status to pharmacological addictions (Holden, 2001) with gambling researchers in particular debating this issue (Shaffer, 1996; Walker, 1989) and arguing that the magnitude of symptoms requires close consideration (Walker, Anjoul, & Milton, 2005).

Along an impulse control disorder spectrum of disorders as described by Hollander and his colleagues (Hollander, 1993; Blaszczynski, 1999a), there are conditions such as Obsessive Compulsive Disorder and Tourette's syndrome, where individuals give in to an urge to relieve anxiety or an impulse to carry out an act. In the behaviorist literature, these conditions are associated with negative reinforcement (escape conditioning). At the other end of the impulse control disorder spectrum (Blaszczynski, 1999b) there are conditions such as Attention

Deficit Hyperactivity Disorder, Sexual Compulsions, or Problem Gambling where individuals perform an inappropriate behaviour because it excites or rewards. In the behaviorist literature, these conditions are associated with positive reinforcement (reward). Thus people can have problems controlling urges because they are anxious, because they crave excitement, or because of a combination of these positive and negative reinforcement schedules. Typically from a pharmacological point of view, the addictions are usually associated with stimulation (Craig & Stitzel, 1990) and avoidance of aversive states (withdrawal or negative emotions). But where do the technological addictions (Griffiths, 1995) sit with respect to the impulse control disorder spectrum?

Indeed this is an interesting question. As there is a convergence in capabilities and functionality between hand-held devices and conventional computers, previous studies of internet use are of relevance. Early theorists suggested that the use of technology was an introverted behaviour (Shotton, 1989) that promoted isolation (Kraut et al., 1998). As classically defined, introverts are individuals whose nervous systems have a greater potential for activation, and that consequently avoid more intense levels of stimulation, whereas extroverts are individuals with a lesser potential for activation, and that consequently seek stimulation to induce optimal levels of arousal (Eysenck, 1994). To suggest that introverts were addicted to computers challenged the conception that addictive behaviour was sensation seeking behaviour (Griffiths 1997; Young, 1996; 1998). For instance, Shaffer (1996) suggested that anything that was exciting could be addictive, whereas Griffiths (1996; 1998; 1999) argued that technology granted access to the object of interest.

Clinical observations suggested that some computer users were social isolates (Shotton, 1989; Young, 1996). Indeed, this was confirmed by a longitudinal study by Kraut et al. (1998). A subsequent reanalysis revealed that the internet served to potentiate existing interests (Kraut et al, 2002). As such, this finding tends to support Griffiths' (1996) position that the internet offers access to objects of interest, rather than Shaffer's (1996) position that anything that is exciting (viz the internet) can be addictive.

The personality traits that predispose people to use the internet are thus of interest (Hamburger, & Ben-Artzi, 2000). A series of studies have observed that people loading on personality traits of disagreeableness are more likely to use the internet (Landers, & Lounsbury, 2006; Wyatt, & Phillips, 2005), shy men used the internet for recreational purposes (Scealy, Phillips, & Stevenson, 2002), lonely men used the internet for entertainment (Matanda, Jenvey, & Phillips, 2004), and people with low self esteem report more problems using the internet (Armstrong, Phillips, & Saling, 2000). Such findings indicate that the technology has much to offer those with low social skill (Amichai-Hamburger, Wainapel, & Fox, 2002). But when we consider messaging behaviour, it is extraverts (Wyatt, & Phillips, 2005) and procrastinators (Baker & Phillips, 2007; Phillips & Reddie, 2007) who are using the internet for email. This suggests that the technology is being used by some as a form of stimulation (Wyatt & Phillips, 2005), or serves for others as a distraction from ongoing tasks (Lavoie, & Pychyl, 2001; Lim, 2002). The suggestion that the internet can serve as a distraction was confirmed by Baker and Phillips (2007), who found that people who responded to messages were likely to be avoiding an important task. In summary, studies of internet use suggest different patterns of use as a function of social skill and interest. And to some extent this is mirrored in our studies of mobile phone use.

Mobile Phone Use

Bianchi and Phillips (2005) administered the Eysenck Personality Questionnaire (Revised) (Eysenck & Eysenck, 1991) and the Coopersmith Self Esteem Inventory (Coopersmith, 1991) to 195 mobile phone users. Details of their average amounts of mobile phone use can be seen in Table 1. Using multiple regression they found that younger age, lower self esteem, and higher levels of extraversion could predict self reported time spent on the mobile phone. As low self esteem tends to be correlated with poor social skill (Kring, Davison, Neale, & Johnson, 2007), the two themes of extraversion and low social skill previously documented in internet usage seem to have emerged in studies of mobile phone use.

Table 1. Means and Standard Deviations, Minimum, Maximum and Skew of Time spent using the mobile phone per week, Mobile Phone Problem Use Scale (MPPUS) score, Social use, Business use, Use of other features and SMS use

DV	Mean	SD	Min	Max	Skew
Time using during the week in minutes	123.32	213.32	2.00	1800.00	4.42
Number of people called on a regular basis	5.47	5.57	0.00	40.00	2.70
MPPUS score	64.86	31.60	27.00	210.00	1.49
Social use as percentage of total use	70.75	32.24	0.00	100.00	-0.83
Business use as percentage of total use	26.19	31.91	0.00	100.00	0.96
Use of other features as percentage of total use	2.99	7.24	0.00	60.00	4.49
SMS use as percentage of total use	31.19	30.85	0.00	98.00	0.629

The number of people phoned is another indicator of mobile phone use (Bianchi & Phillips, 2005). Bianchi and Phillips (2005) found that the number of people phoned could be predicted from gender and extraversion. Multiple regression suggested that male extraverts phone more people. The self reported proportion of use devoted to social and business purposes was also analysed. Young females use their mobile phones more for social reasons. Older males use their mobile phones for business purposes.

Extraverts are by nature sociable and gregarious (Eysenck, 1994), and so it was expected that extraversion could predict *the number of people phoned*, nevertheless it was surprising that extraversion did not predict the proportion of mobile phone *use for social purposes* (Bianchi & Phillips, 2005). Hence there was reason to consider other features of mobile phones (Wen & Mahatanankoon, 2004), and to consider whether the mobile phone was being used for some other reason, such as self-stimulation. For instance, the manipulation of the mobile phone might serve no other function than to generate sensory or tactile sensation. Unfortunately, at this time Bianchi and Phillips (2005) had problems predicting the use of SMS (but see Mahatanankoon, 2007). Bianchi and Phillips (2005) just found that SMS was something used by younger people, and that it was the younger participants that were more interested in other features of mobile phones.

A subsequent study explored the use of the mobile phone for more specific purposes. Butt and Phillips (2008) administered the NEO FFI (Costa & McCrae, 1992) questionnaire to 115 participants. Details of the participants' mobile phone use can be seen in Table 2. The self reported amounts of time spent on the mobile phone could be predicted by the extraversion and agreeableness scales of the NEO FFI. Individuals who were extraverted and low in agreeableness spent more time on their mobile phones.

Table 2. Totals, means and standard deviations along with minimum, maximum and skew for weekly average time (minutes) using mobile phones, the percentage of incoming calls that are perceived as unwanted, and interest in new mobile phone features.

DV	n	M	SD	Min.	Max.	Skew
Calls (mins)	112	149.29	222.38	2.00	1080	2.498
Incoming calls	112	30.13	45.47	1.00	300	3.37
Unwanted (%)	111	13.5	19.45	.00	99	2.232
Outgoing calls	111	22.89	30.53	.00	150	2.290
SMS (mins)	105	104.1	229.0	.00	1710	5.688
Tone/wall (mins)	95	10.53	62.25	.00	600	9.242
Games (mins)	112	27.72	174.53	.00	1800	9.66
Interest in new features[1]	112	3.22	1.10	1	5	-0.54

[1] Measured on 5 point scale, 1 = Very disinterested, 5 = Very interested

The previous study by Bianchi and Phillips (2005) had asked people to report their SMS use as a percent of total use. Butt and Phillips (2008) asked people to report their SMS use in terms of minutes per week. Perhaps this was a more reliable way of expressing usage, as it equates more readily with billing details, because SMS use could now be predicted from personality traits. Higher levels of SMS use were associated with high extraversion, high neuroticism, and lower agreeableness and lower conscientiousness (Butt & Phillips, 2008).

Bianchi and Phillips (2005) found that younger individuals were interested in the other features of mobile phones. Butt and Phillips (2008) asked people about the time they spent changing their ring tone and/or wallpaper. People who reported spending more time altering their ring tones and/or wallpaper were more likely to be extraverted and disagreeable.

In summary, the sorts of personality traits that predict internet use tend to predict the use of mobile phones. A constellation of personality scales associated with low social skill (e.g. low agreeableness, low self esteem) were associated with mobile phone use. In addition, extraversion is also associated with messaging behaviour on the internet (Wyatt & Phillips, 2005) and with mobile phone use.

Even though there is some understanding of the personality traits predicting mobile phone use, this does not explain which features (physical or functional) of mobile phones are important (Choi & Kim, 2004; Griffiths, 2003; Wood, Griffiths, Chappell, & Davies, 2004). This is relevant for regulators, designers and legislators contemplating future innovations (Wen, & Mahatanankoon, 2004). Whether it is location aware advertising that is contemplated in Finland (Häkkilä & Isomursu, 2005), or mobile gaming (e.g. Bridge, 2005; Griffiths, 2003) in South Africa, or laws directed at the use of mobile phones while driving in Australia or the USA, community and user acceptance may influence the success of the innovation (Mahatanankoon, Wen, & Lim, 2005).

Given suggestions that gambling on mobile phones would be a major industry (Bridge, 2005), Phillips, Butt and Blaszczynski (2006) considered the number of individuals playing games on their mobile phones. Game playing was correlated with extraversion and low agreeableness, and we already know that there is a section of the community that reports problems controlling their mobile phone use. Hence it is important to consider the numbers of people actually reporting problems with their mobile phone use when considering implementing mobile gaming.

Bianchi and Phillips (2005) had questioned whether extraverts were using their mobile phones to contact others, or were using their mobile phones to stimulate themselves. In other words, were people using their mobile phones for social purposes to contact people, or was this a secondary objective to mere stimulation (see Mahatanankoon, 2007). Theorists suggest that extraverts are underaroused (Eysenck, 1994), and seek stimulation (Zuckerman, 1979). Although the primary function of a mobile phone is to contact others, it is important to determine whether other features of the mobile phone are sought for purposes of stimulation. Hence follow up studies considered whether mobile phones were being used for purposes of stimulation (Butt & Phillips, 2008), or served to distract (Payne, unpublished study).

Table 3 reports an item analysis of Bianchi and Phillips (2005) Mobile Phone Problem Use Scale. Participants indicated the degree of their agreement with each item on a 10 point rating scale. A percent agreement was calculated by counting the number of people who gave a rating of 6 or higher to an item. As may be seen in Table 3, the larger percentage of the sample agree with question 4 (All my friends own a mobile phone). Other questions with high percentages of agreement also involve question 17 (If I don't have a mobile phone, my friends would find it hard to get in touch with me). These might be expected given that mobile phones are used to contact others, but some of the other questions are of more interest. For instance, a high proportion of people also agree with question 10 (The time I spend on the mobile phone has increased over the last 12 months), and question 11 (I have used my mobile phone to talk to others when I was feeling isolated) and question 2 (I have used my mobile phone to make myself feel better when I was feeling down) implying that their mobile phones can assume a mood regulatory function for some individuals.

Of relevance here is Jacob's (1988) general theory of the addictions. Jacob's general theory of addictions links addictive behaviours to people's states of hyper and hypo arousal and need to escape negative emotions associated with early life experiences that lead to the development of low self esteem and a sense of rejection. According to Jacobs, previous unpleasant experiences are associated with strong persistent negative memories causing negative emotions. To avoid such negative emotions and restore homeostatic balance in arousal levels, people engage in the addictive behaviour. Bianchi and Phillips (2005) found higher levels of mobile phone use were associated with low self esteem, and the present item analysis indicates that the mobile phone is used to ameliorate negative mood. Hence there are parallels between mobile phone use and other behavioural addictions (e.g. gambling). But it is important to consider those items that best predict overall amounts of time spent on the mobile phone.

We performed correlations between each item and their self reported minutes spent on their mobile phones per week. As may be seen in Table 3, those questions that correlate best with time spent on the mobile phone are question 3 (I find myself occupied on my mobile phone when I should be doing other things, and it causes problems), question 6 (I lose sleep due to the time I spend on my mobile phone), question 12 (I have attempted to spend less time on my mobile phone but am unable to) and question 20 (I find myself engaged on the mobile phone for longer periods of time than intended). These questions appear to be assessing poor impulse control. In particular, the item that correlates most strongly with overall use is question 3 that indicates that the mobile phone is distracting them from other tasks, and that this causes problems. This also supports previous suggestions that the technology is distracting people from other obligations (Griffiths, 1995; Phillips, 2006).

Table 3. Item analysis of the Mobile Phone Problem Use Scale.

	Mobile Phone Use Survey	mean	sd	r Lie Scale	r Time on phone	% Agree
1.	I can never spend enough time on my mobile phone	1.62	1.184	-0.031	0.117	1.5
2.	I have used my mobile phone to make myself feel better when I was feeling down	2.67	2.498	-0.293	0.318	19.5
3.	I find myself occupied on my mobile phone when I should be doing other things, and it causes problems	1.81	1.589	-0.167	0.455	5.6
4.	All my friends own a mobile phone	7.78	2.641	-0.142	0.268	78.5
5.	I have tried to hide from others how much time I spend on my mobile phone	1.35	1.094	-0.163	0.163	3.1
6.	I lose sleep due to the time I spend on my mobile phone	1.35	1.321	-0.142	0.442	2.6
7.	I have received mobile phone bills I could not afford to pay	1.88	2.061	-0.214	0.165	7.7
8.	When out of range for some time, I become preoccupied with the thought of missing a call	2.15	1.797	-0.295	0.069	7.2
9.	Sometimes, when I am on the mobile phone and I am doing other things, I get carried away with the conversation and I don't pay attention to what I am doing	2.80	2.391	-0.267	0.224	15.9
10.	The time I spend on the mobile phone has increased over the last 12 months	3.76	2.976	-0.279	0.330	28.2
11.	I have used my mobile phone to talk to others when I was feeling isolated	3.31	2.751	-0.336	0.264	24.1
12.	I have attempted to spend less time on my mobile phone but am unable to	1.84	1.768	-0.102	0.429	4.6
13.	I find it difficult to switch off my mobile phone	3.06	3.056	-0.167	0.325	20.5
14.	I feel anxious if I have not checked for messages or switched on my mobile phone for some time	3.10	2.509	-0.230	0.218	16.9
15.	I have frequent dreams about the mobile phone	1.11	0.495	-0.115	0.053	0.5
16.	My friends and family complain about my use of the mobile phone	1.42	1.139	-0.197	0.355	2.6
17.	If I don't have a mobile phone, my friends would find it hard to get in touch with me	4.35	3.092	-0.258	0.353	32.8
18.	My productivity has decreased as a direct result of the time I spend on the mobile phone	1.42	1.029	-0.175	0.228	1.5
19.	I have aches and pains that are associated with my mobile phone use	1.42	1.267	-0.077	0.151	3.1
20.	I find myself engaged on the mobile phone for longer periods of time than intended	2.48	2.237	-0.257	0.425	14.4
21.	There are times when I would rather use the mobile phone than deal with other more pressing issues	1.68	1.534	-0.202	0.334	4.6
22.	I am often late for appointments because I'm engaged on the mobile phone when I shouldn't be	1.35	1.185	-0.211	0.225	3.1
23.	I become irritable if I have to switch off my mobile phone for meetings, dinner engagements, or at the movies	1.51	1.430	-0.160	0.147	4.1
24.	I have been told that I spend too much time on my mobile phone	1.57	1.687	-0.157	0.399	5.6
25.	More than once I have been in trouble because my mobile phone has gone off during a meeting, lecture, or in a theatre	1.98	2.010	-0.304	0.032	9.2
26.	My friends don't like it when my mobile phone is switched off	3.11	2.692	-0.140	0.184	18.5
27.	I feel lost without my mobile phone	2.99	2.738	-0.178	0.357	17.9

Table 3 also shows the correlation between each item and the Lie Scale of the EPQ. The Lie Scale assesses social desirability (Anastasi & Urbina, 1997), and high positive correlations would imply that participants were telling us what they thought was socially acceptable. Instead, the negative correlations imply that people are being honest in their responses, as is always hoped for in anonymous questionnaires. Hence, people are probably honest in reporting that they receive bills that they cannot pay, but the actual percentage of individuals reporting this is low (7.7%), and the correlation with amounts of self-reported use is low ($r= 0.165$, $n=191$, $p<.05$). In addition, it is likely that some people do experience aches and pains from their mobile phone use (question 19), but while there is a significant correlation with self reported mobile phone use ($r=0.151$, $n=191$, $p<.05$), it is small, and although it only accounts for a small proportion of the variance, it is probably sufficient to justify attempts to replace existing SMS systems involving repetitive button pressing with alternative systems involving gesture (Ancona, Quercini, & Dominici, 2007; Isokoski & Raisamo, 2007) or other alternatives to key pressing such as handwriting (De Stefano, Marcelli, & Santoro, 2007).

The present item analysis confirms the two themes previously seen in internet users. The technology serves for extraverts to contact others, while for those with poorer social skills, it is probably used to regulate mood, or reduce their isolation (Griffiths, 1996; Kraut et al., 2002). There are also indications from item analysis that the mobile phone serves to distract from other activities. Given that Bianchi and Phillips (2005) questioned the use of mobile phones for social purposes, follow up studies considered whether mobile phones were being used for purposes of stimulation (Butt & Phillips, 2008), or served to distract attention away from other tasks (Payne, unpublished study).

For the individual, distraction can be driven externally, and take the form of interruptions. For instance, on the Mobile Phone Problem Use Scale, 9.2% of participants reported that their mobile phone goes off at meetings, lectures or at the theatre (Bianchi & Phillips, 2005). Distraction could also be equated with unwanted phone calls. On average 13.5% of phone calls were reported to be unwanted (Butt & Phillips, 2008), and would be considered to be a distraction from other tasks. These rates of distraction would be likely to change with the implementation of new technologies such as location-aware advertising (Häkkilä & Isomursu, 2005) and SMS friend and dating services (e.g. SMS buddy http://www.mobilinkworld.com/index.php?option=com_wrapper&Itemid=160) (Mahatanankoon, Wen, & Lim, 2005; Wen & Mahatanankoon, 2004).

Alternately, distraction can be driven internally, as the individual seeks out stimulation (Zuckerman, 1979). Butt and Phillips (2008) specifically considered whether people played games on their mobile phones, or devoted more time to ring tones and screen wallpaper. Amounts of time people report spending on ring tones, screen wallpaper, and games may be seen in Table 2. Games accounted for about 20% of mobile phone use. As mobile phones serve to distract during other tasks, it is essential to determine the extent to which this is problematic (Phillips, 2006), for instance distracting people during important tasks such as driving. Driving simulation studies indicate that dual tasking, for example using a mobile phone while driving, can be detrimental to driving performance, and this is a problem use of the mobile phone that is against the law in many jurisdictions.

Problematic Use of Mobile Phones

Although not directly linked to the mechanisms associated with the impulse control disorders or addiction, there are mobile phone behaviours that generate problems that can harm self or others. The use of mobile phones while driving behaviour is considered, because this is a behaviour that is associated with appreciable risk, with the potential to cause harm, and that is against the law in many jurisdictions.

Driving simulation studies indicate that performing two tasks at once, for example using a mobile phone while driving, can be detrimental to driving performance (Alm & Nilsson, 1995; Consiglio, Driscoll, Witte, & Berg, 2003; Hancock, Lesch, & Simmons, 2003). Jurisdictions have banned hand-held telephones while driving because they believe that physical manipulation of the mobile phone, for example, having to hold it up to the ear, would lead to a decrement in driving performance (McLeod, 1977; McLeod & Mierop, 1979), but others consider that is the diversion of attention from the driving task that is a problem (Briem & Hedman, 1995; Lamble, Kauranen, Laakso, & Summala, 1999; McKnight & McKnight, 1993; Strayer & Johnston, 2001). Correlational studies comparing crash data and the use of mobile phones lend weight to the potential dangers of mobile phone use while driving (Laberge-Nadeau, et al., 2003; Reidelmeier, & Tibshirani, 1997; Violanti, 1998; Violanti & Marshall, 1996).

To address distractions while driving, Payne in an unpublished study, asked 76 people about their driving behaviours. Table 4 summarises self reported frequencies of activities during driving. People reported on a 4 point Likert scale the frequency with which they engaged in activities while driving. To summarise findings, the 4 point scale was dichotomized into whether participants agreed or disagreed that they engaged in some other behaviour while driving. Table 4 then reports the percentage of people who agreed with statements. As may be seen in Table 4, people are generally attentive, 82% of them reporting that they focus their attention upon the driving task. People are more likely to report spending time looking at the landscape, and adjusting stereo controls, and talking to other passengers than using mobile phones. Eating and drinking correlate negatively with the lie scale, indicating that those people who are less constrained by social desirability will report eating and drinking while driving. Although not significant, social desirability correlates with the use of hands-free mobile phones.

Table 4. Self Reported Frequency of Behaviours during Driving.

Behaviour	Mean	SD	r Lie scale	r Extraversion	% Agree
Adjusting Mirrors	2.25	0.73	-0.125	0.113	86.8
Adjusting Stereo Controls	2.86	0.80	-0.178	0.043	96.1
Using Hand-held Phone	1.67	0.72	-0.135	0.324**	56.3
Eating and Drinking	2.34	0.83	-0.301**	0.181	84.2
Using Hands-free Phone	2.12	1.12	0.210	0.137	57.9
Talking to Passengers	3.39	0.63	-0.193	0.265*	92.1
Admiring Landscape	2.62	0.67	-0.187	0.078	97.4
Keeping both Hands on Wheel	2.38	0.78	-0.142	-0.007	88.2
Focussing all Attention on Driving	2.00	0.65	-0.157	0.182	81.6

As might be expected, it is the extraverts that report more use of the mobile phone while driving, and extraverts are more likely to report talking to passengers. This replicates previous studies finding that extraverts are more likely to use their mobile phones (Bianchi & Phillips, 2005; Butt & Phillips, 2008), and confirms that sociability is a key predictor of mobile phone use.

Bianchi and Phillips (2005) had questioned whether mobile phones are being used socially or merely for stimulatory purposes. This was important, as it is relevant as to the development of new features on these devices such as games (Yuan, 2003). Extraversion is consistently a predictor of mobile phone use, and emerges as a predictor of handheld mobile phone use as well. The observation that extraverts are reporting more use of handheld mobile phones with driving is also in keeping with previous associations between extraversion and risk taking behaviour (Trimpop, 1994) and sensation seeking (Zuckerman, 1979).

As can be seen in Table 4, a number of things are more likely to distract drivers than mobile phones. Manufacturers are obviously aware of the issue of stereo controls, and tend to design modern cars with stereo controls on the steering wheel. It is likely that socially responsible individuals will be more likely to adopt hands-free technology. As landscape, food, drink, and passengers are more likely to be reported as distractors, one questions why legislation focuses upon mobile phones? There are two likely reasons for this: 1) degree of distraction; 2) documentation of distraction.

The degree of distraction afforded by a mobile phone is liable to be greater than that caused by the landscape or the stereo. The handheld mobile phone is a small complicated device, that is unlikely to be manipulated by feel, and is more likely to require vision for its use (see Baber, 1997; MacKenzie 1992). Although people are less likely to report it as a distractor, when it is used, it is more likely to require direct gaze rather than a glance, and it is more likely to require two hands for its use.

The mobile phone documents the calls that are made, in its billing details. Hence, unlike the other distractors, the mobile phone is most likely to deliver date and location of an offence. As jurisdictions require documentation to be kept (e.g. Australian Communications and Media Authority), the technology associated with mobile phones supports prosecution better than some other distractors (Anandarajan, 2002). It is a curious paradox, that those people who are more likely to use the technology, and are more likely to engage in inappropriate behaviours, are also those for whom the technology is likely to document their offences.

There are other forms of problematic behaviours associated with the use of the mobile phone. The perceived anonymity associated with mobile phone use has been associated with bullying (Charlton, Panting, & Hannan, 2002). As anonymity is associated with reduced helping behaviour (Markey, 2000) technologies to improve helping behaviour are of interest. For instance, Milgram (1965) observed aggressive behaviour was greater when the victim was at a distance, than when they were in physical proximity. More relevantly, Weinberger (1981) observed that face to face interaction was more likely to improve helping behaviour than interaction over the telephone. Although face to face communication may be useful in improving prosocial behaviour, there may be a reluctance to use this mode of communication. For instance, there seems to be a preference for the use of non-interactive communication (SMS) amongst those with lower agreeableness and lower conscientiousness (Butt & Phillips, 2008). Face to face communication might be avoided by those who are unwilling to help. Similarly, there are other indications that people are seeking to control their communications.

Choice of Communication Channel

The Johari window can be used to explain choice of communication channel. As described by Luft and Ingrams (1955) (Luft, 1970), the Johari window (see Table 5) classifies a communication space into areas that are known to the self, and areas that are known to others. Areas that are known to the self are a function of feedback. Areas that are known to others are a function of self-disclosure. As may be seen in Table 5, areas that are known to both the self and others can be considered an open arena of communication. Areas that are known to the self, but not others can be considered a hidden area in which the self attempts to control their self-presentation. Areas that are not known to the self, but are known to others, represent a person's blind spot. To counsellors, this is an area of self-discovery that can be discovered by interaction with others. Areas that are neither known to the self or others, are unknown.

Table 5. The Johari Window as a means of describing a communication space.

	Known to Self	Not Known to Self
Known to Others	1 Open	2 Blind
Not Known to Others	3 Hidden	4 Unknown

The size of the areas within the Johari window are a function of interaction with others, and vary with self-disclosure and feedback. Within the context of phone use, the size of these areas is a function of the messages that the communicator wishes to transmit, and other unintended information that may be incidentally transmitted. The communication space is further divided into areas based upon the ability of the recipient to decode cues from the communicator.

The choice between text, audio, and video is a choice between communication channels that offer more or less incidental cues. Our studies already indicate that some individuals choose a communication channel (text) that delivers fewer cues than other channels (e.g. audio, video) (Butt & Phillips, 2008).

As a communication channel (Welford, 1987) the mobile phone affords the user control over the range of cues provided to others (i.e. textual, aural, and visual), and their willingness to self-disclose. SMS was used by people who in some way wanted to control or limit the amounts of information (e.g. aural, visual) transmitted about themselves (Butt & Phillips, 2008). SMS users were more likely to be neurotic, extraverted, disagreeable, or unconscientious. Disagreeable individuals spent more time on their phone as a "display", while neurotic or unconscientious persons, with their reliance upon SMS seemed reluctant to use the phone to transmit more information about themselves (Argyle, 1984; McKenna & Bargh, 1999; 2000). Similar motivations to control self-presentation (Eastin, Glynn, & Griffiths, 2006) have been observed on the internet for messaging behaviour, and may be successful in understanding messaging behaviour on mobile phones.

To some extent the response to phone calls also potentially offers insight into a person's responsiveness to feedback and self-knowledge. It is the disagreeable extravert that is more likely to report unwanted phone calls (Butt & Phillips, 2008). Butt and Phillips (2008) were puzzled as to why someone would phone a disagreeable extravert, but it is likely that they are phoned for purposes of surveillance, or are phoned to remind them of their interpersonal obligations (Phillips, 2006). In either instance, the disagreeable extravert would perceive the phone call as unwanted because they are then in receipt of feedback that they are untrustworthy or unreliable.

A consideration of the Johari window assists in understanding motivation to choose text over audio, but it remains to be seen whether the same personality traits will explain a choice between audio and video. We suspect high self esteem and traits such as exhibitionism may also contribute to the use of video.

The mobile phone extends the reach and immediacy of interpersonal transactions (Plant, 2000), and as such it greatly enhances the ability to monitor or bully others. This may explain the greater likelihood of reporting unwanted calls or the tendency for the disagreeable to use mobile phones. Plant (2000) observed that the model of the phone, the ring tone and wallpaper, are a display to others of who you are. As such the amounts of use of this communication channel, and the types of channels used may be informative (see Argyle, 1984). For instance, disagreeable extraverts focussed upon the superficial elements of their mobile phone such as ring tone and wallpaper, presumably for self stimulatory purposes (Bianchi & Phillips, 2005) or for the purposes of attracting the attention of others (Welford, 1977).

Implications

Legislators planning new innovations need to know that a proportion of the community already report problems with their mobile phones. For instance, 2.6% of individuals agree that the time they spend on their mobile phones gets them into trouble, another 5.6% agree that they are on their mobile phone when they should be doing other things, and 7.7% of individuals already report problems paying their bills. But this is a minority, and it is difficult to argue that an innovation be blocked just on this basis. Nor is psychological testing mature enough to be able to equitably detect and exclude members of the community who would abuse the technology (but see Schellinck & Schrans, 2006). Hence this should be taken into account in impact statements or harm minimisation plans when considering innovations such as location aware advertising or gambling (Scharl, Dickinger, & Murphy, 2005; Wen & Mahatanankoon, 2004).

Current strategies to curb problem gambling seek to control access by limiting the availability to gaming machines (Abbott & Volberg, 1994). Nevertheless, legalising gambling on mobile phones, as has been done in South Africa, potentially creates millions of new gaming terminals. For instance, would the same people who already have a gambling problem be liable to misuse mobile gaming technology if implemented, or would the new technology create a new cohort of people with problems (but see Abbott, 2005; Grohol, 1999). For instance, Abbott (2005) suggested that current cohorts of gamblers may become inoculated to exposure to electronic gaming machines through a process of familiarity and social adaptation. Furthermore, Grohol (1999) suggested that heavier internet use was a just a fad, similar in scope to that of interest in CB radios in the 1980s. In other words, would the implementation of mobile gaming be neutral, with the percentage of the community with problems remaining stable or would this add to the percentage of the community who go on to develop problems? If the proportions of individuals who develop gambling problems alter, would this occur in a conjunctive or a more disjunctive manner? For instance, would only a fraction (i.e. mobile game players 18%) of an existing fraction (i.e. problem gamblers 2%) be affected by the implementation of mobile gaming, giving a very small minority and low risk (i.e. 0.4%)? Or given the pervasiveness and accessibility of the mobile phone, would the implementation of mobile gaming reach those who currently do not gamble electronically (e.g. 60%) and create the higher rates of problem gambling (e.g. 30%) associated with electronic game machine players, leading to an appreciable increase (e.g. 18%) in problem gamblers.

The implementation of location aware advertising (Scharl, Dickinger, & Murphy, 2005) faces other issues. There is already consumer resistance to advertising over mobile phones (Mahatanankoon, Wen, & Lim, 2005). While the internet has challenged legislators (Smeaton & Griffiths, 2004; Tonneguzzo, 1995; 1997; Watson, Liddel, Moor, & Eshee, 2004), in some jurisdictions legislation is already in place to curb it. For example, the USA has the wire transfer act, to control gambling, and Australia has their SPAM act. Australia has already prosecuted advertising material sent to mobile phones through SMS by way of the SPAM act, and such a precedent may pose a problem for advertisers.

Conclusion

We have addressed mobile phone use from viewpoints of the impulse control disorders and the addictions. People report problems controlling their mobile phone use. These problems are linked to poor self-esteem, and indeed some people report using their mobile phones to regulate their mood (see Jacobs, 1988). Mobile phone use is linked to personality traits (extraversion) that predict sensation seeking, and hence there are clearer parallels between problem mobile phone use and addiction than there are between problem internet use and addiction. Can we say that problem use of a mobile phone is an addiction? From a forensic stance the American courts appear to require: 1) proof of testing of the basic underlying hypothesis or the technique; 2) peer review and publications; 3) known or potential error rate (effect size); 4) existence of an accepted methodology; 5) general acceptance (Freckelton, 2002b). Although the data tend to fit within the framework of a behavioural addiction, and there are peer reviewed publications on this matter, the effect sizes may not suffice (they may not for the behavioural addictions), nor is there general acceptance on the matter. Nevertheless, the behavioural addictions can serve as a useful starting point when conducting research on mobile phone use.

References

Abbott, M. (2005). Do EGMs and problem gambling go together like a horse and carriage. Paper presented at the 15th annual conference of the National Association for Gambling Studies, 10th-12th November, Alice Springs.

Abbott, M.; & Volberg, M. (1994). Gambling and pathological gambling: Growth industry and growth pathology of the 1990s. *Community Mental Health in New Zealand,* **9**(2), 22-31.

Alm, H., & Nilsson, L. (1995). The effects of a mobile telephone task on driver behaviour in a car following situation. *Accident Analysis and Prevention,* **27**, 707–715.

Amichai-Hamburger, Y., Wainapel, G., & Fox, S. (2002). "On the Internet no one knows I'm an introvert": Extroversion, neuroticism, and Internet interaction. *CyberPsychology & Behavior,* **5**, 125-128.

Anandarajan, M., (2002). Internet Abuse in the Workplace. Communications of the Association of Computing Machinery.

Anastasi, A, & Urbina, S. (1997). Psychological testing, 7th edition. New Jersey: Prentice Hall.

Ancona, M., Quercini, G., & Dominici, L. (2007). An improved text entry tool for PDAs. In J.G. Phillips, D. Rogers, & R.P. Ogeil (Eds.), *Proceedings of the 13th Conference of the International Graphonomics Society* (pp. 201-204). Melbourne: Monash University.

Argyle, M. (1984). *The psychology of interpersonal behaviour.* Middlesex, England: Penguin Books.

Armstrong, L., Phillips, J.G., & Saling, L.L. (2000). Potential determinants of heavier *Internet usage. International Journal of Human Computer Studies,* **53**, 537-550.

Baber, C. (1997). *Beyond the desktop: Designing and using interaction devices.* San Diego: Academic.

Baker, J.R. & Phillips, J.G. (2007). Email, decisional styles and rest breaks. *CyberPsychology & Behavior,* **10**(5), 705-708.

Bianchi, A., & Phillips, J.G. (2005). Psychological predictors of problem mobile phone use. *CyberPsychology & Behavior*, **8**(1), 39-51.

Blaszczynski, A. (1999a). Pathological gambling and obsessive compulsive spectrum disorders. *Psychological Reports,* **84**, 107-113.

Blaszczynski, A. (1999b). Pathological gambling: an impulse control, addictive or obsessive-compulsive disorder? *Anuario de Psicología,* **30**(4), 93-109.

Blum, K., Cull, J.G., Braverman, E.R., Comings, D.E. (1996). Reward deficiency syndrome. *American Scientist,* **84**(2), 132-139.

Bridge, P. (2005). CEO Lasseter's Online. *Converging worlds – Technology in the virtual and land-based gaming environments.* Paper presented at National Association of Gambling Studies Conference, Alice Springs.

Briem, V., & Hedman, L. R. (1995). Behavioural effects of mobile telephone use during simulated driving. *Ergonomics,* **38**, 2536–2563.

Butt, S, & Phillips, J.G. (2008). Personality and self reported mobile phone use. *Computers in Human Behavior,* **24**, 346-360.

Charlton, T., Panting, C., & Hannan, A. (2002). Mobile telephone ownership and usage among 10- and 11-year olds: Participation and exclusion. *Emotional and Behavioral Difficulties,* **7**, 152-163.

Choi, D., & Kim, J. (2004). Why people continue to play online games: In search of critical design factors to increase customer loyalty to online contents. *CyberPsychology & Behavior,* **7**, 11-24.

Comings, D.E. (1995). Tourette's syndrome: A behavioral spectrum disorder. In W.J. Weiner & A.E. Lang (eds.), Behavioral Neurology of Movement Disorders (pp. 293-303). New York: Raven Press.

Consiglio, W., Driscoll, P., Witte, M., & Berg, W. P. (2003). Effect of cellular telephone conversations and other potential interference on reaction time in a braking response. *Accident Analysis and Prevention,* **35**, 495–500.

Coopersmith, S., (1991) *Self-Esteem Inventories*, CA: Consulting Psychologists Press, Inc.

Costa, P., & McCrae, R. (1992). *NEO PI-R Professional Manual.* Odessa: Psychological Assessment Resources, Inc.

Craig, C.R., & Stitzel, R.E. (1990). Modern pharmacology (ch. 41, Contemporary drug abuse). Boston: Little Brown & Company.

De Stefano, C., Marcelli, A., & Santoro, A. (2007). On-line cursive recognition by ink matching. In J.G. Phillips, D. Rogers, & R.P. Ogeil (Eds.), *Proceedings of the 13th Conference of the International Graphonomics Society* (pp. 23-37). Melbourne: Monash University.

Eastin, M.S., Glynn, C.J., & Griffiths, R.P. (2006). Self-regulation of communication technology in the workplace. In M. Anandarajan, T. Thompson, & C. Simmers (Eds.), *The Internet and Workplace Transformation* (pp. 28-40). Armonk: ME Sharpe.

Eysenck, H.J. (1994). Personality: Biological foundations. In P.A. Vernon (Ed.), *The neuropsychology of individual differences* (pp. 151-207). San Diego: Academic Press.

Eysenck, H. J., & Eysenck, S. B. G. (1991). *Manual of the Eysenck Personality Scales (EPS Adult).* London: Hodder & Stoughton.

Freckelton, I. (2002*a*). The area of expertise rule. In I. Freckelton & H. Selby (Eds.), *Expert evidence: Law, practice, procedure and advocacy.* (2nd ed.). (pp. 53-75). Sydney: Lawbook Co.

Freckelton, I. (2002*b*). Novel psychological evidence. In I. Freckelton & H. Selby (Eds.), *Expert evidence: Law, practice, procedure and advocacy.* (2nd ed.). (pp. 318-411). Sydney: Lawbook Co.

Griffiths, M. (1995). Technological addictions. *Clinical Psychology Forum,* **76**, 14-19.

Griffiths, M. (1996). Gambling on the Internet: A brief note. *Journal of Gambling Studies,* **12**, 471-473.

Griffiths, M.D. (1997). Psychology of computer use: XLIII. Some comments on 'Addictive use of the Internet' by Young. *Psychological Reports*, **80**, 81-82.

Griffiths M.D. (1998). Internet addiction: Does it really exist? In J. Gackenbach (ed.) *Psychology and the Internet: Interpersonal and transpersonal applications* (pp. 61-75). New York: Academic Press.

Griffiths, M. (1999). Internet addiction: Fact or fiction? *The Psychologist,* **12**(5), 246-250.

Griffiths, M.D. (2003). Internet gambling: Issues, concerns and recommendations. *CyberPsychology & Behavior*, **6**(6), 557-568.

Grohol, J.M. (1999). *Internet addiction guide*. (1999) retrieved April 18, 2002, from http://www.psychcentral.com/netaddiction/

Hamburger, Y. A., & Ben-Artzi, E. (2000). The relationship between extraversion and neuroticism and the different uses of the Internet [Electronic version]. *Computers in Human Behavior,* **16**, 441-449.

Häkkilä, J., & Isomursu, M. (2005). User experiences on location-aware mobile services. *Proceedings of OZCHI2005*, ACM ISBN: 1-59593-222-4,

Hancock, P. A., Lesch, M., & Simmons, L. (2003). The distraction effects of phone use during a crucial driving maneuver. *Accident Analysis and Prevention,* **35**, 501–514

Holden, C. (2001). "Behavioral addictions": Do they exist? *Science,* **294**, 980-982.

Hollander, E. (1993) *Obsessive-compulsive related disorders*. Washington, DC: American Psychiatric Press

Isokoski, P., & Raisamo, R. (2007). Text entry and graphonomics research from the HCI perspective. In J.G. Phillips, D. Rogers, & R.P. Ogeil (Eds.), *Proceedings of the 13th Conference of the International Graphonomics Society* (pp. 197-200). Melbourne: Monash University.

Jacobs, D.F. (1988). Evidence for a common dissociative-like reaction among addicts. *Journal of Gambling Behavior*, **4**(1), 27-37.

Kraut, R., Kiesler, A., Boneva, B., Cummings, J., Helgeson, V., & Crawford, A. (2002). Internet paradox revisited. *Journal of Social Issues,* **58**, 49-74.

Kraut, R., Patterson, M., Lundmark, V., Kiesler, S., Mukopadhyay, T., & Scherlis, W. (1998). Internet Paradox: A social technology that reduces social involvement and psychological well-being? *American Psychologist,* **53**, 1017-1031.

Kring, A.M., Davison, G.C., Neale, J.M., & Johnson, S.L. (2007). *Abnormal psychology* (ch. 8). Hoboken, NJ: John Wiley & Sons.

Laberge-Nadeau, C., Maag, U., Bellavance, F., Lapierre, S. D., Desjardins, D., Messier, S., et al. (2003). Wireless telephones and the risk of road crashes. *Accident Analysis and Prevention,* **35**, 649–660.

Lamble, D., Kauranen, T., Laakso, M., Summala, H. (1999). Cognitive load and detection thresholds in car following situations: Safety implications for mobile (cellular) telephones while driving. *Accident Analysis and Prevention,* **31**, 617–623.

Landers, R.N., & Lounsbury, J.W. (2006). An investigation of Big Five and narrow personality traits in relation to Internet usage. *Computers in Human Behavior,* **22**, 283-293.

Lavoie, J.A.A., & Pychyl, T.A. (2001). Cyberslacking and the procrastination superhighway: A web-based survey of online procrastination, attitudes, and emotion. *Social Science Computer Review,* **19**, 431-444.

Leary, T.F. (1994). *Chaos and cyberculture*. Berkeley, CA: Ronin Publications.

Lim, V.K.G. (2002). The IT way of loafing on the job: cyberloafing, neutralizing and organizational justice. *Journal of Organizational Behavior,* **23***,* 675-694.

Luft, J. (1970). *Group processes: An introduction to group dynamics*. Palo Alto, CA: National Press Books.

Luft, J. & Ingham, H. (1955). *The Johari-window: A graphic model for interpersonal relations*. University of California, Western Training Laboratory.

MacKenzie, I. S. (1992). Fitts' law as a research and design tool in human- computer interaction. *Human-Computer Interaction,* **7**, 91-139.

Mahatanankoon, P. (2007). The effects of personality traits and optimum stimulation level on text-messaging activities and M-commerce intention. *International Journal of Electronic Commerce,* **12**(1), 7-30.

Mahatanankoon, P., Wen, H.J., & Lim, B. (2005). Consumer-based m-commerce: exploring consumer perception of mobile applications. *Computer Standards & Interfaces,* **27**,347-357.

Markey, P.M. (2000). Bystander intervention in computer-mediated communication. *Computer in Human Behavior,* **16**, 183-188.

Matanda, M., Jenvey, V.B. & Phillips, J.G. (2004). Internet usage in adulthood: Loneliness, computer anxiety and education. *Behavior Change,* **21**(2), 103-114

McLeod, P.(1977). A dual task reponse modality effect: Support for multiprocessor models of attention. *Quarterly Journal of Experimental Psychology,* **29**, 651-677.
McLeod, P. & Mierop, J. (1979). How to reduce manual response interference in the multiple task environment. *Ergonomics,* **22**, 469-475.
McKenna, K. & Bargh, J. (1999). Causes and consequences of social interaction on the Internet: A conceptual framework. *Media Psychology*, **1**, 249-270.
McKenna & Bargh 2000
McKnight, A. J., & McKnight, A. S. (1993). The effect of cellular phone use upon driver attention. *Accident Analysis and Prevention,* **25**, 259–265.
Milgram, S. (1965). Some conditions of obedience and disobedience to authority. *Human Relations, 18*, 57-76.
Payne, B. (unpublished). Postgraduate Diploma Thesis. Monash University.
Phillips, J.G. (2006). The Psychology of Internet Use and Misuse. In M. Anandarajan, T. Thompson, & C. Simmers (Eds.), *The Internet and Workplace Transformation* (pp. 41-62). Armonk: ME Sharpe.
Phillips, J.G., Butt, S., & Blaszczynski, A. (2006). Personality and self reported use of mobile phones for games. *CyberPsychology & Behavior,* **9**(6), 753-758.
Phillips, J.G. & Reddie, L. (2007). Decisional style and self-reported email use in the workplace. *Computers in Human Behavior.* **23**, 2414-2428.
Phillips, J.G. & Stelmach, G.E. (1996). Parkinson's disease and other involuntary movement disorders of the basal ganglia. In C. Fredericks & L. Saladin (Ed.) *The motor system: basic principles and pathophysiology* (pp. 424-444). Philadelphia: F.A. Davis.
Plant, S. (2000). *'On The Mobile: The Effect of Mobile Telephones on Social and Individual Life'[WWW* document]. Retrieved September 10,2005, from, http://www.motorola.com
Prensky, M. Digital natives, digital immigrants. *On the Horizon,* **9**, (2001a), 1-6.
Prensky, M. Digital natives, digital immigrants, Part II: Do they really *think* differently?. *On the Horizon, 9,* (2001b), 1-10.
Redish, A.D., Jensen, S., & Johnson, A. (in press). A unified framework for addiction: Vulnerabilities in the decision process. *Behavioral and Brain Sciences*.
Reidelmeier, D. A., & Tibshirani, R. J. (1997). Association between cellular telephone calls and motor vehicle collisions. *New England Journal of Medicine* **336**: 453–458.
Scealy, M., Phillips, J.G. & Stevenson, R. (2002). Shyness and anxiety as predictors of patterns of Internet usage. *CyberPsychology & Behavior,* **6**, 507-515.
Scharl, A., Dickinger, A., & Murphy, J. (2005). Diffusion and success factors of mobile marketing. *Electronic Commerce Research and Applications,* **4,** 159-173.
Schellinck, T. & Schrans, T. (2006). *The successful loyalty data based model to identify problem and at risk gamblers in casinos: Issues and implications*. Paper presented at the 16th conference of the National Association of Gambling Studies, 22nd-25th November, Sydney.
Shaffer, H. J., (1996). Understanding the means and objects of addiction: Technology, the Internet and Gambling, *Journal of Gambling Studies,* **12**, 461-469.
Shaffer, H.J., Hall, M.N., & Vander Bilt, J. (2000). Computer addiction: A critical consideration. *American Journal of Orthopsychiatry*, **70**(2), 162-168.
Shotton, M.(1989). *Computer addiction? A study of computer dependency.* New York: Taylor and Francis.

Smeaton, M. & Griffiths, M.D. (2004). Internet gambling and social responsibility: An exploratory study, *CyberPsychology & Behavior*, **7**, 49-57.

Stein, D.J., Hollander, E., & Cohen, L. (1994). Neuropsychiatry of obsessive compulsive disorder. In E. Hollander, J. Zohar, D. Marazzati, & B. Olivier (Eds.), Current insights in Obsessive Compulsive Disorder (pp. 167-182). Chichester: John Wiley & Sons.

Strayer, D. L., & Johnston, W. A. (2001). Driven to distraction: Dual task studies of simulated driving and conversing on a cellular telephone. *Psychological Science,* **12**, 462–466.

Toneguzzo, S.J. (1995). The Internet: Entrepreneur's dream or regulator's nightmare. In J. O'Connor (Ed.), *High stakes in the nineties*. (pp. 73- 86). Perth: Curtin University.

Toneguzzo, S. (1997). Internet gaming - evidence for parliament of South Australia, Social Development Committee. In G. Coman, B. Evans, & R. Wootton (Eds.), *Responsible gambling: A future winner* (pp.369-389). Adelaide: National Association for Gambling Studies.

Trimpop, R. M. (1994). *The psychology of risk taking behaviour*. Amsterdam:North Holland

Violanti, J. M. (1998). Cellular phones and fatal traffic collisions. *Accident Analysis and Prevention,* **30**, 519–524.

Violanti, J. M., & Marshall, J. R. (1996). Cellular phones and traffic accidents: An epidemiological approach. *Accident Analysis and Prevention,* **28**, 265–270.

Walker, M. B.(1989). Some problems with the concept of "Gambling Addiction": Should theories of addiction be generalized to include excessive gambling, *Journal of Gambling Behavior*, **5**(3), 179-200.

Walker, M., Anjoul, F., & Milton, S. (2005). *A structured clinical interview for pathological gambling (SCIP)*. Paper presented at the 15th conference of the National Association for Gambling Studies, Alice Springs 10th-12th November.

Warden, N., Phillips, J.G., & Ogloff, J.R.P. (2004). Internet Addiction. *Psychiatry, Psychology and Law,* **11**(2), 280-295.

Watson, S., Liddell, P.jr, Moore, R.S., & Eshee, W.D. jr. (2004). The legalization of internet gambling. A consumer protection perspective. *Journal of Public Policy and Marketing.* **23**(2), 209-213.

Weinberger, A. (1981). Responses to old people who ask for help. *Research on Aging,* **3,** 345-368.

Welford, A.T. (1977). Desire for attention. *Australian and New Zealand Journal of Psychiatry,* **11**(3), 157-161.

Welford, A.T. (1987). Ergonomics of human relations. *Ergonomics,* **30**, 3-8.

Wen, H.J., & Mahatanankoon, P. (2004). M-commerce operation modes and applications. *International Journal of Electronic Business*, **2**(3).

Wood, T.A., Griffiths, M.D., Chappell, D. & Davies, M.N.O. (2004). The structural characteristics of video games: A psycho-structural analysis. *CyberPsychology & Behavior* **7**: 1-10.

Wyatt, K. & Phillips, J.G. (2005). Personality as a predictor of workplace Internet use. *Proceedings of OZCHI2005,* ACM ISBN: 1-59593-222-4,

Young, K. (1996). Psychology of computer use: Addictive use of the Internet: A case that breaks the stereotype. *Psychological Reports,* **79**, 899-902.

Young, K. (1998). Internet addiction: The emergence of a new clinical disorder. *Cyberpsychology & Behaviour,* **1**(3), 237-244.

Yuan, M.J. (2003). Develop state-of-the-art mobile games. http://www.javaworld.com/javaworld/jw-11-2003/jw-1107-wireless.html

Zuckerman, M. (1979). *Sensation seeking: Beyond the optimal level of arousal.* New Jersey:Lawrence Erlbaum Associates.

In: Mobile Telephones... ISBN 978-1-60456-436-5
Editors: A.C. Harper and R.V. Buress, pp. 211-229

Chapter 7

SIMULATION OF WIRELESS SENSOR NETWORKS BASED ON A REALISTIC BATTERY MODEL

Marco Aurélio Spohn, Paulo S. Sausen, José R.B. Sousa Angelo Perkusich and Antonio M.N. Lima
Federal University of Campina Grande-FUCG, Brazil

Abstract

A Wireless Sensor Network (WSN) is a power constrained system, because nodes run on limited power batteries. When a node switches between modes (i.e., transmit, receive, and sleep), there are also changes related to the power consumed by the node not only after the node is in a different mode, but also during the transition process. Although the switching energy is considered negligible most of the time, it is strongly dependent on the technology (e.g., the radio) deployed on the nodes. In particular, for low-duty cycle networks such as WSNs, the switching energy may impact on the overall energy consumption. Network simulators usually employ the linear discharge model which is simple, but it does not capture the battery capacity recovery effect. Getting a more accurate measure for the energy consumption of a sensor network may lead to better estimates for the network lifetime. This Chapter explores the application of a more realistic battery model (i.e., the Rakhmatov-Vrudhula battery model) for the simulation of WSNs.

1. Introduction

A Wireless Sensor Network (WSN), sometimes referred to as Wireless Ad Hoc Sensor Network, is a variant of the broader class of Mobile Ad Hoc Networks (MANETs). A WSN comprises a large number of small sized nodes each one having a processing unit, a radio transceiver and antenna for wireless communications, one or more sensor units (e.g., temperature, speed, acoustic, seismic), and a power unit usually equipped with a low capacity battery. Due to its limited power resources, and because batteries cannot be easily replaced, nodes are built out from power saving components.

When applied to WSN, dynamic power management mechanisms provide the means for saving energy and prolonging the network lifetime. The basic approach consists on putting nodes' components to sleep whenever possible. Nodes are brought back to an active state upon the occurrence of some event, or otherwise, following some predefined activation schedule. In the former case, the sensor unit is kept active to react to any expected event. Therefore, usually a node is kept partially on, but the power savings from keeping most units in a sleep state pays off.

When a node switches between modes (i.e., transmit, receive, and sleep), there are also changes to the power consumed by the node not only after the node is in a different mode, but also during the switching process. Although the switching energy is considered negligible most of the time, it is strongly dependent on the technology (e.g., radio) being deployed on the sensor nodes. In particular, for low-duty cycle networks like the sensor networks, the switching energy may impact the overall energy consumption.

Most of the time, the switching energy is neglected to simplify the simulation process. To better understand energy consumption in a WSN, it is necessary to adopt a more realistic battery discharge model. Network simulators usually employ the linear discharge model, which is simple and far from reality, leading to wrong conclusions like, for example, that switching states always implies on battery capacity discharge. Getting a more accurate measure for the energy consumption of a WSN may lead to better estimates for the network lifetime. For example, an error of 6% on the energy consumption of a WSN translates into shortening the network lifetime by two months over an expectancy of three years [1].

Most WSN protocols rely on some sort of network wide search operations. In this case, control packets are usually flooded to the whole network. For example, on demand routing protocols issue *route request* (RREQ) messages when a node is searching for a route to any destination. Receiving nodes broadcast the first copy of every RREQ message until it reaches a node with a valid route to the destination or the destination itself. As another example, a base station (or a sink) may probe a sensor node by flooding the request throughout the whole network, because the location of the node might be unknown. Given the importance of such flooding operations, it is important to understand the energy cost of flooding the whole network considering a more realistic battery model.

This Chapter explores the application of a more realistic battery model for the simulation of WSNs. The rest of this Chapter is organized as follows. Section 2. presents a brief introduction to WSNs. Section 3. presents a brief discussion on battery models, and Section 4. talks about the switching energy. Section 5. explores the application of the Rakhmatov-Vrudhula battery model in some dynamic power management techniques. Section 6. presents an analysis for the blind flooding protocol under the Rakhmatov-Vrudhula battery model.

2. Wireless Sensor Networks

A Wireless Sensor Network (WSN) may well contain hundreds or thousands of small autonomous elements called sensor nodes, and each sensor can feature a large variety of sensors (e.g., temperature, speed, acoustic, seismic). In many cases, nodes are randomly spread over remote areas, making it difficult to perform any maintenance to the nodes. Hence, a node remains live while it has enough battery capacity for its normal operation, and the network lifetime strongly depends on the remaining capacity of the nodes in the network.

A sensor node has a few basic components (see Figure 1) [2,3] as follows: i) a *processing unit* which includes a microprocessor or a micro-controller; ii) a *communication unit*

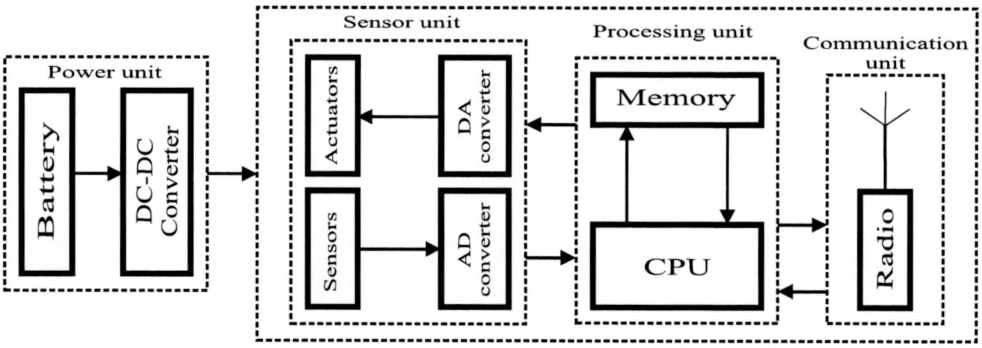

Figure 1. Sensor node: basic structure

which consists of a short range radio for wireless communication; iii) a *sensor unit* that contains a group of sensors and actuators; iv) and a *power unit*, usually a battery, which acts as the power source for all node's components.

3. Battery Models

Given that usually it is unfeasible to replace nodes' batteries after the network has already been deployed, it is paramount to any WSN protocol to be energy efficient. Therefore, energy consumption is usually taken as a performance metric to compare different protocols.

Each battery in a node's power unit can be characterized by two values: (i) open-circuit voltage, V_{oc}, that is the battery's voltage when the battery is totally charged with no discharge electric current; and (ii) the cut-off voltage, V_{co}, that is the battery's voltage when the battery is deemed discharged.

A battery's capacity, C, can be expressed in three different ways [4]: (i) the *theoretical capacity* which is based on the amount of energy stored in the battery, and that is the maximum energy that can be consumed in practice; (ii) the *standard capacity* which is the energy that can be obtained under some conditions specified by the battery manufacturer; and (iii) the *actual capacity* is the amount of energy that the battery delivers under a given load, and is usually used as a metric to judge the battery efficiency. While the actual capacity may exceed the standard capacity, it cannot exceed the theoretical capacity of the battery. In this Chapter, when we mention battery capacity we are actually referring to the actual battery capacity in terms of mA-ms[1].

The battery performance with regards to the discharge pattern depends on two effects [4]: (i) the ***capacity effect*** which depends on the battery *actual capacity* and the discharge electric current; (ii) the ***recovery effect*** due to the battery capacity recovery during idle periods, that is, periods during which the actual electric current is drastically reduced.

Battery models aim at capturing the actual battery characteristics. These models can help predicting the battery's behavior under a large variety of conditions. When integrated into a network simulator, it is possible to analyze protocols under a more realistic scenario,

[1]Read it as mA *times* ms. We have adopted this notation because a sensor node usually requests an electric current of order of mA during very short periods of order of ms.

giving a better picture of a protocol's behavior in terms of energy consumption.

Battery models can be classified according to their characteristics [4]: (i) analytical models; (ii) electrical circuit based models; (ii) stochastic models; and (iv) eletro-chemical models. Some of these models can capture both the capacity effect and the recovery effect. Analytical models are flexible regarding their implementation, and they can be easily configured to any battery specification.

The linear battery model is the simplest analytical model. This model considers the battery as a linear electric current source. This way, the equation $C = C' - It_d$ allows computing the remaining battery capacity, C, where C' is the initial battery capacity, and I is the constant discharge electric current for the period of time t_d. The remaining battery capacity is computed always when the discharge current changes its value.

The Rakhmatov-Vrudhula [5] model is usually used for analyzing the impact of a more realistic battery model on the simulation of a WSN. The next Section presents a brief description of this model.

3.1. Rakhmatov-Vrudhula Battery Model

Rakhmatov and Vrudhula [5] derived the equations for computing the lifetime of a *lithium-ion* battery considering both a constant and a variable load. Two parameters are used for computing the battery lifetime: (i) the α parameter which is related to the battery capacity, and the β parameter with regards to the non-linear battery behavior during charging and discharging periods.

Equation 1 describes the impact of the discharge profile on the battery lifetime, where I_{k-1} is the discharge electric current during the period $k-1$. Function A computes the impact of the battery non-linear behavior, where L is the battery lifetime, t_k is the time duration for period k and t_{k-1} is the time duration for period $k-1$. More details about function A and the algorithm for computing L can be obtained from [5].

$$\alpha = \sum_{k=1}^{n} 2I_{k-1}A(L, t_k, t_{k-1}, \beta) \tag{1}$$

The parameters for computing the battery's lifetime, L, are the following ones: (i) the values for the discharge electric current, $S_I = \{I_0, I_1, \ldots, I_n\}$; (ii) the respective time intervals, $S_t = \{t_0, t_1, \ldots, t_n\}$; and (iii) the α and β parameters. Given these parameters, it is possible to obtain the battery's lifetime or its actual capacity.

A battery's discharge pattern has an important role on the battery's performance. Figure 2 shows the battery's capacity considering the discharge pattern given by $S_I = \{30, 2, 20\}$ mA and $S_t = \{8, 6, 6\}$ ms. This means that the discharge electric current is 30 mA, 2 mA, and 20 mA for the intervals of 8 ms, 6 ms and 6 ms, respectively.

Figure 2 shows the recovery effect (between 8 ms and 14 ms) captured by Rakhmatov-Vrudhula model due to a variation on the discharge current from 30 mA to 2 mA, and the results for the linear model showing that, because it does not capture the recovery effect, it actually computes a reduction to the battery capacity. Note that for a battery to recover its capacity it does not imply reducing its consumption to zero, but it certainly requires a reduction on the consumption.

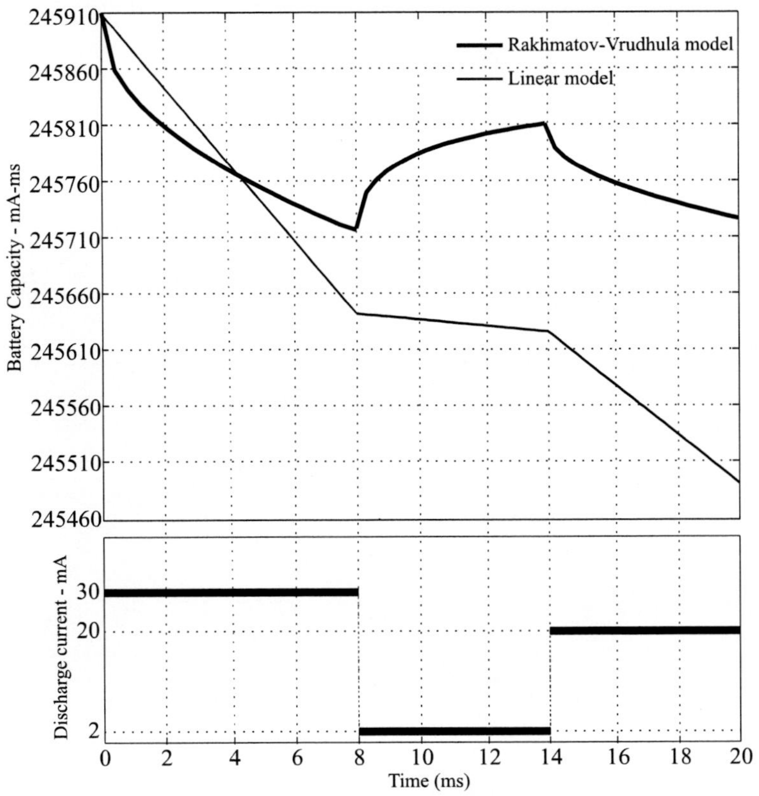

Figure 2. Battery capacity as a function of its discharge electric current.

Figure 3 shows the difference between the energy consumption for the linear model and the *Rakhmatov-Vrudhula* model. The linear model presents an energy consumption 50 times larger than the *Rakhmatov-Vrudhula* model considering the same discharge profile with a 10 s duration. This huge difference is due to the fact that the linear model cannot capture the battery capacity effect and the recovery effect.

Handy and Timmermann [6] present an algorithm for integrating the Rakhmatov-Vrudhula battery model into simulation environments for wireless mobile networks. The proposed mechanism estimate the lifetime for battery operated devices following an interactive approach without requiring the knowledge about the whole past of the battery discharge pattern. This Chapter presents results considering $\alpha = 245910$ and $\beta = 4034$ which were estimated based on an alkaline battery [7] usually employed in WSN devices.

4. Switching Energy

Switching Energy (*SE*) refers to the energy related to the transitioning of a node to a different state (e.g., transitioning from a transmission state to a sleeping state). Depending on the type of the transition, it can consume or it can recover some battery capacity. In this Chapter, the parameter *SE* is represented in mA-ms units (see Section 3.).

Figure 4 shows that when a node transitions from the receive mode to the sleeping

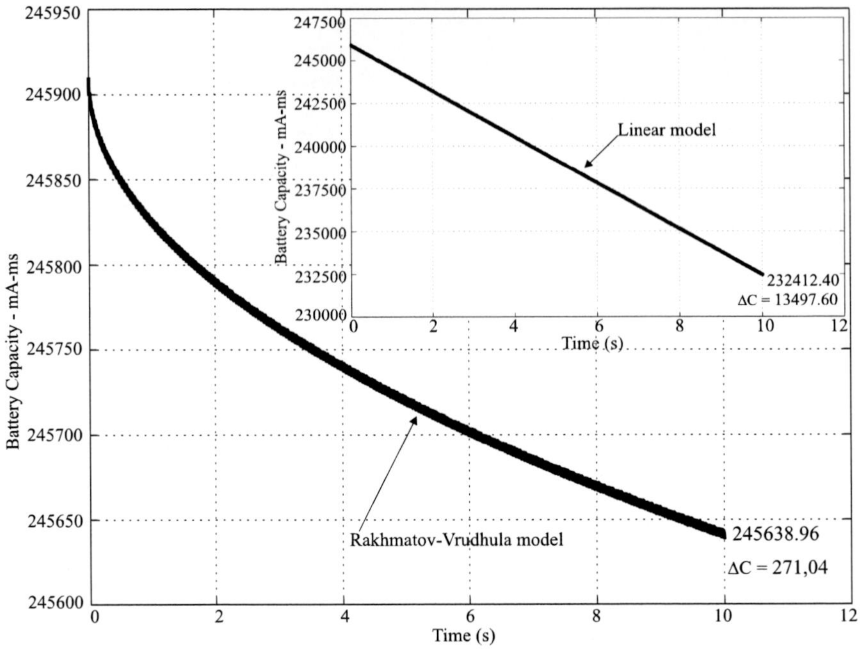

Figure 3. Linear model versus the Rakhmatov-Vrudhula model.

Table 1. Transition times (in μs) for the radios TR1000™ [8] and CC1000™ [9]

Radio	Data Rate/Modulation	Tx→Sleep	Rx→Sleep	Sleep→Tx	Sleep→Rx
TR1000	19.2 Kbps/OOK	10	10	16	20
TR1000	115.2 Kbps/ASK	10	10	16	20
CC1000	38.4 Kbps/OOK	10	10	850	850

mode the battery actually recovers some energy (i.e., ***Switching Time 1***, ST_1), and when the node transitions back to the receiving mode (after a sleeping period) the node spends some energy during the process (shown as ***Switching Time 2***, ST_2). During the sleep state, the node acquires most of its recovered battery capacity. It is important to mention that it is possible to observe the battery capacity recovery due to the Rakhmatov-Vrudhula battery model. The adoption of a simpler battery model (e.g., the linear model) can lead to wrong conclusions when analyzing the transitioning phase of a node. If that is the case, one can conclude that the switching energy always implies the reduction of the battery capacity.

Ruzzelli et al. [1] take into account the switching energy when evaluating the performance of WSN in terms of energy consumption. In most cases the switching energy is neglected to simplify the simulations. However, in energy constrained networks like the WSNs, it becomes relevant to consider all the gains due to the battery capacity recovery effect when evaluating the network lifetime.

Table 1 presents the switching time results of each operational mode for the radios TR1000™ [8] and CC1000™ [9]. These radios integrate a great part of the sensor devices

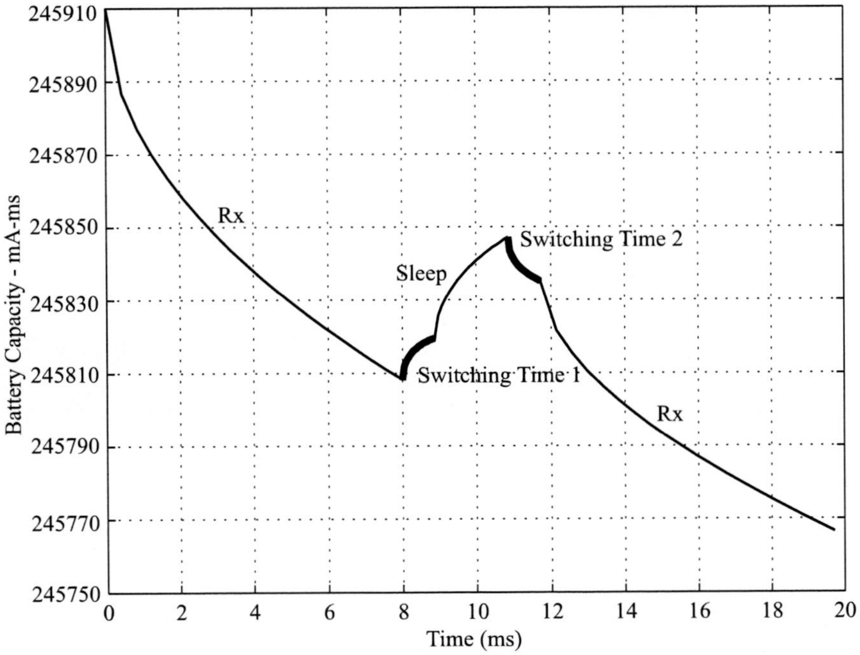

Figure 4. Analyzing the switching energy.

Table 2. Current (in mA) for each operational state for the radios TR1000™ [8] and CC1000™ [9].

Radio	Data Rate/Modulation	I_{Tx}	I_{Rx}	I_{Sleep}
TR1000	19.2 Kbps / OOK	12.00	4.50	0.005
TR1000	115.2 Kbps / ASK	12.00	4.80	0.005
CC1000	38.4 Kbps / FSK	27.00	10.00	0.001

(e.g., Berkeley Mica e Mica2 Motes) actually deployed in WSNs. The current values corresponding to each one of these states are presented in Table 2.

5. Dynamic Power Management

This section presents an analysis of two dynamic power management mechanisms that take advantage of the switching energy. Analyzing such protocols is only possible due to a more realistic battery model (i.e.,the Rakhmatov-Vrudhula battery model) which captures the battery capacity recovery effect.

5.1. Dynamic Power Management with Mode Switching (DPM-MS)

Dynamic Power Management with Mode Switching (DPM-MS) [10] is a power management technique which schedules state transitions to a sleep state based on the current node's battery capacity. As the battery capacity degrades, upon reaching some predefined energy level, the scheduling to the sleep state takes place after a transmission in order to improve the battery capacity recovery.

When a node subsystem is switched to a sleep state, there is a huge reduction on the current drawn from the battery by that component. Hence, it allows us to take advantage of the battery recovery effect. In a sensor node, the radio is the subsystem that, in average, draws more energy from the battery [11]. For the radios under consideration in this text, transmission (i.e., Tx) is the mode that consumes more energy (see Table 2). Therefore, DPM-MS adopts switching to the sleep state from a transmission state. By doing that, it is possible to take more advantage of the battery recovery effect (as shown in Figure 5), because the gap between the current drawn from the battery between a transmission state and the sleep state is larger than considering the other state (i.e., between Rx and Sleep, as shown in Table 2).

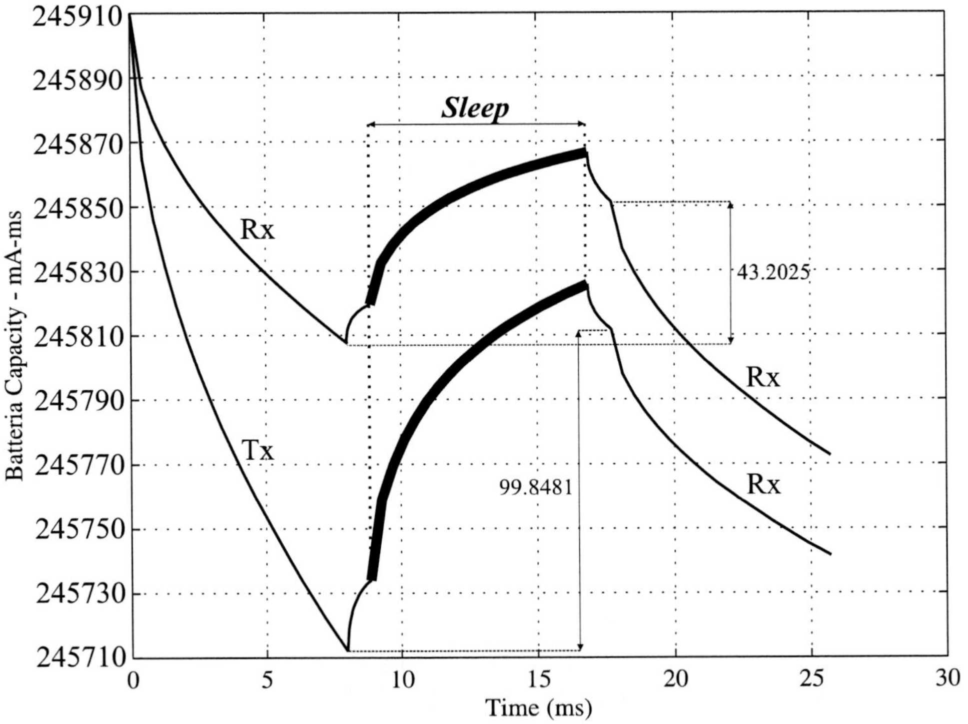

Figure 5. Battery capacity recovery.

The results presented in this Section are based on the parameters for a CC1000 radio [9], considering packets of 40 Bytes. This radio is used on the Mica Motes [11], and the nominal transmission speed is 38.4 Kbps.

To define the DPM interval, Sausen et al. [10] considered five sleep times. Figure 6 and Table 3 present the values for these intervals, the recovered energy (i.e., $E(Sleep_k)$) for each sleep interval, the energy recovered or consumed during the first and second transitions (i.e., $SE(ST_1)$ and $SE(ST_2)$ respectively), and the total energy recovered or consumed during the sleep state (i.e., $E(DPM)$). Based on these results, the 8 ms interval (i.e., $Sleep_4$) was adopted as the default DPM interval. Any other interval could be con-

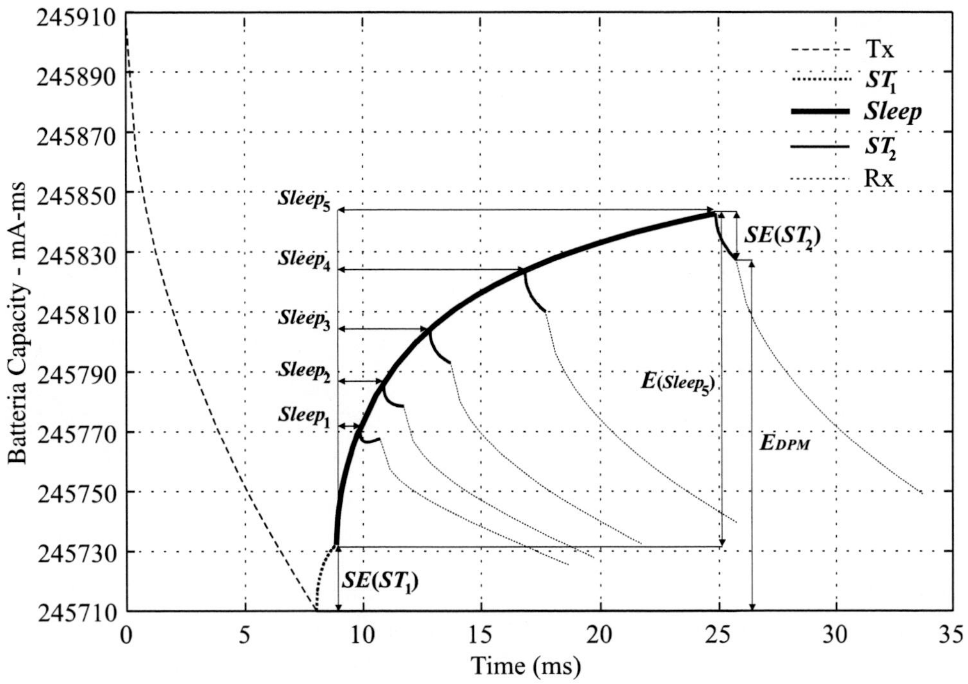

Figure 6. Sleep times for DPM-MS.

Table 3. Energy (in mA-ms) for the *sleep* intervals shown in Figure 6.

k	$Sleep_k$ ms	$SE(ST_1)$	$E(Sleep_k)$	$SE(ST_2)$	$E(DPM)$
1	1	22.0725	38.8088	-3.3500	57.5313
2	2	22.0725	53.8740	-7.6376	68.3088
3	4	22.0725	71.9832	-11.2339	82.8218
4	8	22.0725	91.5731	-13.7975	99.8481
5	16	22.0725	110.4824	-15.3337	117.2212

sidered, but there is a tradeoff between the energy recovered and the unavailability of the communication unit (i.e., while the radio is in the sleep state the node cannot communicate with other nodes). An 8 ms sleep interval allows recovering 99.84 mA-ms for the period equivalent to a transmission time of a 40 Bytes packet. For example, if the $Sleep_5$ interval were adopted, the energy recovered increases by just 17.4% at the expense of doubling the inactive period.

5.1.1. Computing the Switching Energy

Two radio models (i.e., TR1000 and CC1000) are considered for computing the Switching Energy (i.e., *SE*), and two *discharge patterns* (DP) as follows:

$$DP_1 = Tx \rightarrow Sleep \rightarrow Tx$$
$$DP_2 = Tx \rightarrow Sleep \rightarrow Rx$$

Let P_A and P_B be the power consumed during mode A and B respectively, and $T_{A\to B}$ and $T_{B\to A}$ be the transition times between these two modes respectively.

Both radio manufacturers do not provide all the necessary information for computing the switching energy. To circumvent this problem, the following assumptions are employed [10, 12, 13]:

- **Assumption 1** The energy consumed when the node switches from mode A to mode B is the energy consumed for the mode A during the transition time, that is,

$$SE_{A\to B} = P_A T_{B\to A} + P_B T_{A\to B}$$

- **Assumption 2** The energy consumed when the node switches from mode A to mode B is the energy consumed for the mode B during the transition time, that is,

$$SE_{A\to B} = P_A T_{A\to B} + P_B T_{B\to A}$$

- **Assumption 3** The energy consumed when the node switches from mode A to mode B is the average between the energy consumption for modes A and B during the transition time, that is,

$$SE_{A\to B} = \frac{(P_A + P_B)(T_{A\to B} + T_{B\to A})}{2}$$

Table 4 presents the results when applying these three assumptions to the Radio CC1000 with an 8 ms sleep time, FSK modulation, and at a 38.4 Kbps rate. Positive values represent battery capacity recovery, while negative values translate into battery capacity lost. For this particular radio, an 8 ms DPM interval provides actual energy gain (see E_{DPM}) for all the three assumptions.

Table 4. Battery capacity recovered/lost in mA-ms for the CC1000 radio (38.4 Kbps/FSK).

Assumption	DP	E($Sleep_4$)	$SE(ST_1)$	$SE(ST_2)$	$SE(ST_1 + ST_2)$	$E(DPM)$
1	1	111.9600	-3.3736	-61.3872	-64.7608	47.1992
	2	111.9600	-3.3736	-30.0408	-33.4143	78.5456
2	1	64.5739	54.1369	-4.3573	49.7797	114.3536
	2	64.5739	54.1369	-0.9573	53.1797	117.7536
3	1	91.5731	22.0725	-29.5626	-7.4901	84.0830
	2	91.5731	22.0725	-13.7975	8.2750	99.8481

Tables 5 and 6 present the results for a TR1000 radio at 19.2 Kbps. The simulations takes into account the fact that this radio allows both OOK and ASK modulation. There is a small difference between the results for these two cases for the battery capacity recovery considering an 8 ms sleep period.

Table 5. Battery capacity recovered/spent (in mA-ms) for a TR1000 radio at 19.2 Kbps and OOK modulation

Assumption	DP	$E(Sleep_4)$	$SE(ST_1)$	$SE(ST_2)$	$SE(ST_1+ST_2)$	$E(DPM)$
1	1	54.4623	3.3343	-4.3741	-1.0398	53.4225
	2	54.4623	3.3343	-2.7674	0.5669	55.0292
2	1	53.5734	4.2601	-3.4483	0.8119	54.3853
	2	53.5734	4.2601	-1.9411	2.3191	55.8925
3	1	55.1581	2.6570	-2.7710	-0.1139	55.0440
	2	55.1581	2.6570	-1.6738	0.9832	56.1413

Table 6. Battery capacity recovered/spent (in mA-ms) for the TR1000 Radio at 115.2 Kbps and ASK modulation

Assumption	DP	$E(Sleep_4)$	$SE(ST_1)$	$SE(ST_2)$	$SE(ST_1+ST_2)$	$E(DPM)$
1	1	54.4623	3.3343	-4.3741	-1.0398	53.4225
	2	54.4623	3.3343	-2.8522	0.4820	54.9443
2	1	53.5734	4.2601	-3.4483	0.8119	54.3853
	2	53.5734	4.2601	-2.0011	2.2590	55.8324
3	1	55.1580	2.6570	-2.7710	-0.1140	55.0440
	2	55.1580	2.6570	-1.7257	0.9313	56.0893

5.1.2. Performance Analysis

A scenario to drain the battery capacity is used to account for the gains due to the energy recovery effect. Only transmissions are considered, because these are the most energy costly operations. Figure 7 shows that the recovered energy allows DPM-MS transmit twice as many packets transmitted without DPM-MS. This is accomplished with a very short sleep period (i.e., 8 ms), and it certainly can be improved by letting the battery rest for a longer period of time (Figure 6 shows that longer sleep times recover more battery capacity).

The Rakhmatov-Vrudhula battery model has proved efficient to provide a better insight on the non-linear behavior of battery capacity, and also on the impact of appropriately managing state transitions in order to take advantage of the recovery effect. By just considering a standard linear battery model there is no way to fully understand the efficiency of any DPM technique.

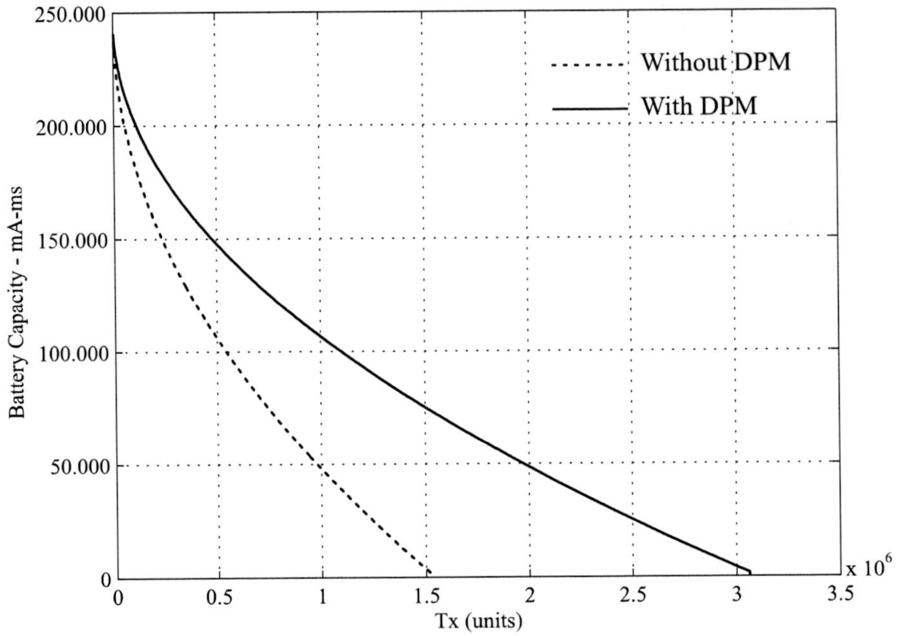

Figure 7. Transmissions achieved with and without DPM.

5.2. Dynamic Power Management with Scheduled Switching Modes (DPM-SSM)

Dynamic Power Management with Scheduled Switching Modes (DPM-SSM) [14] is another power management technique which evolved from DPM-MS. DPM-SSM provides several DPM modes which are triggered depending on the battery remaining capacity.

A Differential Hybrid Petri Nets (DHPN) [15] model for a sensor node with DPM-SSM is shown in Figure 8. There are three possible DPM modes (see Table 7), represented by places P_{10}, P_{13}, and P_{16} respectively. The transition to a particular DPM mode depends on the current battery capacity which is represented by place P_{21}. The three DPM modes differ by the number of transmissions (represented by places P_{11}, P_{14} and P_{17}) taking place (i.e., P_{18}) before the node transitions to a sleeping state. The processing unit is either active or inactive.

Table 7. DPM modes

DPM Mode	Battery Capacity	Transition after
DPM1	$65\% < C \leq 80\ \%$	6 transmissions
DPM2	$50\% < C \leq 65\ \%$	3 transmissions
DPM3	$C \leq 50\ \%$	1 transmission

The parameters for the sensor node (see Table 8) were obtained from the specification for the *Berkeley Mote* [11].

Table 8. Sensor node parameters

Unit	Specification
Computational	Active = 8 mA, Inactive = 15 μA
Communication	Tx = 27 mA, Rx = 10 mA, Sleep = 1 μA
Power source	C=245910 mA-ms, V_{oc}=1.5V, V_{co}=0.8V

The interaction between discrete events dynamic and continuous dynamics is captured in the DHPN model as follows. Each state relates to a set of discrete events and their corresponding continuous evolutions while the system is in the current state. The continuous evolutions are enabled through differential transitions which activate the corresponding function f_j. Each one of the f_j ($j = 1 \dots 6$) functions (see Table 9) execute the following: (i) updates the discharge pattern (i.e., the sets S_I and S_t; and (ii) starts the algorithm for computing the battery's lifetime for the new battery discharge pattern. Therefore, while discrete events are being handled, the battery discharge pattern is updated and the corresponding battery lifetime is computed.

Table 9. Differential transitions with respective functions and continuous calculations

Transition	Function	Continuous calculation
T_{22}	f_1	Tx Operation
T_{23}	f_2	Rx Operation
T_{24}	f_3	Transitory (Sleep→ Rx)
T_{25}	f_4	Transitory (Sleep→ Tx)
T_{26}	f_5	Transitory (Tx→ Sleep)
T_{27}	f_6	Sleep state

For example, the model shown in Figure 8 presents an initial marking $M_0(t)$=$[1\,1\,0\,0\,0\,0\,0\,0\,0\,0\,0\,0\,0\,0\,0\,0\,0\,0\,0\,0\,0\,0\,245910]^T$, which means that the sensor and communication subsystem are actives. $M_0(t)$ enables, for example, the following sequence of events $T_1 T_{22}$. The differential transition T_{22}, which on its turn activate function f_1. The resulting S_I and S_t sets are $S_I = \{27\}$ mA and $S_t = \{8\}$ ms; that is, the system is currently drawing from the battery an electric current of 27 mA (see Table 8) for the time period of 8 ms.

Assuming a fully charged battery, a node could start with all its units active without any restriction; that is, no DPM is initially employed. However, when the battery capacity is below 80% its full capacity ($H_{P_{21},T_{13}}$: $65\% < x_1 \leq 80\%$ is True), a transition T_{13} is fired and a token is placed at place P_{10} enabling the DPM1 mode. The same procedure is repeated once the battery capacity is below its 65% and 50% full capacity, leading to transitions T_{16} and T_{19} respectively.

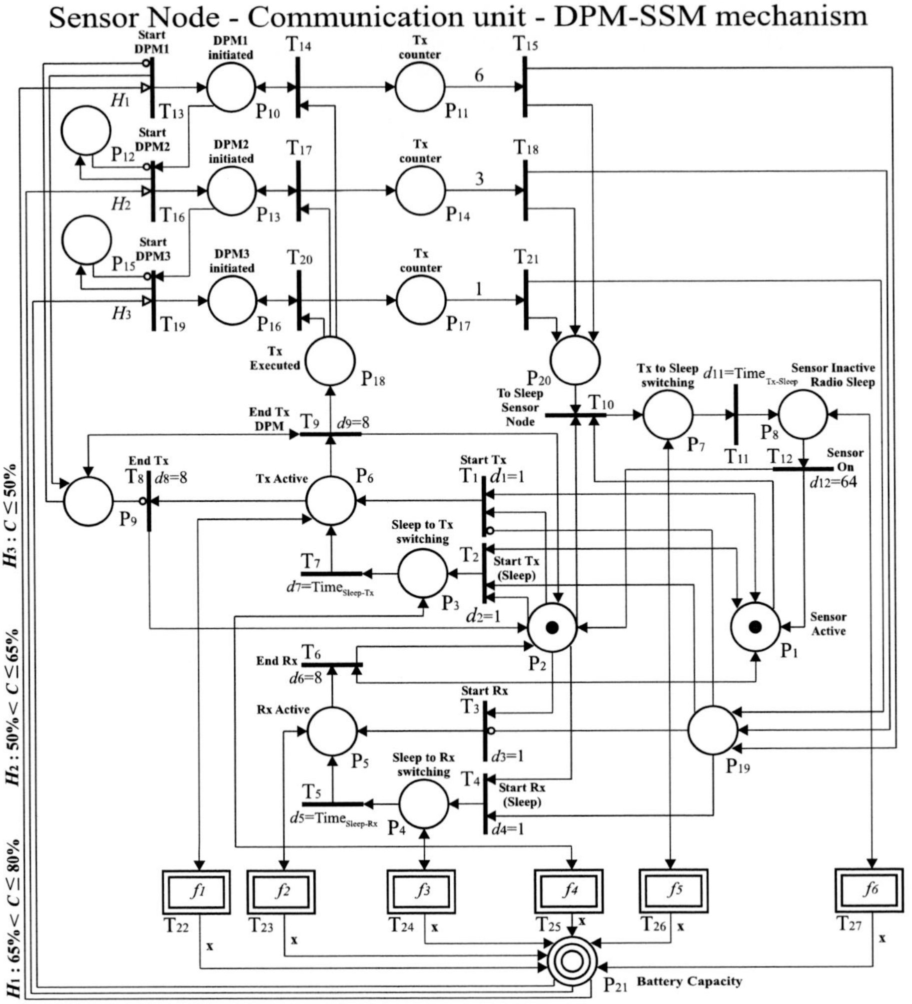

Figure 8. DHPN model for a sensor node using the DPM-SSM technique.

5.2.1. Performance Analysis

The marking $M(t)$=[$1\,1\,0\,0\,0\,0\,0\,0\,1\,1\,5\,0\,0\,0\,0\,0\,0\,0\,0\,0\,1\,9\,6\,5\,0\,3$]^T is used as an initial evaluation for the DHPN model as shown in Figure 8. This marking presents the sensor node currently operating on DPM1 mode and that 5 transmissions have already taken place (i.e., $m_D(P_{11})(t) = 5$). Among the many possible sequence of events starting from marking $M(t)$, the following one is taken into account: $T_1T_{22}T_9T_{14}T_{15}T_{10}T_{26}T_{11}T_{27}T_{12}T_4T_{24}T_5T_{23}T_6$.

This sequence results in a transition to an inactive state and it also triggers the differential transition T_{27} which is responsible for updating the battery capacity while the node remains in a sleep state. The sensor node remains in this state for 64 ms. Following that, the transition T_{12} brings the node back to an active state. Figure 9 shows the battery capacity for the given scenario.

To account for the gains due to the energy recovery effect, a scenario to drain the battery

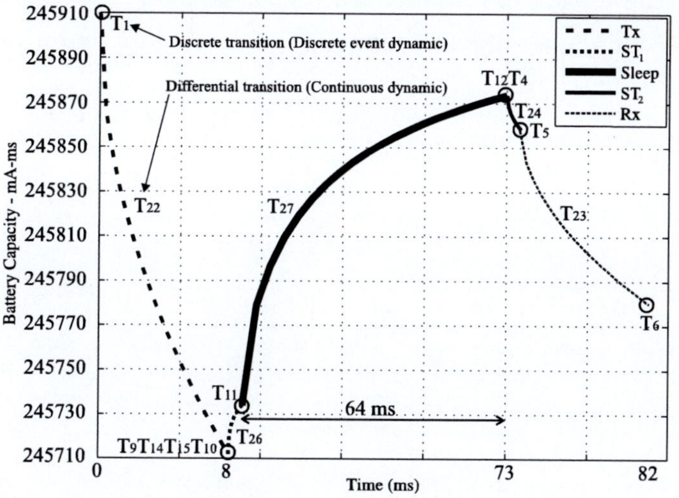

Figure 9. Battery capacity for a given scenario under DPM1 mode.

capacity is considered. In this case, only transmissions are considered, because these are the most energy costly operations. The number of transmissions achieved with and without DPM is taken into account. Figure 10 and Table 10 present the simulation results, showing that the recovered energy allows DPM-SSM transmit many more packets compared to without having any DPM technique. Larger sleeping intervals increase the recovered energy allowing an increased number of packet transmissions. Table 10 shows in more details the exact number of transmissions achieved in all cases. Even for very short sleeping periods (i.e., 8 ms), DPM-SSM can almost duplicate the number of total transmissions, and it can go up to 5.9 times the number of total transmissions when applying a 64 ms sleeping period.

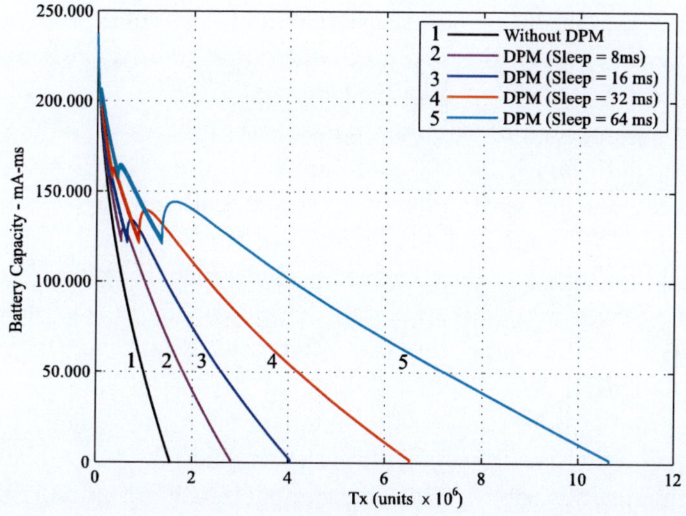

Figure 10. Transmissions achieved with and without DPM.

Table 10. Number of transmissions (Tx) achieved with and without DPM

Mode	Total number of Transmissions (Tx)	Gain (%)
No DPM (Ref.)	1,542,700	—
DPM (Sleep = 8 ms)	2,796,400	81.27
DPM (Sleep = 16 ms)	4,039,000	161.81
DPM (Sleep = 32 ms)	6,514,700	322.29
DPM (Sleep = 64 ms)	10,640,300	589.72

6. Blind Flooding Analysis

Most WSN protocols resort to the broadcasting of control messages. For example, on demand routing protocols apply blind flooding when searching for a route to any given destination (even though *expanding ring search* is usually employed to reduce the impact of flooding to the whole network, blind flooding is employed after expanding the ring to some extent). Another typical scenario is when a base station probes a sensor node or a sub-set of sensor nodes for any given information. Considering that the base station does not know the current location of the target nodes, flooding the whole network is the only way to possibly reach the targets. Given that, this Section presents an analysis regarding the energy cost of blind flooding over WSNs using the Rakhmatov-Vrudhula battery model.

In order to perform the performance analysis, the Rakhmatov-Vrudhula model is incorporated into the energy model of the network simulator *ns-2* [16], because this simulator provides only the linear model.

The battery capacity discharge for all nodes in the network is then computed. Nodes are modeled with the same characteristics as of the *Mica Motes* [11]: ATMEL ATMega128L processor, active mode current of 8 mA, inactive mode current of 15μ A, CC1000 Radio (Tx mode = 27 mA, Rx mode = 10 mA, inactive mode = 10 mA, and sleep mode = 1 μA), and power supply composed of two alkaline batteries.

The network is composed of 50 nodes spread over an area of $1500m$ X $300m$. Simulations are repeated 10 times, for 900 s of simulation time. IEEE 802.11 is used as the Phy/MAC layer protocol, and the random waypoint model with no pause time and constant speed of 1 m/s is used for modeling node mobility. Blind flooding is implemented by the FLOOD protocol [17]. To reduce the chance of packet collision, re-broadcasts are randomly jittered by 10 ms. A sensor node is randomly selected as the broadcast source, sending 2 packets of 512 Bytes per second during the whole simulation time. Table 11 summarizes the simulation parameters.

Table 12 presents the simulation results regarding the battery capacity consumption for both battery models under consideration. It is shown that the linear model presents results far from a more realistic scenario. The linear model does not provide useful insight on the real battery consumption of a node, showing a consumption roughly 400 times larger than the Rakhmatov-Vrudhula model. The linear model is easy to implement, but does not provide meaningful results. On the other hand, the Rakhmatov-Vrudhula model provides

Table 11. Simulation parameters

Parameter	Specification
Processor Unit	Active = 8 mA, Inactive = 15 μA
Number of nodes	50 nodes
Terrain	1500m X 300m
Radio Range	250 m
Simulation time	900 s
Mobility model	random waypoint mode
Pause time	none
Speed	1 m/s
Radio	Tx = 27 mA, Rx = 10 mA, Sleep = 1 μA
Battery Capacity	C=245910 mA-ms, V_{oc}=1.5V, V_{co}=0.8V

realistic results because it captures the non-linear battery behavior at the expense of being more compute intensive. Anyway, only when having a realistic battery model we are able to have a better idea about the network lifetime resulting from running any protocol in a WSN.

Table 12. Simulation Results: Linear Model (LM) *versus* Rakhmatov Model (RM) in Joule.

Scenario 1 (One Traffic Source)				Scenario 2 (Five Traffic Source)			
Sample	**RM Consumption**	**LM Consumption**	**Difference**	**Sample**	**RM Consumption**	**LM Consumption**	**Difference**
1	0,3727	144,3248	387,25	1	0,3762	145.4163	386,54
2	0,3712	144,3619	388,88	2	0,3764	145.4380	386,35
3	0,3718	144,3605	388,23	3	0,3763	145.4311	386,48
4	0,3726	144,2200	387,10	4	0,3760	145.3353	386,54
5	0,3728	144,3562	387,24	5	0,3764	145.3964	386,29
6	0,3728	144,3574	387,24	6	0,3764	145.4362	386,41
7	0,3727	144,3003	387,16	7	0,3762	145.4049	386,50
8	0,3726	144,2987	387,15	8	0,3763	145.4312	386,47
9	0,3727	144,3175	387,18	9	0,3763	145.4141	386,48
10	0,3727	144,2897	387,15	10	0,3764	145.4328	386,40
Average	**0,3725**	**144,3187**	**387,46**	**Average**	**0,3763**	**145,4136**	**386,45**
STDEV	**0,001**	**0,04**	**0,60**	**STDEV**	**0,000**	**0,03**	**0,08**

References

[1] A. G. Ruzzeli, G. M. P. O'Hare, T. Tynan, P. C. , and J. M. Havingat, "Protocol assessment issues in low duty cycle sensor networks: The switching energy," *SUTC '06: Proceedings of the IEEE International Conference on Sensor Networks*, pp. 136–143, 2006.

[2] V. Raghunathan, C. Schurgers, S. Park, and M. B. Srivastava, "Energy-aware wireless microsensor networks," *IEEE Signal Processing Magazine,*, vol. 19, no. 2, pp. 40–50, March 2002.

[3] I. F. Akyildiz, W. Su, Y. Sankarasubramaniam, and E. Cayirci, "A Survey on Sensor Networks," *IEEE Communications Magazine*, pp. 102–114, August 2002.

[4] K. Lahiri, A. Raghunathan, S. Dey, and D. Panigrahi, "Battery-driven system design: a new frontier in low power design," *IEEE Comp. Society - VLSID,*, pp. 1–7, 2002.

[5] D. Rakhmatov and S. Vrudhula, "An analytical high-level battery model for use in energy management of portable electronic systems," *Proc. of ICCAD,*, pp. 1–6, 2001.

[6] M. Handy and D. Timmermann, "Simulation of mobile wireless networks with accurate modelling of non-linear battery effects," *IASTED-ASM,*, pp. 532–537, 2003.

[7] Panasonic, "Panasonic industrial alkaline batteries data sheet:," *http://www.panasonic.com/industrial/battery/oem/chem/alk/index.html*, 2007.

[8] R. M. Inc., "Rf monolithics, inc.:tr1000 datasheet," *http://www.rfm.com*, 2007.

[9] Chipcon, "Chipcon products: Cc1000 datasheet," *http://www.chipcon.com*, 2007.

[10] P. S. Sausen, J. R. B. Sousa, M. A. Spohn, A. Perkusich, and A. M. N. Lima, "Exploring the switching energy effect in a dynamic power management technique for wireless sensor networks," in *Proceedings of the IEEE IECON*, 2007.

[11] Crossbow, "Crossbow technology inc. data sheet:," *http://www.xbow.com*, 2007.

[12] F. Jin, H. Choi, and S. Subramaniam, "Hardware-aware communication protocols in low energy wireless sensor networks," *IEEE - MILCOM 2003*, vol. 1, no. 1, pp. 676–681, 2003.

[13] S. Park and M. B. Srivastava, "Dynamic battery state aware approaches for improving battery utilization," *Proceedings of CASES '02*, pp. 225–231, 2002.

[14] P. S. Sausen, J. R. B. Sousa, M. A. Spohn, A. Perkusich, and A. M. N. Lima, "Dynamic power management with scheduled switching modes in wireless sensor networks," *MASCOTS'07 - Modeling, Analysis, and Simulation of Computer and telecommunication Systems.*, pp. 1–8, 2007.

[15] J. R. B. Sousa, A. M. N. Lima, and A. Perkusich, "Modeling and analyzing power consumption in sensor networks nodes based on differential hybrid petri nets," in *The 31st Annual Conference of the IEEE Industrial Electronics Society*, 2005, pp. 389–394.

[16] NS2, "The network simulator," *http://www.isi.edu/nsnam/ns/*, 2007.

[17] T. Kunz, "Multicasting in mobile ad-hoc networks: Achieving high packet delivery ratios," in *The Proceding of the Anual Conference of the Centre for Advanced Studies on Collaborative Reseach*, 2003, pp. 1–15.

In: Mobile Telephones…
Editors: A.C. Harper and R.V. Buress, pp. 231-240
ISBN: 978-1-60456-436-5

Chapter 8

A New Method for Managing Emergency Calls

Landoni Giovanni*
Department of Cardiothoracic Anesthesia and Intensive Care
Istituto Scientifico San Raffaele, Milano, Via Olgettina 60, 20132 Milano, Italy
Fochi Oliviero†
Anesthesiology and Intensive Care Unit 1, Ospedale Niguarda Ca' Granda, DEA EAS
Piazza Ospedale Maggiore 3, 20162 Milano, Italy

Saturday, 4 p.m. You and your friend Joe are walking downtown when Joe gasps, grabs his chest, and, before you can reach him, falls to the ground. You shout for help, draw your cell phone and dial 911. Ten minutes later an ambulance arrives, some paramedics surround Joe and start cardiopulmonary resuscitation. They load him on the ambulance and take off. When you arrive at the hospital the doctors tell you there was nothing they could do, your friend was already dead when he entered the emergency room.

Abstract

The "chain of survival" is a series of steps linking a person whose life is in danger to medical service providers. It currently functions appropriately in relation to the resources available. We suggest a simple improvement to the survival chain: this chapter presents a method for managing emergency calls received from a user/citizen, which can be used in a cellular telecommunication system.

A database should be instituted, on a voluntary basis, for trained personnel. Such database should comprise resuscitation specialists, physicians, but also anyone who has undergone recognized training (e.g.: Basic Life Support, Advanced Life Support, Advanced Trauma Life Support). Upon receiving a medical emergency call, an off-duty operator can be selected from the database (in a cooperative, non-competitive manner with the dedicated emergency services), based on current position of her/his cellular phone. If the call includes information as to the type of emergency (e.g.: trauma, cardiac arrest) the selection of the

* E-mail address: landoni.giovanni@hsr.it. Phone: ++390226434524; Fax ++390226437178.
† E-mail address: oliviero.fochi@ospedaleniguarda.it. Phone: ++393474207940.

operator can take into account a predefined compatibility with the specific emergency. Consequently, the operator who is both closest to and best prepared for the emergency is contacted via cellular phone. If the operator is available she/he can (again on a voluntary basis) precede or join the mobile unit on site, managing the emergency according to her/his training standard, possibly in cooperation with the ambulance personnel and even up to hospital admission.

A simple software system able to identify the nearest volunteers in the existing database is needed, and the spread of Basic Life Support training among the population, along with the distribution of semi-automatic defibrillators in public places, could improve first aid in cardiorespiratory circulatory arrest, trauma or catastrophes.

Introduction

Out-of-hospital emergencies are a burden for society with sudden, out-of-hospital cardiac arrests claiming approximately 1000 lives each day in the United States alone. [1] Road accidents are the fourth cause of death in the United States and the main cause of death in people between 1 and 44 years of age. As opposed to what happens inside the hospitals, emergencies occurring on the territory usually involve a much younger population, and subsequent deaths or disabilities heavily impact a country's labor force, yielding very high social costs. Resources have been dedicated to implement cardiac resuscitation and a dedicated emergency call number is widespread in all Western nations. In medical emergencies, the most important factor that determines one's probability to survive without sequelae, is often time. It is known, for example, that one third of those suffering a heart attack die before reaching a hospital, and most of them die within one hour after the symptoms occur. In such cases, the rapid intervention of a person with even only basic training is vital, since she/he could be able to administer first aid on the spot, which undoubtedly increases the probability of survival. Basic life support training to millions of people and the availability of automated external defibrillators could improve the survival after out-of-hospital ventricular fibrillation from 17 to 33% when cardiopulmonary resuscitation is started within 4 min after collapse [2] and up to 74% for patients who received their first defibrillation within 3 min after collapse. [3]

Similarly, the first hour following a road accident or any traumatic accident is now considered the "golden hour": most of what can be done to save the patient's life must be initiated within an hour from the traumatic event. This concept was first introduced in 1977 by Dr. James K. Styner, after a plane crash in which his wife had lost her life, and has since then become the object of many trauma care training programs. [4]

Whatever the cause of the medical emergency, many countries have implemented a "chain of survival" which includes a first call for help, a rapid activation of first aid, transportation of the patient to an advanced facility (emergency room) or mobilization of an advanced rescue unit, and prosecution of care in a hospital (e.g. operating room, intensive care unit). [5,6] In the specific setting of cardiorespiratory arrest, the ''chain of survival'' presupposes rapid access to an emergency medical system, early cardiopulmonary resuscitation, defibrillation and advanced resuscitative care with appropriate management during transport to the hospital

In some countries (e.g. United States) different emergency services, such as police, fire department, paramedics, are managed by means of a single number. In others (e.g. Italy) police, medical emergency center, fire departments are contacted via different numbers.

According to the current system for managing emergency calls, when a call is received a telephone operator has the task of recording information relating to the nature of the call, its urgency, the location of the call and so on, so as to be able to send out as soon as possible a mobile emergency unit. The operator, depending on the gravity and priority of the emergency call received, organizes the mobilization of a first-aid vehicle, an advanced life support vehicle, or a helicopter. Obviously, depending on the location from where the call originates or – should it be different – the location where the emergency is happening, the most closely situated vehicle is alerted.

The method described hitherto, which is currently the most used worldwide, functions satisfactorily in relation to the resources available. These resources are obviously limited and cannot provide suitable coverage for the whole territory, in particular with the rapidity that an emergency often requires.

The greater cost which must be borne in order to achieve larger coverage arises mainly from the need to have highly specialized personnel present in the mobile rescue units. Since such an increase cannot achieved without incurring excessively high costs, it is necessary to envisage the intervention of qualified personnel within few minutes from the trauma or cardiovascular arrest.

The object of the present chapter is to describe a method for managing emergency calls which is able to overcome some of the drawback mentioned above, allowing the emergency center to be put in contact with a specialized operator available in the vicinity of the emergency zone (in a cooperative, non-competitive manner with the dedicated emergency services). [7]

Such an innovation could be applied not only to health emergencies requiring first aid, but also to other types of interventions, such as those performed by police, fire departments, and so forth.

Description

Following an emergency situation a user makes an emergency call to the emergency center, dialing a telephone number (e.g. 9-1-1). The call location may or may not coincide with the location of the emergency. The emergency center is able to record the identification data of the emergency call (location and type of emergency), and it then transmits these data to the mobile unit center, which, for example, is a first-aid center. The latter mobilizes one or more mobile rescue units, which is/are required at the location of the emergency.

At this step of the rescue "chain", our invention introduces a list or database of operators who are not part of the personnel within the emergency center. The database is managed directly by the emergency center or by an additional management center. The list includes for example physicians with specific training, medical doctors, qualified personnel or also anyone who has undergone recognized training, such as BLS (Basic Life Support), ALS (Advanced Life Support), ATLS (Advanced Trauma Life Support), and is able to act in emergency situations. The operators voluntarily participate to this database in their off-duty time.

After receiving the emergency call, the emergency center selects one or more operators from the database, depending on her/his present position with respect to the location of the emergency. For example, a predetermined relation may require that the selected operator be within the same cell (few hundred meters) of the location of the emergency. The identification data of the call (location and type of emergency) are then transmitted to the selected operator, preferably via a text message. The operator (who is off-duty and may be at home, driving, on a bus…) has the possibility to confirm or deny her/his availability, again on a voluntary basis.

In case the operator does confirm her/his availability to intervene, the emergency center transmits this information to the mobile unit mobilized beforehand. The mobile unit is made aware of the fact that a medical operator able to provide professional assistance to the rescue operations will be present on the scene. It is obvious that, depending on the location of the selected operator, she/he will reach the location of the emergency before, at the same time as, or after the mobile unit. The information contained in the message sent to the operator could also include the expected time of arrival of the mobile unit, in order to allow the operator to decide whether to intervene or to wait, based on her/his personal expertise.

GSM Cellular Communication System

The territory, in accordance with radio mobile technology, is divided into a large number of adjacent cells which are repeated in a geometrically uniform manner and each cell has a main radio station which irradiates it. It should be noted that in densely populated areas the splitting technique is used: a large-sized cell is divided into a certain number (typically three) of smaller cells. It is thus possible to pass gradually, with successive splitting steps, from large cells used in rural areas (with a radius of several kilometers) to cells having a radius of a few hundred meters for the more dense coverage required in large urban centers. Other types of communication systems could also be used, such as the satellite system GPS.

From here on reference will be made to a radio mobile communication system of the GSM type.

Operator Selection

Database

What must be understood is that personnel serving as "emergency operators" must not be on duty. All those who volunteer to participate must be willing to be contacted in their free time, even at home, or when driving for example.

One vital step in applying this innovation will be the inclusion/exclusion criteria for participation to the database of operators. No doubt that physicians of critical areas – such as resuscitation, critical care, anesthesiology, surgery, cardiology, emergency medicine - will be the most desired, but also the least numerous. Medical doctors whose field of specialization does not deal with emergencies (general medicine, gynecology, ophtalmology are some examples) would also be included, as far as they have undergone some recognized specific training. As for nurses, they should be included, provided their daily work involves critical areas (emergency departments, intensive care units, operating rooms) or they have undergone specific emergency training. Professional paramedics are not available in many countries, but they should be included when possible.

The most difficult decision is perhaps the inclusion/exclusion of non-healthcare workers with Basic Life Support training. These operators might not offer a significant help to "conventional" rescue units, and might even represent an obstacle to rescuers. However some considerations must be made. It is now accepted that in many cases of cardiocirculatory arrest the therapeutic actions which are most effective in increasing survival and returning the patient to a normal life, are external chest compressions, artificial ventilation (such as mouth-to-mouth) and early defibrillation. [8] These maneuvers, however, must be put into action within five minutes from a cardiovascular arrest, and lose effectiveness with time. They are very useful in order to buy time until more advanced rescue arrives. An enormous number of people can be instructed to resuscitate a patient through modern Basic Life Support (BLS) and similar programs, and can learn how to effectively operate a semiautomatic defibrillator. As BLS courses become increasingly popular and semiautomatic defibrillators are distributed in densely populated areas, it will be less and less likely to be caught by a heart attack in a room where neither a defibrillator nor somebody who is able to operate it are present. Therefore non-healthcare workers should perhaps be included, and specifically associated with the emergency "cardiocirculatory arrest". Their ability to offer help in other situations may be more questionable.

Location

It is possible to associate with any cell an operator, together with his/her momentary position, and it is likewise possible to associate the location of the emergency with any cell. In the application of the present invention it is possible to envisage the possibility that third parties (for example a mobile telephone company) are able to offer the service, providing in real time, upon request by the emergency center, the location of each operator present in the database.

Using a GSM cellular communication system of the type described above, the proximity relation therefore corresponds to a comparison between the cell which represents the location of the selected operator and the cell which represents the location of the emergency.

The selection may be performed by imposing as a rule that the cell of the emergency should be adjacent to the cell in which the operator is momentarily located, or that the two cells should coincide. Depending on the area covered by the cell grid, be it a rural or urban zone, the proximity relation varies. In the case of an urban area with a cell of a few hundred meters, the above mentioned rule may prove restrictive, and it is consequently necessary to ascertain the extension of the cell within the territory. The rule may then impose that the emergency cell can be reached from the cell of the call by means of a minimum travel path which passes through a predetermined number of cells, while in the case of rural areas the rule may state that the cell containing the emergency should correspond to the call cell.

Compatibility

The method here described has a particular advantageous application if the identification data of the call furthermore comprise the type of emergency call and the list or database associates each operator with an area of expertise. Using these additional data of the emergency call, the step of selecting the operator is performed in such a way that the type of operator has a predefined compatibility with the type of emergency. Consequently, the operators who are both closest to the zone of the emergency and who, in terms of specialization characteristics, are the best prepared to deal with the particular emergency situation are chosen.

The type of call may comprise, for example, a heart attack, a trauma resulting from a road accident, a respiratory arrest, and other similar acute states, while the type of operator represents her/his specialization, for example resuscitation, critical care, surgery, cardiology, pneumology. As a result, it follows that the predefined compatibility relation requires that one of the following rules be satisfied: heart attack corresponds to cardiology, resuscitation or critical care; trauma corresponds to surgery, resuscitation or critical care; respiratory arrest corresponds to pneumology, resuscitation or critical care.

Contacting an Operator

When selection is made, the call data previously recorded by the emergency center must be transmitted to the selected operators. This can be achieved via a phone call or a SMS, and both means have advantages and disadvantages. A direct call on the operator's mobile phone is probably the quickest way to reach her/him and to acknowledge her/his availability. The SMS text, though, allows an operator not to be disturbed should she/he not be available, and would probably put less pressure on an operator who is not willing to come into action for whatever reason. Furthermore, the location and type of emergency would remain written on the operator's cell phone and more easily remembered. Contacting an operator via SMS text would require awaiting for confirmation of her/his availability, and not receiving a response within a predetermined time (e.g.: one minute) would have to be considered as a negative response.

Practical, Ethical and Legal Issues

It is undeniable that selection from the database by the emergency center could represent a very stressing event, and, if repeated, might even interfere with the operator's private life. Some rewards and guarantees must therefore be granted to all those who accept to be a part of the database of operators. First of all, adhesion to the database must be completely voluntary, and privacy must be respected. The names and data of the operators must be treated according to local laws on privacy. The operator should not face any economical cost for her/his participation: the mobile phone and the calls to the emergency center should be free of charge.

Some thought must be given to how to reward the operators. This can be done in different ways: one example could be a single payment at the moment of subscription and a "prize" for every time an operator is contacted and confirms her/his availability.

Intervention in every single case should also be entirely voluntary: an operator should not worry about the ethical and legal consequences of not intervening when her/his help is required. It is a physician's duty to intervene when she/he witnesses a medical emergency, but it should probably be different when she/he *is noticed* that a medical emergency is taking place.

It is also necessary to be very cautious with the legal consequences of acting in an emergent situation. It is always possible to cause some degree of damage to a patient, and very few would subscribe for participation to the database if this meant increasing their likelihood of encountering legal liabilities. Some ad hoc exceptions to current laws should be therefore applied in case an operator acts in an emergency case after being contacted by the emergency service, depending on different local realities and cultural background.

It is hard to foresee how the medical population, or more in general healthcare workers, will consider the possibility to enter a program which will expose them to unpredictable and difficult situation in their free time. Nevertheless, it is imaginable that physicians, nurses, paramedics, and even laymen, might welcome the chance to turn all the effort they put into training, learning skills, into an actual contribution to saving human lives.

Step by Step Description

With reference to the flow chart shown in Figure 1, this innovation envisages as a first step following occurrence of the emergency the same first step that is currently needed to initiate rescue: a call to an emergency number by means of a commonly used communication device. In step two the emergency center records the data of the emergency call, namely the location of the event and the type of emergency, and immediately mobilizes a first-aid mobile unit. At the same time - or immediately after - the mobile rescue unit is alerted, an operator listed in the preformed database is selected depending on the same identification data of the emergency call, so that the location and type of operators have the predetermined proximity and affinity relation with the location and type of the emergency (step three). As a fourth step, the identification data recorded by the emergency center are then transmitted to the chosen operators. It is essential to check the availability of the selected operators (step five), and this may be achieved, for example, with a direct phone call to the mobile number of the operators, or by sending her/him an SMS text containing the identification data of the emergency, and awaiting for a reply by means of SMS of phone call.

In case of failure to communicate with the operators, for any reason (failure to locate the operator, no reply, non-availability of the operator: option NO in step 5), the steps of selecting and communicating with an operator can be repeated for a predetermined number of times, for example five. In other words, if the selected operators cannot be reached, selection of new operators is performed, whereby the requirements or rules described above must still be satisfied.

In the case where communication with the operator selected from the list has been confirmed (option YES in step 5), the chosen operator moves to the location of the event (step 6) so as to provide first aid and to assist the mobile unit that is about to reach the scene or has already arrived.

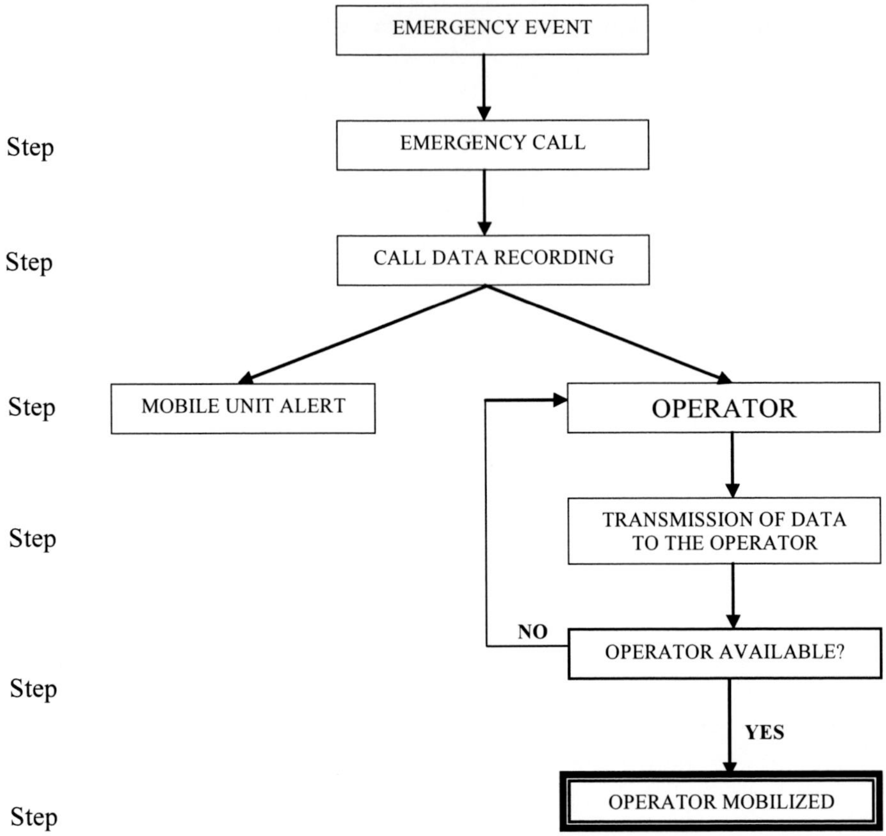

Figure 1. Flow diagram of the proposed method for managing emergency calls.

Conclusion

Available methods for managing territorial medical emergencies are nowadays very effective. Nevertheless, resources are limited, and such methods are far from being perfect. It is impossible, for example, to cover the whole territory with mobile rescue units, or to reach the spot where the emergency is taking place in such short times as those which are often required by a cardiocirculatory arrest or a trauma. It is furthermore impossible to have on-board emergency specialists for every mobile unit: in some countries many rescue operators have undergone only minimal training, or are not even professional personnel, but volunteers that may not have any relation to the medical world (lawyers, taxi drivers, waiters…). While their work is very valuable, it may not be sufficient in numerous situations, as they do not have the ability or even the legal possibility to put into action some lifesaving maneuvers, administer drugs, deal with medical-legal issues, and so on.

As it can be appreciated from the description above, this innovation is able to overcome some drawbacks of the current management of medical rescue on the territory. Trained operators would be widely available for rapid intervention, at a much lower cost than providing every mobile rescue unit with specialized personnel. Furthermore, allowing some time for the database to grow with more and more volunteers, operators will become increasingly widespread across the city/state/country, so that more people in need of medical help can be reached in reasonably short times.

Throughout this chapter, emergencies are always intended as *medical* emergencies, but it is not hard to imagine how this method could be adapted to function in other types of emergencies. As an example, whenever a theft is being witnessed by a citizen who calls the emergency number, the call data could be redirected to an off-duty policeman, or to anyone with a specific training, with the same modalities and rules to localize and communicate with the operator as described above.

With the method according to this invention it is indeed possible to contact in real time a medical operator present in proximity of an emergency in order to reduce the intervention time and increase the probability of surviving of a person whose life is at risk. This is feasible wherever mobile phones and/or beepers are widespread, medical personnel are plentiful and a dedicated emergency call system already exists and functions properly. A simple software system is needed which has to be able to identify the nearest volunteers among those included in the existing database.

Saturday, 4 p.m. You and your friend Joe are walking downtown when Joe gasps, grabs his chest, and, before you can reach him, falls to the ground. You shout for help, draw your cell phone and dial 911. One minute later a man arrives running and says he's a doctor. He starts cardiopulmonary resuscitation on Joe. When the paramedics arrive they speak to the man who is resuscitating your friend. They load Joe on the ambulance, while the doctor keeps working with them. When you arrive at the hospital the doctors tell you that Joe is alive and awake. Good thing help came so fast.

Appendix

For the latest guidelines on cardiopulmonary resuscitation, see: 2005 American Heart Association Guidelines for Cardiopulmonary Resuscitation and Emergency Cardiovascular Care. *Circulation*. 2005;112:IV1-203.

For informations about Basic Life Support and Defibrillation courses, see: www.erc.edu (Europe) or www.ilcor.org (United States)

About the Authors

Giovanni Landoni, MD

Anesthesiology and Intensive Care Specialist. Head of Research of the Department of Anesthesiology and Senior Anesthesiologist (Istituto Scientifico San Raffaele, Milano, Italia.).

Assistant Professor (Università Vita-Salute San Raffele, Milano, Italia).

Invited speaker in numerous national and international congresses.

Member of the Editorial Board of the "Journal of Cardiothoracic and Vascular Anesthesia", reviewer for numerous medical journals.

Vice President and Web Master of I.T.A.C.T.A. (Italian Association of Cardiothoracic Anaesthesiologists) www.itacta.org

Authors of more than 100 publications, half of them on Indexed Journals.

Numerous stages abroad (USA, Spain, the Netherland, India, Uganda)

Extra curriculum.

Played football in the Ugandan Premiere League (2001-2002).

President of the no-profit organization "Associazione Progetto Calcutta – O.N.L.U.S." www.calcuttaproject.org (2003-2004).

Third position. International Championship of Mathematics, Lombardia, Italy (1995)

Oliviero Fochi, MD

Resident, Anesthesiology and Intensive Care Unit 1, Ospedale Niguarda Ca' Granda, DEA EAS, Milano, Italy and 1st School of Anesthesiology and Intensive Care, Università degli Studi di Milano, Italy.

Author of more than 20 publications, one third of which on indexed journals.

Stage in Salvador de Bahia, Brasil.

BLSD instructor.

ATLS provider.

Extra curriculum.

Musician, participated in numerous concerts and music festivals.

References

[1] State-specific mortality from sudden cardiac death--United States, 1999. *MMWR Morb Mortal Wkly Rep*. 2002;51:123-6.

[2] Lombardi G, Gallagher J, Gennis P. Outcome of out-of-hospital cardiac arrest in New York City. The Pre-Hospital Arrest Survival Evaluation (PHASE) Study. *Jama*. 1994;271:678-83.

[3] Valenzuela TD, Roe DJ, Nichol G, Clark LL, Spaite DW, Hardman RG. Outcomes of rapid defibrillation by security officers after cardiac arrest in casinos. *N Engl J Med*. 2000;343:1206-9.

[4] Carmont MR. The Advanced Trauma Life Support course: a history of its development and review of related literature. *Postgrad Med J*. 2005;81:87-91.

[5] Weisfeldt ML, Becker LB. Resuscitation after cardiac arrest: a 3-phase time-sensitive model. *Jama*. 2002;288:3035-8.

[6] Cobb LA, Fahrenbruch CE, Walsh TR, Copass MK, Olsufka M, Breskin M, Hallstrom AP. Influence of cardiopulmonary resuscitation prior to defibrillation in patients with out-of-hospital ventricular fibrillation. *Jama*. 1999;281:1182-8.

[7] Landoni G, Biselli C, Maj G, Zangrillo A. Faster rings in the survival chain: Mobile phones could improve the response to the dedicated emergency call system. *Resuscitation*. 2007.

[8] 2005 American Heart Association Guidelines for Cardiopulmonary Resuscitation and Emergency Cardiovascular Care. *Circulation*. 2005;112:IV1-203.

In: Mobile Telephones…
Editors: A.C. Harper and R.V. Buress, pp. 241-251
ISBN: 978-1-60456-436-5

Chapter 9

Acute Exposure to a Standing Wave GSM-like Mobile Phone Signal in Rats Treated with Subconvulsive Doses of Picrotoxin: Effects on Cerebral Activity

Elena Lopez-Martín[1,*], Jose Luis Relova-Quinteiro[2,†] Julio Claudio Bregains[3,‡] and Francisco Jose Ares-Pena[3,‡]
[1]Morphological Sciences Department, Faculty of Medicine, University of Santiago de Compostela, Campus Norte, 15782, Santiago de Compostela, Spain
[2]Physiology Department, Faculty of Medicine, University of Santiago de Compostela, Campus Norte, 15782, Santiago de Compostela, Spain
[3]Applied Physics Department, Faculty of Medicine, University of Santiago de Compostela, Campus Sur, 15782, Santiago de Compostela, Spain

Abstract

The spectacular advances in wireless communications technology and the increased use of mobile phones in the past decades have led to much of the research into the effects of electromagnetic fields on humans and animals being centred on the study of the biological effects of mobile phones. Among the numerous studies carried out in this field, many involve the design of experimental radiofrequency systems that are manageable in the laboratory and that can be applied to biological models of interest for the study and measurement of the effects of this type of non-ionizing radiation on humans and animals, particularly on their nervous systems.

The present study assessed the validity of an experimental model of standing wave radiation of GSM radio frequency (similar to that emitted by mobile phones) applied to the biological study of the effects on cerebral activity of rats exposed to acute doses of this radiation after having been converted into a model of seizure-proneness by injection of low doses of picrotoxin. Various studies have taken into account physical parameters (frequency,

* cmelena@usc.es
† fsrelova@usc.es
‡ fajulio@usc.es; faares@usc.es

polarization, modulation, pattern and uniformity of the field, type of transmission and radiation from the equipment, as well as the cavity dimensions and material) and biological parameters (dielectric properties of the tissue under study, orientation relative to polarization, and also age, sex, weight, application of anaesthetics, drugs and electrodes), as well as possible artefacts in the calculation of the rate of energy absorption. Analysis of the behaviour of the rats during exposure to the radiation and of postmortem morphological effects, indicated by the c-Fos protein, with specific absorption rates (SAR) always below the thermic levels at 900 MHz (as outlined in the prevailing legislation) and application of different potentials, revealed an important degree of modification of neuronal activity in the brain. The validity of this and other experimental models currently used in the study of the effects of mobile phones on the nervous system is discussed in light of the results. Finally, a personal view is offered of the future of experimental research into mobile phone signals and of the methodology of biological tests used to detect possible adverse effects.

1. Introduction

In the last two decades, the use of wireless devices that employ radiofrequency (RF) radiation has increased dramatically. Since there is a tremendous amount of electrical activity involved in neural transmission, it has been suggested that the nervous system might be sensitive to exposure to electromagnetic (EM) radiation [D` Andrea et al 2003]. The proximity of the mobile terminal to the user's head and the relatively long periods of use have led to numerous experimental studies on the possible effects of microwave fields on the central nervous system (CNS) [Kuster et al, 1997]

The most widely used mobile phone system, Global System for Mobile communication (GSM), is generally operated at carrier frequencies in the near-gigahertz range (890–960 MHz). The carrier signal is pulsed at 217 HZ with a burst of 577 μs pulsed width, which evokes electromagnetic fields at both RF and extremely low frequency (ELF) ranges. Mobile phone microwaves have travelling wave properties that impinge on the human head and therefore on the brain. However, the effects on brain physiology in humans are not clear, and some studies have shown that the microwaves emitted by digital mobile phones do not have any effect -at least under normal resting conditions- on the spectral power of conventional EEG bands or during cognitive processing [Hietanen et al; 2000; Krause et al 2000; Haarala C. et al 2003]. This contrasts with the results of previous studies, which have shown increases in the power or synchrony of high frequency bands [Relova et al., 2005; Croft et al., 2007; Vecchio; et al, 2007], and that exposure affects regional cerebral blood flow [Huber et al 2005].

The experimental set-up for controlled radiation of laboratory animals must be designed with care to replicate the actual exposure conditions for human users of mobile phone signals [Hossman and Herman; 2003]. Electromagnetic parameters such as polarization, modulation, field pattern and uniformity, type of transmission and radiating equipment, chamber material and dimensions of the set-up are factors that affect microwave and RF absorption in tissues and that must be controlled. [Michaelson, et al., 2006]. Experimental radiating devices are the principal tools used in the study of biochemical, morphological and neuropathological effects of microwaves in tissues of the nervous system of small animals (Morrissey et al., 1999; Lopez-Martín, et al., 2006a; Trastoy-Rios et al., 2006 a, b; Koester et al., 2007).

Epilepsy sufferers (and potential sufferers) are of particular interest in studies of electromagnetic pollution and there is interest in studying the possible bioelectrical interactions within the brain tissue of such people, as mobile phones may be hazardous to their health. It has been shown that exposure of rats to RF GSM signals at 900 MHz of an intensity similar to the signals emitted by mobile phones triggers seizure and c–fos activation [López–Martín et al., 2006b] after administration of acute subconvulsive doses of the GABA antagonist picrotoxin [Nutt et al., 1982].

In the present study experimental animals were exposed to a standing wave GSM-like signal generated within a cage. The different levels of power of the signal were determined with the aid of an electromagnetic simulator, so that well-defined conditions for RF exposure may be considered for exposure of small animals, to allow determination of the average and peak specific absorption ratio (SAR). In this model of seizure-proneness (achieved with subconvulsive doses of picrotoxin), the effects of the radiation were assessed by observation of clinical signs and post-mortem by immunochemical testing with a marker of neuronal activation (c–Fos) [Morgan et al., 1991; Willoughby et al., 1995] and analysis of morphological effects on two anatomically different areas of cerebral cortex: frontal and parietal cortex (related to the motor system).

2. Materials and Methods

2.1. Animals and Experimental Design

Forty-eight adult male Sprague-Dawley rats (200-250 g), previously housed under standard conditions (12/12 h light/dark cycle, 22°C) and provided with food and water *ad libitum*, were divided equally into six groups (see Table 1). Animals in groups 1 and 2 were injected with picrotoxin, then exposed for 2 hours to 900 MHz GSM-type radiation in the radiation chamber, i.e. to radiation pulsed with a TDM period [Trastoy-Rios et al., 2006 a, Lopez-Martín et al., 2006b]; group 3 animals were injected with picrotoxin then immobilized for 2 hours in the radiation chamber without radiation; groups 4 and 5 were exposed for 2 hours to radiation (LP and HP) without prior injection of picrotoxin; and group 6 animals were immobilized for 2 hours in the chamber without radiation or prior injection of picrotoxin. Picrotoxin, when used, was injected intraperitoneally at a subconvulsive dose of 2 mg kg^{-1} [Nutt et al., 1982; López-Martín *et al.*, 2006a]. The animals in all groups were treated one at a time, and the receiving antenna and video camera were present in the chamber in all experiments. Radiation was emitted at an average power level of 1 W or 0.25 W, i.e., within the average and peak levels of power (0.25 W and 2 W, respectively) typical of mobile phone use.

The validity of the numerical phantom for the rats was tested in a previous experiment in which six rats of average weight 195.5 g (range 192-202 g) were irradiated as above, without injection of picrotoxin. In an additional experiment, carried out to estimate the extent to which errors in rat placement may have affected measurements, a single rat (approx. 200g) was placed successively in each of six positions differing from the standard position by a slight translation or rotation [López-Martín et al., 2007].

All experiments were carried out in accordance with the research protocols established by the Animal Care Committee of the University of Santiago de Compostela.

Table 1. Description of experimental groups.

Group	Treatment
1	Picrotoxin treated (PT) and radiated at low power (LP)
2	Picrotoxin treated (PT) and radiated at high power (HP)
3	PT and non-radiated (NR)
4	No picrotoxin treatment (NPT) and radiated at low power (LP)
5	NPT and radiated at high power (HP)
6	NPT and NR

2.2. Experimental Set-up and Specific Abpsortion Rates (SAR)

The experimental set-up allows radiation of small laboratory animals with RF signals of certain pre-established values of frequency, intensity and modulation. In the set-up the experimental animal is immobilized in a methacrylate holder (Figure 1), which is then placed-together with a transmitting antenna- inside a metallic chamber that is large enough to minimize any stress to the animal; the chamber currently used is 150 cm long, 70 cm high and 46 cm wide (internal measurements). The chamber may also contain a receiving antenna and video camera [Trastoy-Rios et al., 2006 a,b, López-Martín et al., 2006a]. Both the receiver and emitter antennas are connected to a whole signal generator and measurement system, which includes a signal generator with its amplifier, a spectrum analyzer, two power measurers and two directional couplers.

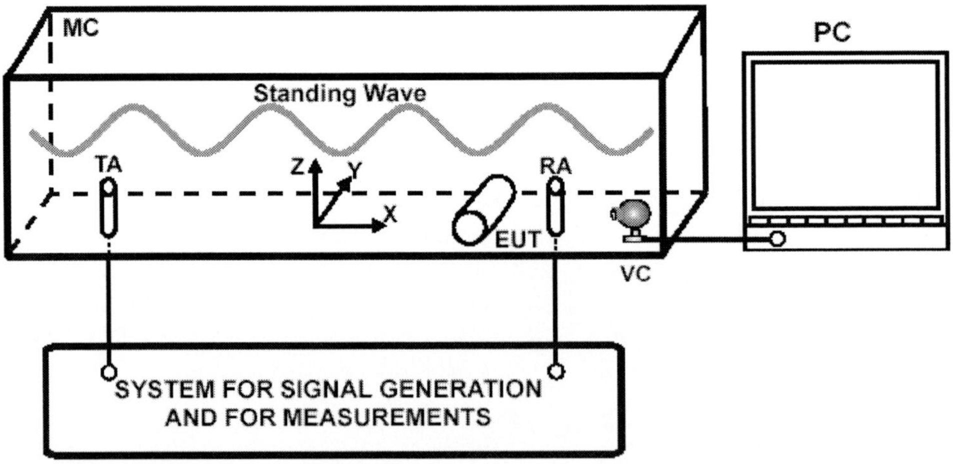

Figure 1. Diagram of the experimental set-up. System for signal generation and for measurements: The signal generator, spectrum analyzer and power meters are from Agilent (Models E4438C, E4407B and E4418B, respectively), the linear power amplifier from Aethercomm, and the directional couplers from Narda (Model 3282B-30). MC, metallic cage, TA, transmitting antenna; RA, receiving antenna. The origin of coordinates is located at the centre of the floor of the chamber, EUT: an element under test (with the rat enclosed within), VC: videocamera, PC: personal computer.

Animals in groups 1, 2, 4 and 5 were irradiated for 2 hours with 900 MHz GSM modulated signals of radiation pulsed with TDM period of 4615μs and a slot width of 577 μs; the power level was chosen to be somewhat lower than the average power radiated by a GSM mobile phone.

To study the possible effects of microwave fields on the CNS, eight rats (previously treated with picrotoxin or untreated) were placed, one at a time, within the EUT (Element Under Test) chamber and the power absorbed by each animal was determined by experimental measurement of incident, transmitted and reflected powers in the system (see Fig. 1).

The positions of the animal holder and receiving antenna (if used) were determined in relation to the field distribution inside the chamber [Trastoy-Rios et al., 2006 a,b].

The power absorbed by the experimental rat was used to correct local SAR predictions calculated by numerical simulation. The value of SAR obtained from the simulation was normalized to the experimental absorbed power value for each animal. Specifically, the local SAR, SAR_E, was calculated by use of the following expression

$$SAR_E = SAR_S \times (P_E/W_E)/\,(P_S/W_S) = SAR_S \times (WBMSAR_E)/\,(WBMSAR_S),$$

where SAR_S is the corresponding simulated SAR, $WBMSAR_E$ and $WBMSAR_S$ are respectively the experimental and simulated whole-body mean SARs, P_E and P_S the corresponding absorbed powers, and W_E and W_S the weights of the experimental and simulated rats. The estimated specific absorption ratio SAR_E for the experimental animals is obtained by adjusting the simulated values to the actual weights and absorbed powers of the rats [Trastoy-Rios et al., 2006a,b, Lopez-Martín et al., 2007]. The SARs are proportional to whole-body mean SAR, at least over the small weight range (200-260 g) used in the present study.

2.3. Behavioural and Clinical

All animals were observed and videotaped continuously for two hours after they received the intraperitoneal injection of picrotoxin. Behavioural modifications, myoclonic jerks and occurrence and latency of convulsive seizures were recorded. These were in all cases followed by full recovery.

2.4. Histology and Statistical Analysis

Immediately after they were injected ip with 2 mg/kg of picrotoxin (Sigma), the animals were immobilized in methacrylate tubes in the radiation cage. After 2 h (this period was chosen because it is used in most studies [Fritz et al., 1997] and was considered as the maximum allowable continuous acute exposure per day) in the radiation cage the animals were then removed and sedated with pentobarbital before being transferred to the histochemistry laboratory. After a further 1 h they were given an overdose of pentobarbital and prefixed by transcardial perfusion with physiological saline followed by formaldehyde in phosphate buffer (pH 7.4).

After the animals were killed the brains were excised and transverse 50 μm sections were cut with a vibratome. The sections were processed for c-Fos immunochemistry.

The free-floating sections were pretreated for 1 hour with normal rabbit serum and Triton X-100 (respectively 10% and 0.25% in 0.2% potassium-phosphate buffered saline (KPBS)) and then exposed overnight at room temperature (approximately 25ºC) to polyclonal sheep anti-Fos antibody (from Cambridge Research Biochemicals, Billingham) (1:1000 in KPBS), rinsed with KPBS, incubated for 1 hour with biotinylated rabbit anti-sheep antibody (from Vector, Burlingame, CA, USA) (1:500 in KPBS), rinsed thrice with KPBS, incubated for 30 min with an avidin-biotin-peroxidase complex prepared following the instructions given by the manufacturer (DAKO, Glostrup, Denmark), and labelled with 3,3-diaminobenzidine (DAB) in imidazole-HCL-buffered H_2O_2 solution (DAKO).

Morphological changes in the brains were examined and the different groups of rats (exposed and non-exposed) were compared. The brain sections were observed under a light microscope, and each pimple was counted as a single activated nucleus of a neuron. For each rat, three or four sections were taken from each of the following areas for examination of c-Fos activity in two different transversal anatomical brain rat levels, and c-Fos positive neurons were counted in cortical areas (frontal and parietal motor) (see Figure 2). In each area, c-Fos-positive cells were counted in a $0.30 \times 0.24\ mm^2$ field, magnified x 20, in a Nikon Eclipse E200 microscope interfaced to a computer on which morphometric software (Kappa, Monrovia, CA, USA) was run. Counts per field were expressed as means ± SEM for individual animals or experiments per group.

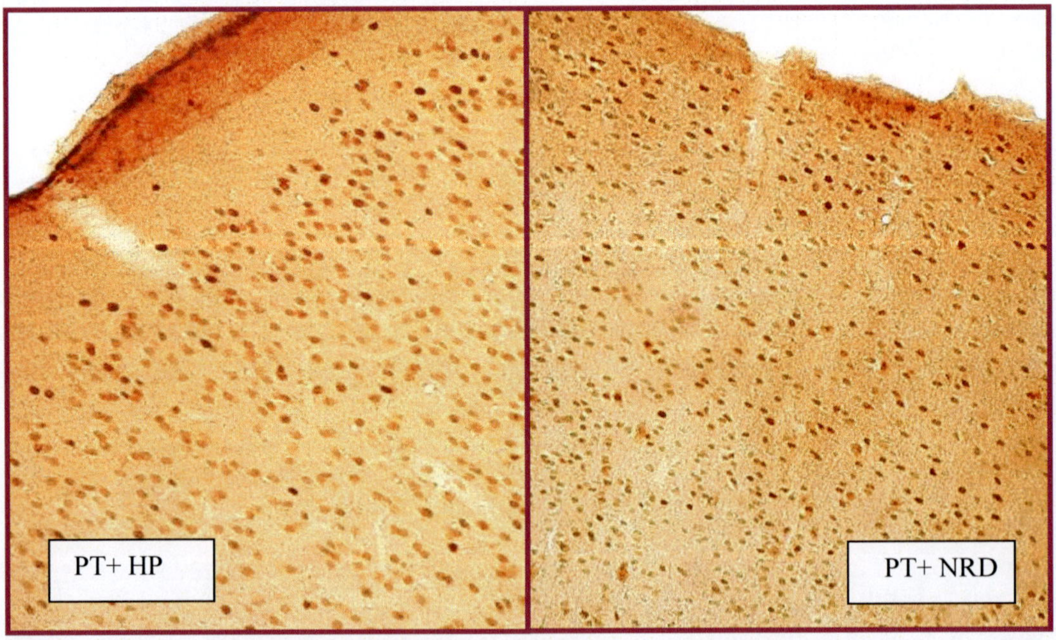

Figure 2. Photomicrographs of transverse sections through the parietal cortex of picrotoxin-treated (PT) and irradiated rats on left hand side) and picrotoxin-treated (PT) and non-irradiated (on right hand side), showing c-Fos-positive neurons as DAB dots × *[20X]*.

Table 2. Density of Fos-positive nuclei (mean ±SEM) in frontal and parietal cortex in rats with or without treatment of picrotoxin and exposed to signal 900 MHz GSM (low or high power or not irradiated).

Group	Treatment	Radiation	Motor cortex	
			Frontal	Parietal
1	PT	LP	$118 \pm 7^{no\ 2}$*	$136 \pm 6^{no\ 2}$*
2	PT	HP	$132 \pm 6^{no\ 1}$*	$146 \pm 6^{no\ 1}$*
3	PT	NR	64 ± 7*	80±6*
4	NPT	LP	$20 \pm 6^{no\ 6}$*	$24 \pm 7^{no\ 6}$*
5	NPT	HP	52± 6 *	61±7*
6	NPT	NR	30 ± 5^{4}*	32 ± 6^{4}*

* Indicates a significant difference within each column (Two–way ANOVA and post hoc Bonferoni test; $p < 0.05$).

[1-6] Indicate no significant differences between the groups within each column.

The significance of between-group differences was estimated by two-way analysis of variance. A natural logarithm transformation was applied to the data, in order to achieve normality and homocedasticity. Differences between groups were considered significant within the range $0.001 < p < 0.05$.

3. Results

3.1. Exposure to Electromagnetic Radiation

Numerical simulation of the system was made by considering the cage with the antennas (see Fig. 1) and a generic rat inside the cage (numerical model of a 200 g weight female rat, obtained from MRI scanning with 1.15 mm thick slices and with 60 different specified tissues [SPEAG–Schmid & Partner Engineering AG, 2006]).

The experimental SAR values then were estimated by interpolation of the above-mentioned estimated values. The mean power absorbed by each group of radiated rats (normal and epileptic), the corresponding incident power, absorbed power, efficiency, weights and the interpolated SARs are shown in Table 3.

3.2. Behaviour during Exposure

After 5 or 10 minutes of exposure, the irradiated picrotoxin-treated rats (group 1 and 2) began to exhibit myoclonic jerks and then suffered intermittent seizures during 20-30 min, whereas non-irradiated picrotoxin-treated rats (group 3) exhibited bursts of locomotor activity for between 3 and 5 min; in both cases, the animals remained immobile but alert for the remainder of their time in the radiation chamber. Rats that were not treated with picrotoxin (groups 4, 5 and 6) initially showed stress (largely attributable to immobilization), but did not exhibit any abnormal activity or signs of seizure.

Table 3. Power absorbed by rats irradiated with 0.25 or 1W at 900 MHz GSM radiation, with (groups PR) or without (groups NPT) prior treatment with picrotoxin, together with rat weights and both peak and mean SARs in both brain and whole body (mean values for 8 rats in all cases)

Group	Absorbed Power	Incident Power	Efficiency = 100 Abs. Power / Inc. Power	Mean weight	Mean SAR (experim.) in Brain	Mean SAR (experim.) in Body	Peak SAR averaged in 1 g Brain	Peak SAR averaged in 1 g Body
	mW		%	g	W/kg			
1	46.10	225.00	20.49	262.65	0.1556	0.1530	0.1729	0.8476
2	187.37	666.88	28.10	209.95	0.7912	0.7780	0.8790	4.3438
3	62.01	217.08	28.57	229.80	0.2392	0.2352	0.2658	1.3032
4	189.41	716.67	26.43	211.83	0.7927	0.7795	0.8807	4.3521

Whole-body mean SAR was calculated directly from the measured absorbed power, and other SARs were calculated by means of the equation given in Section 2.2.

3.3. Morphological Analysis

In picrotoxin-treated animals, c-Fos positive neurons appeared systematically in cortical areas (mainly in layers 2-4), and in some animals more or less scattered c-Fos immunoreactivity was observed in subcortical areas (hippocampus and thalamus). In the irradiated rats treated with picrotoxin, there were almost double the number of C-Fos neurons in the cortical areas (frontal and parietal) than in the same areas in the non irradiated rats treated with picrotoxin. In the rats that did not receive picrotoxin, the counts were lower than in the picrotoxin-treated animals, and were higher in animals exposed to high power radiation than in those exposed to low power radiation. Analysis of variation revealed that both picrotoxin and radiation significantly affected the number of c-Fos positive neurons in both cortical areas ($p< 0.001$).

The significant differences were apparent in each of the areas, in the comparison of the different strengths of radiation and in the comparison between picrotoxin-treated and non treated animals ($p< 0.001$)

Comparison of the counts in both cortical areas in the three groups of animals treated with picrotoxin but exposed to different strengths of radiation with the groups that received similar radiation exposure but no picrotoxin, revealed significant differences in all cases ($p< 0.001$).

However, comparison of the cell counts in both cortical areas in the animals treated with picrotoxin did not reveal any differences between animals exposed to high and low power radiation ($p= 0.281$ or $p=0.695$). Comparison of these two groups with animals not exposed to radiation did, however, reveal significant differences ($p< 0.001$).

Finally, in animals not treated with picrotoxin the counts of positive c-Fos cells in both cortical areas (frontal and parietal) were significantly higher in the animals exposed to high power radiation than in the other two groups ($p< 0.005$ and $p< 0.17$), but there were no significant differences in the cell counts corresponding to non irradiated animals and those exposed to low power radiation ($p= 0.533$ or $p= 0.855$).

3.4. Commentary

In this biological experimental model the signal standing-wave (which stimulates simulates nine-hundred MHz GSM radiation at low and high power) triggered a marked increase in neuronal excitability in seizure-prone rats, as manifested by behavioural indicators and neuronal c-Fos expression.

The metallic walls of the cavity enforce a stationary wave within, which implies a steady spatial distribution of the field with maxima and minima that could affect the whole body of the exposed rats. In this sense, the exposure is roughly similar to the human exposure to radiation generated by a mobile phone inside a room, because the metallic materials that could be in the structure of walls, or even the furniture and windows, would lead to similar framework. The effects of the cerebral nervous system and electric field spatial distribution generated inside a room near a mobile base station [Blas, et al., 2007] or the standing wave model applied to GSM radiation have not yet been investigated.

3.5. Conclusion

The effects on the cerebral activity in risk groups in which there is occurrence or possibility of seizure when the nervous system is exposed to continuous action of standing wave must be studied in further detail at the cellular level in humans.

References

Blas, J.; Bahillo, A.; Lago, F.A.; Fernandez, P.; Lorenzo, R.M.; Abril E.J.; Mazuelas, S.; Compatibilidad electromagnética Variablidad especial del campo eléctrico en un recinto próximo a una estación base GSM. Universidad de la Laguna Tenerife (2007) Ayala Alfonso A.J. *Proceedings of XXII Simposium Nacional de la Unión Científica Internacional de Radio,* 978-84-690-7500-5 URSI 25-29.

Croft, R.J.; Hamblin, D.L.; Spong, J.; Wood, A.W.; R. McKenzie, R.J.; Stough. *The effect of mobile phone electromagnetic field on the alpha rhythm of human electroencephalogram. Bioelectromagnetics* (2008) 29 (1), 1-10.

D' Andrea, A. J.; Chou, C.K.; Johnston, S.; Adair, R.E.; *Microwave effects on the nervous system. Bioelectromagnetics* (2003) 6:S107-S147.

Fritze, K., Wiessner, C., Kuster, N., Sommer, C., Gass, P., Hermann, D.M. Kiessling, M.; Hossmann, K.A. *Effect of global system for mobile communication microwave exposure on the genomic response of the rat brain. Neuroscience* (1997) 81 (3), 627–639.

Haarala, C.; Björnberg, L.; Ek, M.; Laine, M.; Revonsuo, A.; Koivisto, M.; Hämäläinen, H. *Effect of 902 MHz electromagnetic field emitted by mobile phones on human cognitive function: A replication study. Bioelectromagnetics* (2003) 24: 283-288.

Hietanen, M.; Kovala, T.; Hamalainen, A.M.; Human brain activity during exposure to radiofrequency fields emitted by cellular phones. *Scand J Work Environ Health.* (2000) 26:87-92.

Hossmann, K.A., Hermann, D.M., *Effects of electromagnetic radiation of mobile phones on the central nervous system. Bioelectromagnetics* (2003) 24, 49-62.

Huber, R., Treyer, V., Schuderer, J., Berthold, T., Buck, A., Kuster, N., Landolt, H.P., Achermann, P. Exposure to pulse-modulated radio frequency electromagnetic fields affects regional cerebral blood flow. *European Journal of Neuroscience* (2005) 21 (4), 1000-1006.

Koester P., Sakowski J., Baumman W., Glock H.W., Gimsa J.A. A new exposure system for the in vitro detection of GHz field effects on neuronal networks. *Bioelectrochemistry* (2007); 70 (1).

Krause, C.M.; Sillanmaki, L.; Koivisto, M.; Haggqvist, A.; Saarela, C.; Revonsuo, A.; Laine, M.; Hamalainen, H.; Effects of electromagnetic field emitted by cellular phones on the EEG during a memory task. *Neuroreport* (2000) 11:761-764.

Kuster, N.;. Kastle, R.; Schmid T.; Dosimetric evaluation of handheld mobile communication equipment with know precision. *IEICE Trans Commun* (1997); E 80B:645-652.

López-Martín, E., Brégains, J.C.; Jorge Barreiro, F.J.; Moreno, E.; Ares, F. Pulsed modulated fields of GSM radiation triggers c-Fos activation on brain of rats treated with subconvulsive doses of picrotoxin; Bioeffects of electromagnetic Fields. Carlo Lemos (2006a) *Proceedings of International conference of electromagnetic fields health and environment* 2; 101-106.

López-Martín, E., Relova-Quinteiro, J.L.,Gallego-Gómez, R., Peleteiro-Fernández M., Jorge–Barreiro F.J., Ares-Pena, F.J., GSM radiation triggers seizures and increases cerebral c-Fos positivity in rats pretreated with subconvulsive doses of picrotoxin *Neurosci. Letters* (2006 b) 398 (1-2),139-44.

López Martín, E.; Brégains, J.C.; Trastoy Ríos, A.; Jorge Barreiro, F.J.; Moreno, E.; Sebastián Franco, J.L.; Moreno Piquero, E.; Ares Pena F.J.; Aplicaciones biomédicas I: "Cálculo de SAR en el cerebro de animales pequeños expuestos a la radiación de un sistema experimental GSM. Efectos Biológicos", Universidad de la Laguna Tenerife (2007) Ayala Alfonso A.J. *Proceedings of XXII Simposium Nacional de la Unión Científica Internacional de Radio,* 978-84-690-7500-5 URSI 25-29.

Michaelson, S.M.; Elson, C.E.; Anderson, L.E.; Interaction of non-modulated and pulsed modulated radiofrequency fields with living matter: Experimental results. Frank S.Barnes and Ben Greenebaum. *Handbook of biological and medical aspects of electromagnetic field.* Edited CRC press by USA. (2006) 3, 53-147.

Morgan, J.I., Curran, T., Proto-oncogene transcription factors and epilepsy *.Trends. Pharmacol Science* (1991) 12 (9), 343-349.

Morrissey, R.W., Raney, S., Heasley, E., Rathinavelu, P., Dauphinee, M., Fallon, J.H., IRIDIUM exposure increases c-fos expression in the mouse brain only at levels which likely result in tissue heating. *Neuroscience* (1999) 92, 1539-1546.

Nutt, D.J., Cowen, P.J.., Batts, C.C. D.G. Grahame-Smith, D.G. Green, A.R., Repeated administration of subconvulsant doses of GABA antagonist drugs (I).Effect on seizure threshold (kindling). *Psychopharmacology* (1982) 76, 84–87.

Relova, J.L., Peleteiro, M., Pertega, S., Vila, J.A., López Martín, E., Ares F., Human studies. Changes in the electroencephalographic spectrum of epileptic patients induced by mobile phone. Dublin, Ireland. Proceedings of Bioelectromagnetics 2005: A joint meeting of The Bioelectromagnetics Society and European Bioelectromagnetics *Association,* 391-394.

SPEAG–Schmid & Partner Engineering AG. 2006. *Reference manual for the SEMCAD simulation platform for electromagnetic compatibility, antenna design and dosimetry.* Available from: http://www.semcad.com

Trastoy-Rios, A.; Lopez-Martin, E.; Jorge-Barreiro, F.J.; Sebastian-Franco, J.L. Muñoz-San Martin, S.; Ribas-Ozonas, B.; Moreno-Piquero, E.; Ares-Pena, F.J.; "Experimental setup for exposures to a 900 MHZ-GSM radiation: application to small mammalian animals", Madeira, *Proceedings of International conference on electromagnetic fields, health and environment,* (2006a); 972-8822-04-9.

Trastoy-Rios, A.; Lopez-Martin, E.; Jorge-Barreiro, F.J.; Sebastian-Franco, J.L. Muñoz-San Martin, S.; Moreno-Piquero, E.; Ares-Pena, F.J.; "Experimental setup for radiation of small animals", Manchester, *Proceedings of 9th European Conference on Wireless Technology,* (2006 b); 2-9600551-5-2; 235-237.

Vecchio, F; Babiloni,, C.; Ferri, F; Curcio, G.; Fini, R; del Percio, C.; Rossini, P.; Mobile phone emission modulates interhemispheric functional coupling of EEG alpha rhythms. *European Journal of Neuroscience* (2007) 25: 1908-1913.

Willoughby, J.O., Mackenzie, L., Medvedev, A. Hiscock, J., Distribution of fos-positive neurons in cortical and subcortical structures after picrotoxin-induced convulsions varies with seizure type. *Brain Research* (1995) 683, 73–85.

In: Mobile Telephones...
Editors: A.C. Harper and R.V. Buress, pp. 253-264
ISBN: 978-1-60456-436-5

Chapter 10

THE SUITABILITY OF S-CURVES IN FORECASTING THE MOBILE PHONE MARKET

George P. Boretos *

67 Dexamenis St., 14452, Metamorfosi, Athens, Greece

Abstract

One of the most popular and fastest growing technologies of our times is that of mobile phones.

Accurate forecast of the size of the mobile phone market is essential for both regulators, and corporate players in order to make the right decisions, and embrace effective strategies in a continuously changing market place. S-shaped logistic curves have been used successfully in measuring technological diffusion for the Internet, computers and other areas of technological innovation. S-curves work better when applied to populations with natural barriers that will eventually decelerate their growth as they move closer to their upper limit. In the case of mobile phones there are indeed barriers such as the number of people in a certain region, the number of people with sufficient income to finance mobile use, and finally the number of people that are neither too young nor too old to use mobile phones.

Recently, in 2005, this methodology was applied to actual mobile accounts for years up to 2004, in order to produce forecasts concerning the future of the mobile phone business. These projections concerned the whole World and selected regions such as China.

In order to better illustrate the suitability of S-curves in describing the dynamics of the mobile phone market, we test the forecasts made back in 2005 with actual data for recent years. In the case of World mobile accounts, the emergence of a second growth wave after the completion of a first one is confirmed, thus leading to a new estimate of approximately 3.5 billion accounts in the near future. On the other hand, China's projections with limited data up to year 2001 lead to relatively accurate estimates of mobile accounts with less than 8% error for 2006 which is considerably smaller than the expected statistical error.

---

* Tel./Fax: +30210.28.17.795; Mobile phone : +306936.681.105; e-mail : georgeboretos@yahoo.gr

Introduction

During the last thirty years many studies [1,2,3,4,5,6,7,8,9,10,11,12,13] brought to the attention of the scientific and business community the notion that simple models from population ecology, such as the s-shaped logistic curves and logistic substitution, could be easily used to describe various growth processes from other field areas such as energy consumption, economic recession, creativity, transportation, product sales, stock prices and the diffusion of new technologies.

One of the most popular and fastest growing technologies of our times is that of mobile phones. Accurate forecast of the size of the mobile phone market is essential for both regulators and corporate players in order to make the right decisions and embrace effective strategies in a continuously changing market place.

Recently, in 2005, s-curves (the alternative name for logistic curves) were applied to actual mobile account data for years up to 2004, in order to produce forecasts concerning the future of the mobile phone business [14]. These projections concerned the whole World and selected regions such as China. The bottom line of these forecasts was that the number of active mobile accounts around the globe will continue to grow in the next few years and will probably reach its peak by the end of the decade. Mobile subscriptions grow following two successive s-curves, one that was almost completed by 2004 and a second one that was just emerging. Although the uncertainty of the estimation for the first wave was small it was evident that we had to wait for a few more years before more accurate projections could be drawn for the second one.

In order to better illustrate the suitability of S-curves in describing the dynamics of the mobile phone market, we test the forecasts made back in 2005 with actual data for recent years. In the case of World mobile subscriptions, the emergence of the second growth wave is confirmed and more accurately quantified, thus leading to a new estimate of approximately 3.5 billion subscriptions in the near future. On the other hand China's projections with limited data up to year 2001 led to relatively accurate estimates of mobile subscribers with less than 8% error for 2006 which is considerably smaller than the expected statistical error.

The Use of S-curves in Forecasting Mobile Phone Business

The model of natural growth or simply the well known s-curve is described [15] by the following equation:

$$N(t) = \frac{K}{1 + \exp[-\frac{\ln(81)}{\Delta t}(t - t_m)]} \tag{1}$$

The curve's name comes from its S-like shape as can be seen in Figure 1. N(t) represents the total number of members of the population over time. N'(t), the first derivative of N, depicts the growth of the population over time. K, the *capacity* of the population, indicates how large the population N will become. Δt, the *characteristic duration* of the curve, shows how quickly the population will approach its maximum capacity. More specifically, Δt is defined as the time needed for N to grow from 10% to 90% of K. Finally, t_m is the *midpoint* of the curve i.e. the point of time where the population will reach 50% of K.

In other words the three parameters K, Δt, and tm depict how big the population will become, how quickly the growth process will evolve and when exactly in time it will take place.

This is exactly where the core benefit of using logistic fits in forecasting lies. With no detailed knowledge of a certain population, we can make projections concerning its growth potential based on just three parameters. Moreover, s-curve models give us the opportunity to distinguish among different periods of time with varying attributes concerning the size and evolution patterns of the population. For instance, the behavior of the logistic growth changes completely around the midpoint tm. Before that point, population growth increases continuously over time, but after that growth decelerates gradually as the population moves towards saturation. This kind of observation helps us determine underlying trends about the population and foresee changing behavior throughout the growth process.

Another important issue concerning natural growth is that it usually involves two or more s-curves that may coexist, or more often, be successive. During the transition period between two successive growth waves, instabilities may occur as the system fluctuates upwards and downwards until it finds a new direction posed by the second s-curve [16]. Very often we will have to wait for the new growth process to emerge and summon additional data in order to confirm and quantify the new trend.

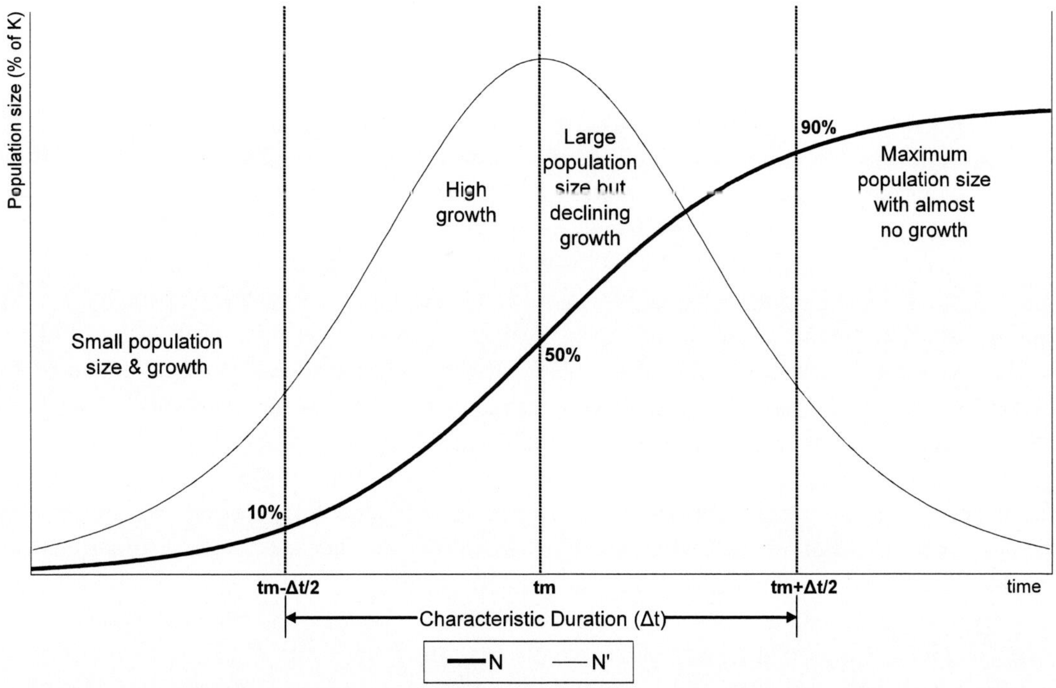

Figure 1. S-curve and successive periods of growth, saturation, and decline.

As a general rule, S-curves work better when applied to populations with natural barriers that will eventually decelerate their growth as they move closer to their upper limit. Since we live in a world full of barriers this is true for most of the natural processes regardless of their particular nature thus explaining the successful application of s-curve models in so many diverse fields such as economy, business, stock exchange, energy and the diffusion of new technologies.

In the case of mobile phones there are indeed barriers such as the number of people in a certain region, the number of people with sufficient income to finance mobile use, and finally the number of people that are neither too young nor too old to use mobile phones. Therefore we can use s-curves for modeling that process.

Recently, in 2005, s-curve modeling was applied to actual mobile subscriptions data for years up to 2004, in order to produce forecasts concerning the future of mobile phone business. These projections concerned the whole World and selected regions such as China [14].

In the following paragraphs we will examine the cases of World and China mobile accounts in order to ascertain the suitability of s-curves in making accurate forecasts and unveiling underlying trends in the mobile phone market.

All fits are produced by using Loglet Lab [15]. The software uses the bootstrap method in order to produce min-max ranges for each estimated model parameter thus giving us insights on the uncertainties embedded in the model.

Forecasting World's Mobile Subscriptions

In the article 'The Future of the Mobile Phone Business' [14], the logistic fit for World's mobile subscriptions, based on available actual data from 1992-2004, resulted in two s-curves as can be seen in Figure 2. The parameters of the original fit can be seen in Table 1. According to that the initial growth curve was almost completed by 2004 reaching 1.2 billion subscriptions, with a characteristic duration of 5.3 years, and year 2000 as its midpoint. The accuracy of the fit was satisfactory for all parameters. A second growth curve emerged after 2002 with a lower ceiling of just 0.7 billion subscriptions, characteristic duration of 2.7 years and 2004 as its midpoint. The uncertainty of the second s-curve and especially for the parameters K and Δt was significant since the growth process was at its earlier stage and still seeking direction. It was mentioned explicitly at the 2005 study that the emergence of the second growth curve had to be further confirmed over the following years and in that event the existing logistic fit would probably underestimate the total number of mobile subscribers in the world. Moreover it was mentioned that further growth could be expected by multiple subscriptions by the same user, for instance one subscription for personal use and one for professional use. Let us see what happens when we put all the above estimations and hypotheses on test based on the new logistic fits.

By using actual data for two more years, 2005 and 2006 [17,18], we can produce a new logistic fit for world mobile accounts. As we can see from Table 1 and Figure 2 the new estimate for the first growth curve is very close to the original one. Parameters Δt, and tm are exactly the same with the previous logistic fit and parameter K slightly above the expected statistical error. The second growth curve though proved significantly bigger than it was originally estimated. It is expected to reach approximately 2.4 billion subscriptions instead of just 0.7 billion of the original estimation. As it was mentioned before an underestimate was expected for the second s-curve as it was initiated just a few years before the first fit was made, and so we had to wait for a little while before we could more accurately quantify it. This is exactly what happened. After two more years the new growth wave obtained a new direction towards a considerably higher ceiling with a characteristic duration of 5 years and a midpoint at year 2005. The expected statistical error for parameters K, and Δt of the second s-curve has been significantly decreased as compared to the 2005 estimate.

It is interesting that the two growth waves of mobile subscriptions have almost the same characteristic duration of 5 years and that the second one succeeds the first s-curve after it is almost completed as can be seen in Figure 3. Combined the two s-curves result to a total number of 3.5 billion mobile subscriptions worldwide by the year 2010 i.e. a penetration of 51% over the total population of the world, according to United Nation's population estimates [19]. According to the 2005 study only 37% of total world population is likely to use a mobile phone in the near future. The remaining 63% consists of either very old (over 70 years old) or too young (less than 15 years old) or from very poor people (less than 2$ per day income) that cannot finance the use of a mobile phone. But if there are strong barriers to mobile usage, how can the predicted penetration of 51% be achieved? A possible explanation to that may be the existence of multiple mobile subscriptions for each user. When mobile phones started to penetrate the population during the 90's even the ownership of a single mobile phone was considered a luxurious possession that only a few could afford. As time passed by, the use of mobile phones became a standard issue that most people could afford in many countries. As mobile penetration increased many users found out that they needed more than one subscription, usually to accommodate for personal and professional use.

Table 1. World's Active Mobile Accounts. Estimated parameters and accuracy of the logistic fits

Parameters		World 2005 estimate				World 2007 estimate				% diff.[d]
		Value[a]	Min[b]	Max[b]	Error[c]	Value[a]	Min[b]	Max[b]	Error[c]	
1st S-Curve	K1	1,244	1,164	1,309	5.8%	1,161	1,084	1,222	6.0%	-6.7%
	Δt1	5.3	5.1	5.5	3.8%	5.3	5.1	5.5	3.8%	0.0%
	tm1	2000	2000	2000	0.0%	2000	2000	2000	0.0%	0.0%
2nd S-Curve	K2	709	504	973	33.1%	2,394	2,159	2,696	11.2%	237.7%
	Δt2	2.7	1.9	3.7	33.3%	5.0	4.5	5.6	11.0%	85.2%
	tm2	2004	2003	2004	0.0%	2005	2005	2006	0.0%	0.1%

[a] Value: Parameter Value as estimated from the logistic fit. K in millions, Δt and tm in years.

[b] Min, Max: Min-Max range of the parameter as estimated by the bootstrap method of Loglet Lab with 90% Confidence level.

[c] Error: The ratio of the average distance of the parameter value from its estimated Min and Max to the parameter value.

[d] % Difference of 2007-2005 estimates for Parameter Values.

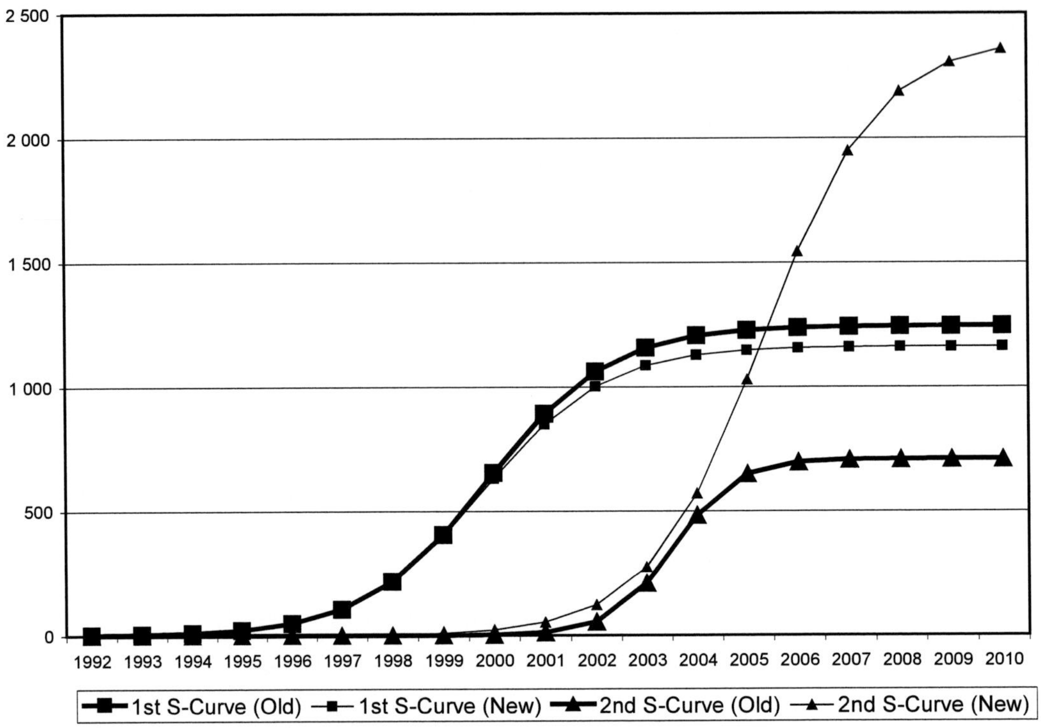

Figure 2. World's Active Mobile Accounts (in millions). Comparison of Old (2005) and New (2007) logistic fits.

Table 2. Mobile penetration and ownership for selected countries (2006 figures)

Country	Population (in millions)	Mobile acc.[a]	Mobile pen.[b]	Mobile Owners (millions)	Mobile Ownership[c]	Number of mobile accounts per user
Brazil	189	100	53%	85	45%	1.2
China	1,324	461	35%	397	30%	1.2
France	61	52	85%	50	82%	1.0
Germany	83	84	102%	73	88%	1.2
India	1,120	166	15%	101	9%	1.6
Indonesia	225	64	28%	63	28%	1.0
Italy	58	72	123%	53	91%	1.4
Mexico	108	57	53%	51	47%	1.1
Russia	144	120	84%	112	78%	1.1
Spain	43	46	106%	37	86%	1.2
UK	60	70	116%	51	85%	1.4
Total	3,415	1,291	38%	1,073	31%	1.2

[a] Mobile accounts in millions, reproduced with the kind permission of ITU [21].

[b] Mobile penetration. % of Mobile accounts over population, reproduced with the kind permission of ITU [21].

[c] % of population that own at least one mobile account, source Wireless World Forum-World Mobile Penetration Rates [20].

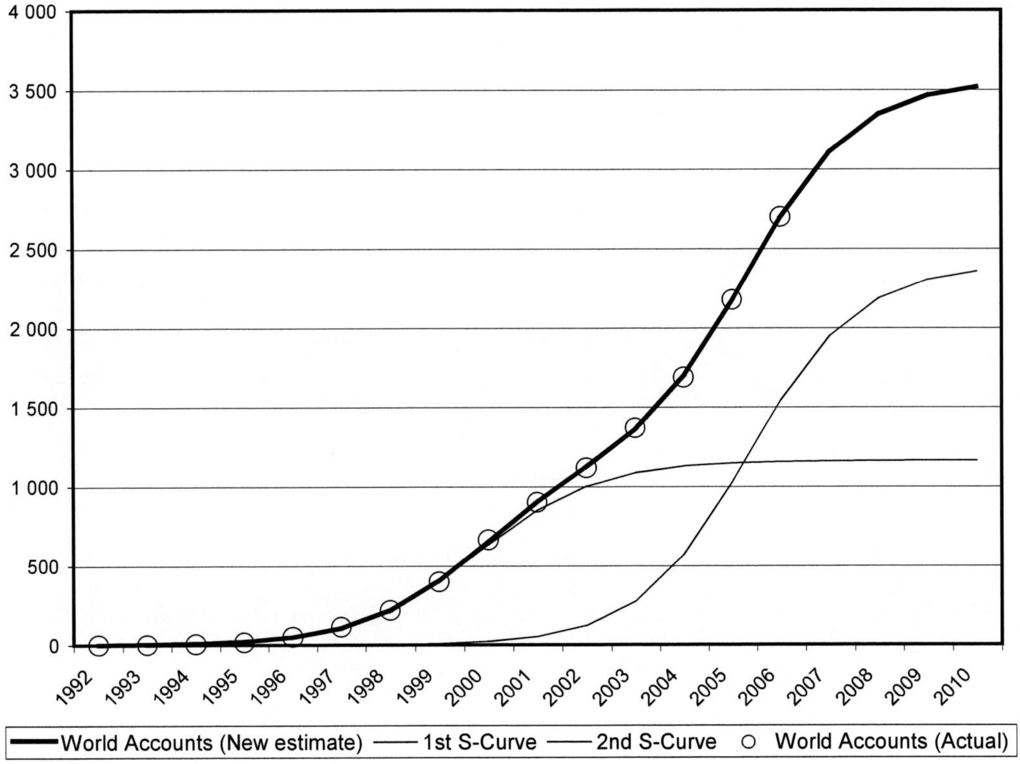

Figure 3. World's Active Mobile Accounts (in millions) evolve in two successive s-curves.

Table 2 presents both mobile usage [20] and mobile subscriptions [21] as a percentage over the population for a group of 11 selected countries. This group including a mixture of more or less developed countries represents 52% of population [19], and 48% of mobile subscriptions around the globe [18]. Although the total number of mobile subscriptions reaches 1.3 billion accounts, mobile users are only 1.1 billion. These figures result to an average of 1.2 mobile accounts for each user, verifying the trend of multiple subscriptions. Therefore the 51% level of penetration can indeed be achieved even without compromising the 37% usage barrier if the number of subscriptions per user rises above current level reaching 1.4 accounts per user in the near future. Of course the target penetration of 51% could be achieved even at lower mobile accounts per user levels, if the percent of people living in poverty around the World decreased in the near future thus bringing the 37% usage barrier at a higher level.

Another example that further supports the above hypothesis is that of Greece. Greek mobile market, although small in size, is one of the most advanced in the world in terms of mobile penetration. The 2006 level of mobile subscriptions as a percentage of population was approximately 100% [21] i.e. one mobile account per person. A recent study [22] revealed though that the actual mobile ownership in the Greek population remains considerably less than 100%. More specifically the study shows that the percentage of population more than 15 years old that owns a mobile phone was only 78% at 2006, again an indication of multiple subscriptions per user. The evolution of mobile users as a percentage of total population is shown at Figure 4. The estimated parameters K, Δt, and tm of a logistic fit to this series are shown in Table 3. The capacity K of the s-curve depicts that eventually 84% of all Greek population over 15 years old will own a mobile phone. This figure is very close to the 85% barrier after the exclusion of population over 70 years old and the very poor [19, 23]. Again in this highly developed mobile market the notion of multiple accounts per mobile user as well as the inherent barriers to mobile penetration of age and income level is verified.

Table 3. Greece's mobile ownership in population over 15 years old (% of population). Estimated parameters and accuracy of the logistic fit

	Greece (mobile ownership in population over 15 years old)			
Parameters	**Value[a]**	**Min[b]**	**Max[b]**	**Error[c]**
K	84%	80%	87%	4.2%
Δt	10.8	9.0	12.5	16.2%
tm	2000	2000	2000	0.0%

[a] Value: Parameter Value as estimated from the logistic fit. Δt and tm in years.

[b] Min, Max: Min-Max range of the parameter as estimated by the bootstrap method of Loglet Lab with 90% Confidence level.

[c] Error: The ratio of the average distance of the parameter value from its estimated Min and Max to the parameter value.

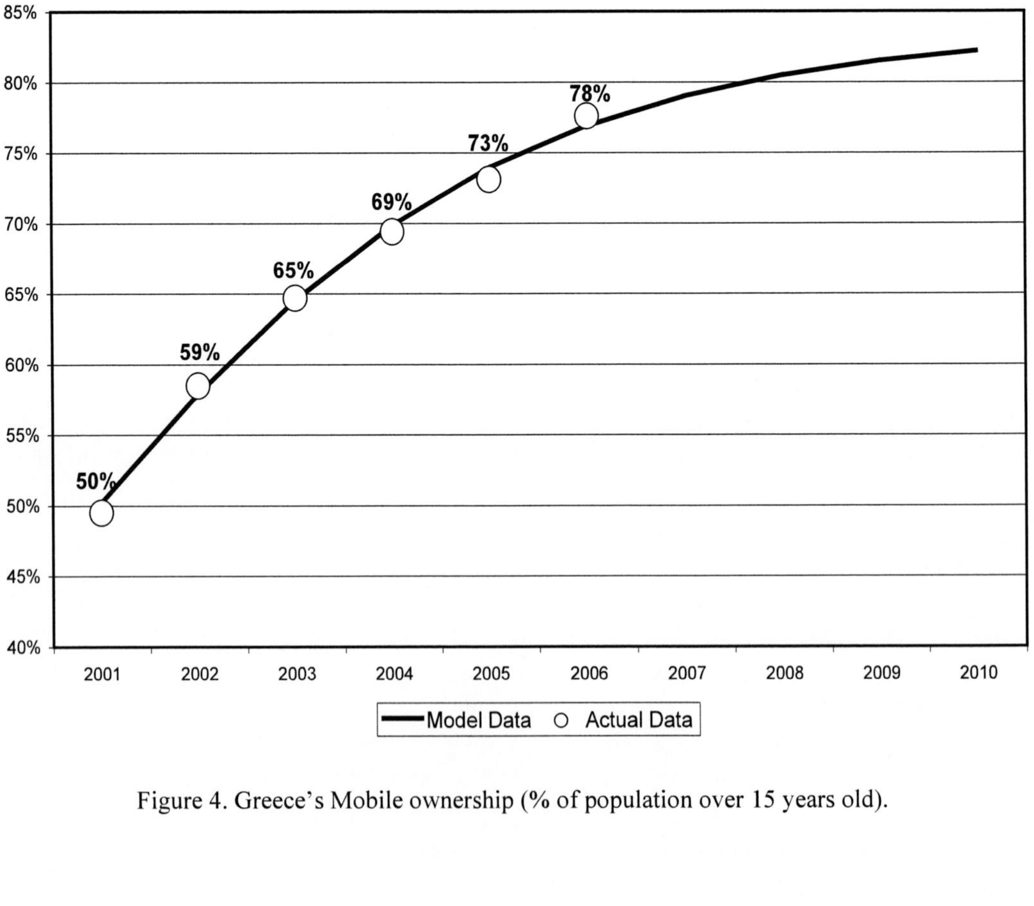

Figure 4. Greece's Mobile ownership (% of population over 15 years old).

Forecasting China's Mobile Subscriptions

Another interesting case of a logistic fit is the evolution of the mobile market in China in terms of mobile subscriptions. In the 2005 study, the original logistic fit was based in actual data available only for a limited period of time from 1992-2001. That fact in conjunction with the significant economic and social reform that still takes place in China was expected to produce less reliable estimates especially for the capacity K of mobile subscriptions as can be seen in Table 4. By using recent data for years 2002-2006 [21] we can see that the original fit was much more accurate than it was expected. As can be seen in Figure 5 actual data after 2001 fell relatively close to the estimation produced from the logistic, despite the fact that the maximum statistical error was significant as can be seen from the enveloping min-max logistic fits of Figure 5.

The new logistic fit with all data for years from 1992-2006 resulted in a new set of parameters very close to the original one. For the year 2006 the original estimation was 499 million mobile subscriptions when the actual number reached 461 million. This is a satisfactory estimate with less than 8% error considering the fact that it was made five years in advance in the rapidly changing technological, economic, and social environment in China.

Based on the new logistic fit, the growth process has already passed its midpoint at year 2003, its characteristic duration is 7.8 years and it is moving rapidly towards its maximum capacity K of 531 million subscriptions. We expect that China's mobile subscriptions will reach 521 million by the year 2010 with a mobile penetration of 39%, according to United Nation's population projections [19], which is very close to the original 40% estimation [14].

Table 4. China's Active Mobile Accounts.
Estimated parameters and accuracy of the logistic fits

Parameters		China (1992-2001 data/old fit)				China (1992-2006 data/new fit)				% diff.[d]
		Value[a]	Min[b]	Max[b]	Error[c]	Value[a]	Min[b]	Max[b]	Error[c]	
K		544	420	698	25.6%	531	498	555	5.3%	-2.4%
Δt		6.4	6.1	6.7	4.7%	7.8	7.2	8.3	7.1%	21.9%
tm		2002	2002	2003	0.0%	2003	2003	2003	0.0%	0.0%
Most recent data	Year	2006				2006				
	Actual	461				461				
	Forecasted	499				451				
	Error	-7.6%				2.3%				

[a] Value: Parameter Value as estimated from the logistic fit. K in millions, Δt and tm in years.

[b] Min, Max: Min-Max range of the parameter as estimated by the bootstrap method of Loglet Lab with 90% Confidence level.

[c] Error: The ratio of the average distance of the parameter value from its estimated Min and Max to the parameter value.

[d] % Difference of 2007-2005 estimates for Parameter Values.

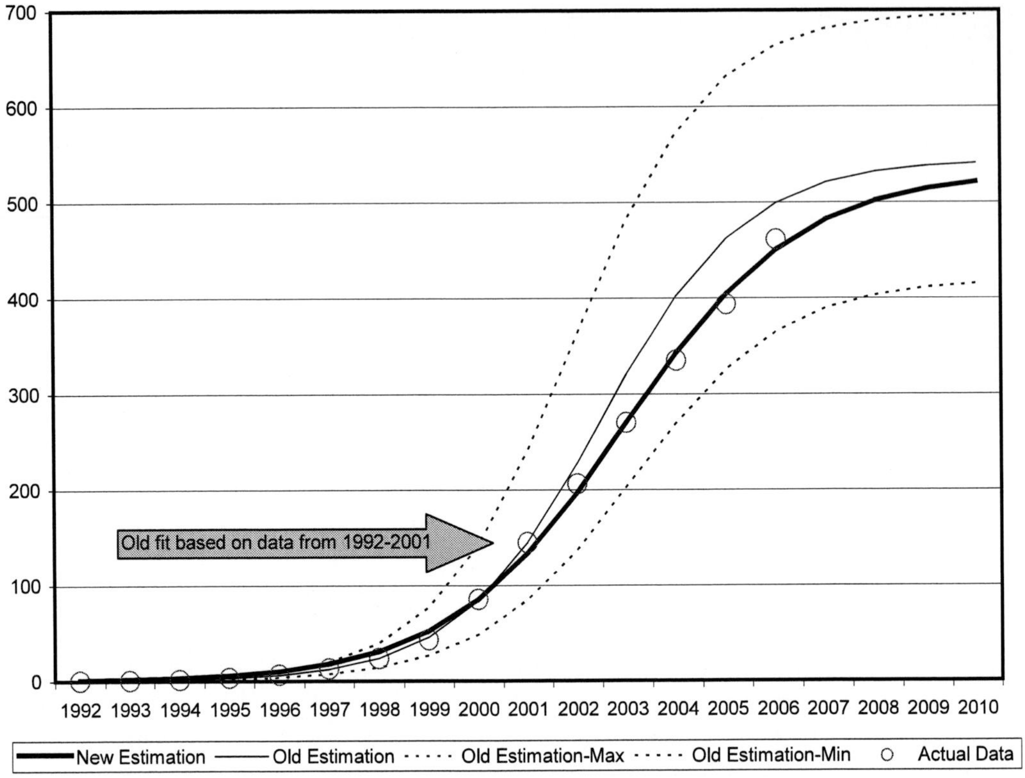

Figure 5. China's Active Mobile Accounts (in millions). Comparison of Old (2005) and New (2007) logistic fits.

Conclusion

Mobile phones, as many other processes, can be effectively modeled and forecasted by using one or more s-curves from population ecology. Furthermore, logistic curves provide us with not only satisfactory projections of the number of mobile accounts, but even more they unveil underlying growth trends in the mobile industry, even before they are massively unleashed and experienced.

A very important finding from the comparison of the 2005 study and the new logistic fits with just a few more years of actual data is that we have to keep monitoring our forecasts as new trends emerge, and market circumstances change continuously. On one hand we have the case of China's mobile market estimates which, although made with limited data up to only year 2001, were effectively accurate with less than 8% error for the year 2006 forecast. That unveils the significant forecasting capacity of s-curves especially in a fast moving market with many varying parameters over time. On the other hand, in our analysis we found that although the first s-curve for World's mobile subscriptions was estimated with accuracy, the second one, as it was anticipated, had to be readjusted for new data before we had a reliable estimation. The important fact is that the increased uncertainty for the second growth wave was known in advance thus giving us an early warning, and highlighting the need for careful monitoring in the following years.

S-curves have proven to be a valuable instrument in modeling mobile accounts as they are for many other processes as well. Yet again they are just an instrument and must be used wisely and carefully by both the scientific and business community when making forecasts. It is always a matter of good judgment when to use s-curves and when to not, when to trust the resulted forecasts or wait for confirmation and readjust accordingly, and when to use 'high level' approximations or go into more detailed formulations of the actual process. For instance, we can just model the total number of mobile subscriptions, or we could divide the population in different market segments based on attributes such as age, income level, geography or other criteria. This way we could deliver even more accurate forecasts and highlight the potential of different market segments. Subscription forecasts could then be related to specific pricing policies for different target groups and therefore market value forecasts could be delivered as well. S-curve parameters K, Δt, and tm could be linked with actual parameters or resources thus creating a control model. For instance we could link investments made in product or network development or even marketing expenses to the speed of deployment of the logistic curve as this is measured by its characteristic duration Δt. The list of different approaches and future applications is endless and one should decide where to focus in order to increase accuracy and detail of information or take a good 'overview' of a growth process with less effort.

Whatever the forecasting need may be regarding the mobile phone industry or other areas, s-curves in conjunction with practical experience can be a valuable aid for both regulators and corporate players in making timely and cost effective decisions, and embracing effective strategies in a continuously changing market place.

References

[1] Marchetti, C., and Nakicenovic, N.: The dynamics of energy systems and the logistic substitution model, *International Institute for Applied Systems Analysis, Laxenburg, Austria,* IIASA Research Report RR-79-13 (1979).

[2] Marchetti, C.: Recession 1983, *Technological Forecasting and Social Change,* **24** (4), 331-342 (1983).

[3] Marchetti, C.: Infrastructures for movement, *Technological Forecasting and Social Change,* **32** (4), 373-393 (1987).

[4] Marchetti, C.: A personal memoir: From terawatts to witches: My life with logistics at IIASA, *Technological Forecasting and Social Change,* **37** (4), 409-414 (1990).

[5] Marchetti, C.: Millenarian cycles in the dynamics of the Catholic Church: A system analysis, *Technological Forecasting and Social Change,* **46** (2), 189-196, (1994).

[6] Marchetti, C.: Longevity and life expectancy, *Technological Forecasting and Social Change,* **55** (3), 281-299, (1997).

[7] Modis, T.: Competition and forecasts for Nobel prize awards, *Technological Forecasting and Social Change*, **34** (2), 95-102 (1988).

[8] Modis, T.: *Predictions-Society's Telltale Signature Reveals the Past and Forecasts the Future*, Simon & Schuster, New York, 1992.

[9] Modis, T.: Technological substitutions in the computer industry, *Technological Forecasting and Social Change,* **43** (2), 157-167 (1993).

[10] Modis, T.: Genetic re-engineering of corporations, *Technological Forecasting and Social Change,* **56** (2), 107-118 (1997).
[11] Modis, T.: *Conquering Uncertainty,* McGraw-Hill, New York, 1998.
[12] Modis, T.: *An S-Shaped trail to Wall Street,* Growth Dynamics, Geneva, 1999.
[13] Modis, T.: Forecasting the growth of complexity and change, *Technological Forecasting and Social Change,* **69** (4), 377-404 (2002).
[14] Boretos, G.P.: The future of the mobile phone business, *Technological Forecasting and Social Change,* **74** (3), 331-340 (2007).
[15] Meyer, P.S., Yung, J.W., and Ausubel, J.H.: A Primer on Logistic Growth and Substitution: The Mathematics of the Loglet Lab Software, *Technological Forecasting and Social Change*, **61** (3), 247-271 (1999).
[16] Modis, T.: Fractal aspects of natural growth, *Technological Forecasting and Social Change*, **47**, 63-73 (1994).
[17] © GSM Association 1999 – 2007, (http://www.gsmworld.com/news/statistics/pdf/gsma_stats_q4_05.pdf)
[18] © GSM Association 1999 – 2007, (http://www.gsmworld.com/news/statistics/pdf/gsma_stats_q4_06.pdf)
[19] United Nations, Department of Economic and Social Affairs, Population Division: World Population Prospects-the 2006 revision, (http://esa.un.org/unpp/index.asp?panel=2-data extracted 10 Nov 2007)
[20] Wireless World Forum, World Mobile Penetration Rates, (http://www.w2forum.com/)
[21] ITU ICT Eye, reproduced with the kind permission of ITU (International Telecommunication Union), (http://www.itu.int/ITU-D/icteye/Indicators/Indicators.aspx)
[22] Greek Research & Technology Network, National survey on new technologies and Information Society in Greece, Funded by the operational Program 'Information Society' under the 3rd Community Support Framework of Greece, (http://www.ebusinessforum.gr/information/statistics/grnet_statistics/index.php?downid=1519&parent=1272&language=el#1272)
[23] 2005 World Development Indicators, The World Bank, p.64-69

In: Mobile Telephones…
Editors: A.C. Harper and R.V. Buress, pp. 265-274
ISBN: 978-1-60456-436-5

Chapter 11

A NOVEL TECHNOLOGY FOR MOBILE TELEPHONE NETWORKS AND SECURITY

P.P. Yupapin[1,*] and W. Suwancharoen[2]
[1]Faculty of Science, King Mongkut's Institute of Technology Ladkrabang
Bangkok, Thailand
[2]Faculty of Engineering, Rajamangala University of Technology Isan
Sakon-nakon, Thailand

Abstract

We propose a new system of an optical wireless link for telephone networks, especially, for mobile telephone applications. The message is kept in secret under a signal scramble and coding by using nonlinear effects of light in a micro ring resonator. The message is multiplexed with the chaotic signals and transmitted to the end users via an optical wireless link. The original message can be performed the secret by using the synchronous encryption before transmission via the optical wireless link. By these techniques, the messages can be kept in secret and retrieved by the specific user by the synchronous decryption, filtering and cancellation the scramble codes. The potential of using such a proposed device and system to perform the mobile telephone networks security is discussed in details.

Keywords: Future communication, Telephone network security, Networks security

1. Introduction

Mobile telephone technology has been widely used with the very large demand for two decades. The variety of improvement in related technology has seen. However, the problem of keeping telephone message in secret has become the issue of present research works, where the message security has become the issue of the present investigation. Suchat et al [1] have reported that message could be secured via telephone link by using the well known technique known as quantum cryptography. However, such a technique may

* E-mail address: kypreech@kmitl.ac.th

cause the problem of the system implementation for long distance link due to the quantum key generation and regeneration reliable in the telephone networks. Recently, Yupapin et al [2] have shown that the nonlinear property of light known as Kerr effects nonlinear type in fiber optic has shown the promising application, where the relationship between light intensity and nonlinear refractive index of the device known as a micro ring resonator has shown the good potential of application. They have shown that the nonlinear penalty of light circulation can become benefit to the system when it is made incorporating in a nonlinear micro ring resonator, where the practical parameters are micro ring radius, nonlinear material refractive index and coupling coefficient. For instance, the ring parameters such as radius, coupling coefficient, linear and nonlinear refractive indices can be used to produce, i.e. fabricate the practical device. The fabrication of a nonlinear ring device with a radius of 10 micron has been reported [3, 4]. The device scale has been decreased into the range of few microns which is suitable to implement incorporating the electronic devices, especially, the mobile telephone hand set. The more reliable system and technique can be easily implemented into the telephone networks. One of the applications is that the chaotic signals can be generated in the ring device [5]. Therefore, the message can be changed or modulated and encrypted with the random signals by controlling the input of the micro ring [6], which means the decoded signals is difficult to intercept without the secret code. Further, the micro ring device is easy to fabricated and implemented within the telephone hand set, which will be a large demand of applications. This paper, we begin with the chaotic signal generation part which is a device in a micron scale having the nonlinear property which is a key command of the required function. The modulation or encoding signal with the chaotic signals which the signals can be scrambled with the chaotic signals is generated by the external electronic process before return into the same transmission unit. The next process where the multiplexed signals are required to transmit to the networks via communication systems is explained. The signals are then received by the receiver and then de-multiplexed to obtain the chaotic signals (or digital codes) before the second security is processed by a chaotic filter. Finally, the original signal is retrieved by the cancellation technique. The possibility of using such a proposed system in the near future is discussed. Fig. 1 is a schematic diagram of the proposed system for telephone link security which is described as following. The message is in the form of electrical signal which is modulated and transmitted via a laser diode. The signal then becomes a chaotic signal which can be encrypted then multiplexed via either wire or wireless link. The de-multiplex part can be processed by the synchronous decryption, where the chaotic signal is filtered and cancelled. Finally, the message is retrieved by the specific clients at the end user.

2. Operating Principles

From Fig. 1, after the original message (Gaussian pulse) is entered into a system via a micro ring device, the chaotic signals generation and cancellation are performed, where the signal encryption and filtering are also described in details.

2.1. Chaotic Signal Generation

A simple device schematic diagram is as shown in Fig. 2, when light from a monochromatic light source is launched into a ring resonator with a constant light field amplitude (E_0) and random phase modulation (ϕ_0), which results in temporal coherence degradation. Hence, the input light field (E_{in}) can be expressed as

$$E_{in}(t) = E_0 \exp^{j\phi_0(t)}. \qquad (1)$$

The equation (2) is given by [6]

$$\left|\frac{E_{out}(t)}{E_{in}(t)}\right|^2 = (1-\gamma)\left[1 - \frac{(1-(1-\gamma)x^2)\kappa}{(1-x\sqrt{1-\gamma}\sqrt{1-\kappa})^2 + 4x\sqrt{1-\gamma}\sqrt{1-\kappa}\sin^2(\frac{\phi}{2})}\right] \qquad (2)$$

The optical fields E1 and E2 represent the right and left hands circulations in a ring resonator, respectively. A close examination of equation (2) indicates that a ring resonator in the particular case is very similar to a Fabry-Perot cavity, which has an input and output mirror with a field reflectivity, 1-κ, and a fully reflecting mirror. Where n_0 and n_2 are the linear and nonlinear refractive indices, the coupling coefficient is κ. Where $x = \exp^{-\frac{\alpha L}{2}}$ represents a roundtrip losses coefficient, $\phi_0 = kLn_0$ and $\phi_{NL} = kLn_2|E_1|^2$ are the linear and nonlinear phase shifts respectively, $k = 2\pi/\lambda$ is the wave propagation number in vacuum.

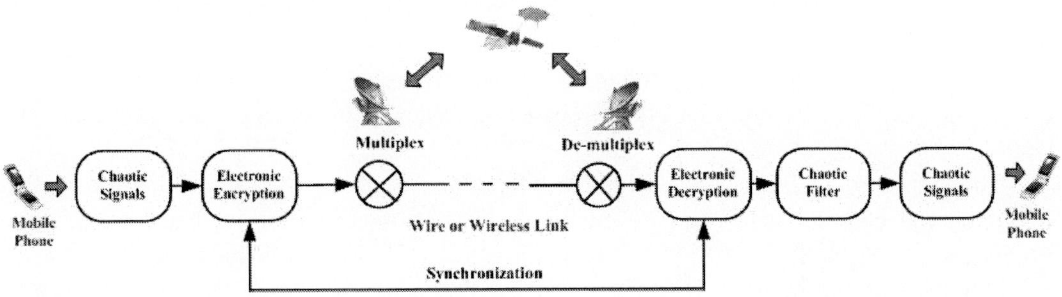

Figure 1. A schematic diagram of the proposed telephone networks security via an optical wireless link.

This nonlinear behavior of light traveling in a single ring resonator (SRR) is described. When the parameters of the system are fixed to λ_0 = 1.55 μm, n_0 = 3.34, A_{eff} = 50 μm^2, where the waveguide ring resonator loss is α = 0.5 $dBmm^{-1}$. The practical bending loss of the waveguide fabricated by InGaAsP/InP is confirmed by reference [3], where the propagation loss as low as 1.3±.02 $dBmm^{-1}$ at 1.55 μm, where the fractional coupler intensity loss is γ = 0.1, and R_1 = 10 μm. The coupling coefficient of the fiber coupler is fixed to κ = 0.0225. The nonlinear refractive index used in $n_2 = 2.2 \times 10^{-15}$ m^2W^{-1}, and the data of 10,000 iterations of roundtrips inside the optical micro ring is plotted. We assume that $\phi_L = 0$ for simplicity, however, the change in phase is slightly altered the optical output [6], which means the dispersion can be neglected when the resonant output is occurred. The chaotic signals are generated by using Eq. (2).

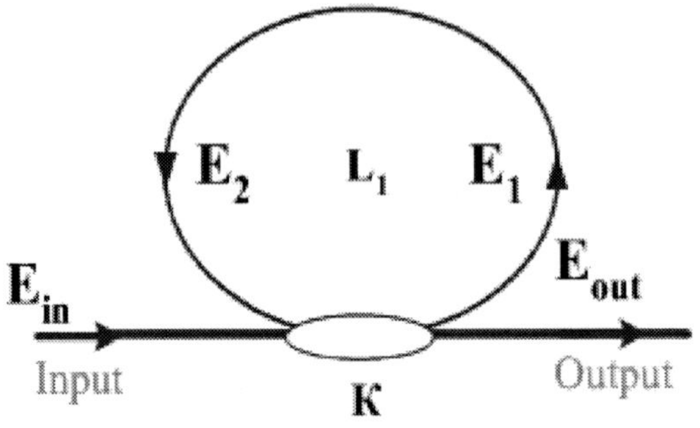

Figure 2. A schematic diagram of a micro ring resonator.

2.2. Electronic Encryption

The quantitatively present logic coding can be expressed by Eq. (3).

$$u(v) = \begin{cases} 0........v\langle 0 \\ 1........v \geq 0 \end{cases} \tag{3}$$

For instance, when $u(v)$ represents the logic states, v is the signal voltage. The quantization and re-quantization processes the similar transfer characteristics. We assume that the quantizing involved is infinite, which means that the system input signal is never clipped by saturation of the quantizing. In this case, the corresponding transfer functions relating the quantizing output to its input can be expressed analytically in terms of the quantizing step size.

The signal quantization can be further understood by using the approximation method in which the chaotic signal can be encoded. The quantizing plots of the various input powers are ranged from 1.8 -1.82 mW as shown in Fig. 3. These plots show the improvement of the approximation method as shown in Fig. 3(a)–(c), and deterioration until or least-square method in Fig. 3(d) is introduced. The signal processing of the output using the approximation method is plotted. Where (a) shows the relationship between the input and output signals, (b) the red line (straight line) is the power reduction with the threshold power of 3.5 mW, (c) the signal after threshold power condition uses, and (d) the output signal when the approximation method is employed. The logical code with the logic state "1" or "0" is generated. From the previous description, after the chaotic behaviors of the device is characterized, the next step is that the random coding can be generated by controlling

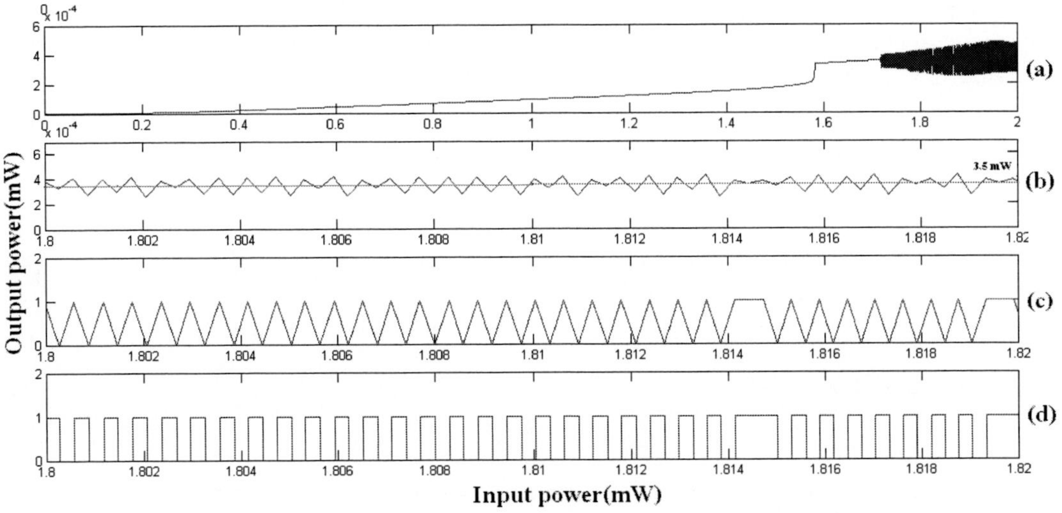

Figure 3. Shows the chaotic signal and coding using the approximation method.

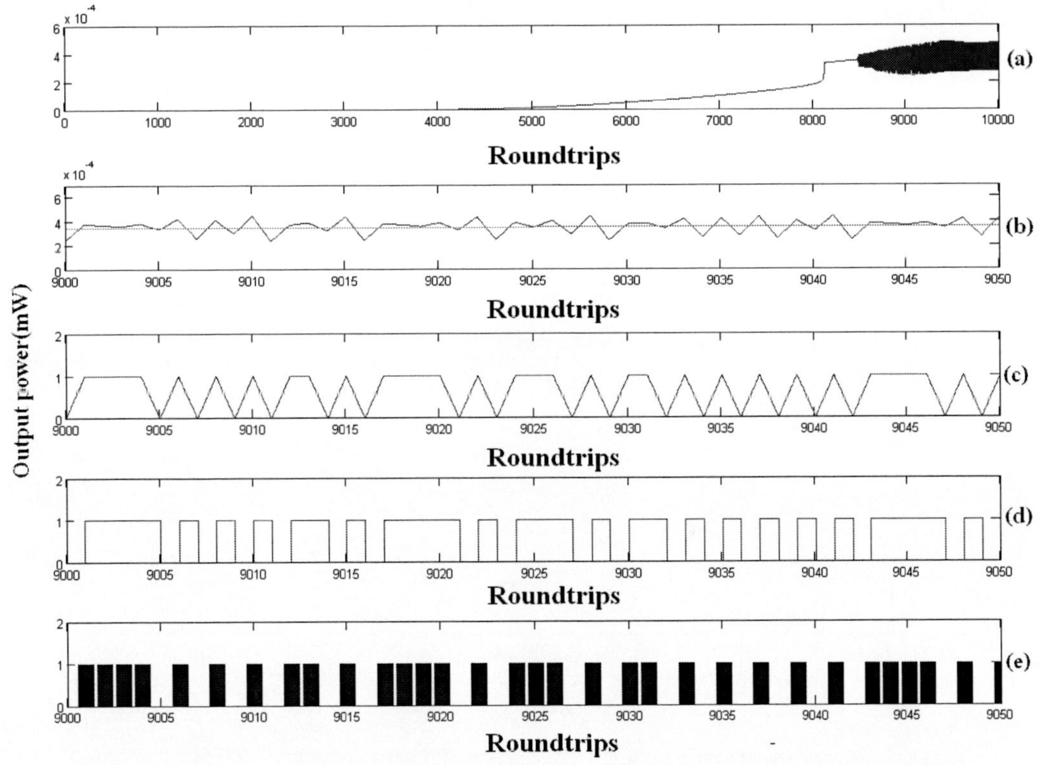

Figure 4. Chaotic codes: [0111101010101101011110101110101101010101010111110101].

The chaotic coding generation can be processed as the followings: (i) the chaotic signals can be generated within a fiber ring resonator by controlling the optical input power into a ring resonator, which can be specified by the roundtrips number, i.e. circulation time, (ii) to start the chaotic coding with the threshold power, when it is marked before averaging by using the least-square method, (iii) the clipping signals is introduced, (iv) the chaotic code generation is completed by using the approximation and sampling methods. The first chaotic code generation is as shown in Fig. 4, (a) the relationship between the output signals and roundtrips, where the chaotic behavior occurs when the roundtrips is 10,000 roundtrips, and the optical power is 0.5 mW, (b) the threshold power is 3.5 mW, and the encoding roundtrips are ranged from 9000 – 9050, (c) the clipping signals, (d) the clipping signal using the least-square method, (e) the chaotic codes using the approximation method, where the logic codes is configured as the following : [01111010101011010111101011101011010101010101110101], which there are 50 logic codes, and a roundtrip time is found to be 29 x 10^{-12} sec.

2.3. Chaotic Filter

The chaotic filter is performed by using a serial connection of two micro ring devices, which one is the transmission unit, the other is within the receiver unit. Similarly, Eq. (4) from Eq. (2) is given by [6].

$$E_{out1} = (E_{in1})\sqrt{(1-\gamma)\left[1-\frac{(1-(1-\gamma)x^2)\kappa}{(1-x\sqrt{1-\gamma}\sqrt{1-\kappa})^2+4x\sqrt{1-\gamma}\sqrt{1-\kappa}\sin^2(\frac{\phi}{2})}\right]} \quad (4)$$

When the $E_{out1} = E_{in2}$ and $\kappa_1 = \kappa_2$ and the Eq. (4) is expressed as

$$E_{out}(t) = (E_{in2})\sqrt{(1-\gamma)\left[1-\frac{(1-(1-\gamma)x^2)\kappa}{(1-x\sqrt{1-\gamma}\sqrt{1-\kappa})^2+4x\sqrt{1-\gamma}\sqrt{1-\kappa}\sin^2(\frac{\phi}{2})}\right]} \quad (5)$$

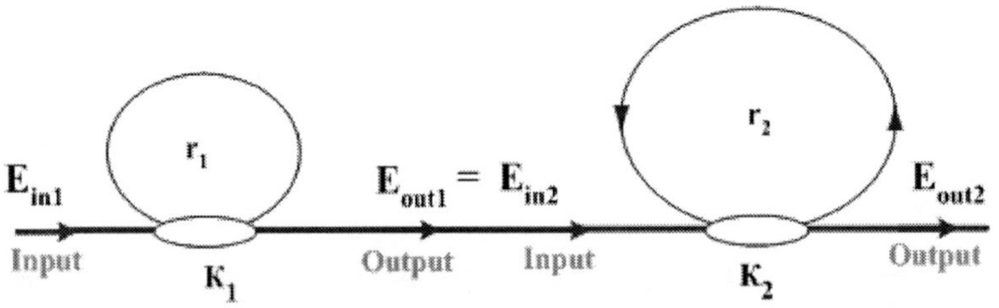

Figure 5. Shows a schematic diagram of a serial connection of two micro ring resonators.

In the Fig. 6, the nonlinear behavior of light in a micro ring resonator can be employed as the band pass and band stop filters as following details. The band stop filter occurs when the roundtrips of the travelling light in the ring device are ranged from 9,050-11,500 as shown in Fig. 6(a). The two side band filters occurred with the roundtrips between 7,550-8,550 and 11,500-12,500. The band pass filter is as shown in Fig, 6(b) with the roundtrips between 8,250-11,550, where the side band stop filters are occurred between 8,000-8,500 and 11,500-

11,800 roundtrips. The band pass filter has also seen in Fig. 6(c), when the roundtrips are ranged from 9,550-11,500 roundtrips.

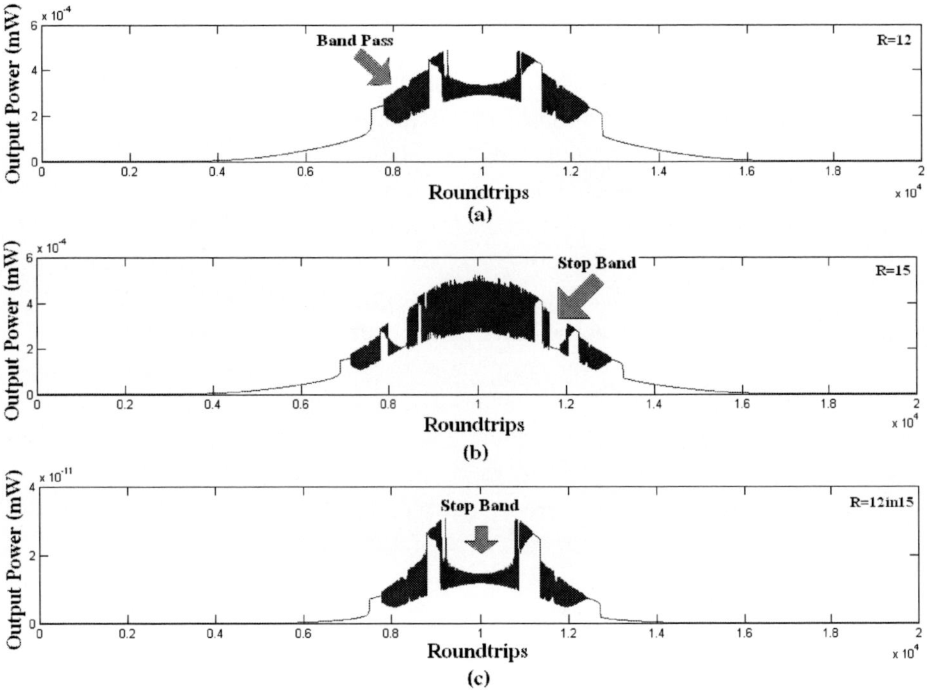

Figure 6. Shows the chaotic output and input the micro-ring at the roundtrips 20,000 (a) 12 μm, (b) 15 μm, and (c) R= 12 μm series into the R=15 μm

2.4. Chaotic Cancellation

After the signals are multiplexed with the generated chaotic noise, then the chaotic cancellation is required by the individual user. To retrieve the signals from the chaotic noise, we propose to use the add/drop device with the appropriate parameters. This is given in details as followings. The two complementary optical circuits of ring-resonator add/drop filters can be given by the equation as (6) where $\beta = kn_{eff}$ the propagation is constant, $k = 2\pi/\lambda$ is the vacuum wave number, n_{eff} is the effective refractive index of the waveguide and the circumference of the ring is $L = 2\pi R$, here R is the radius of the ring [6].

$$\left|\frac{E_t}{E_i}\right|^2 = \frac{(1-\kappa_1) - 2\sqrt{1-\kappa_1}\cdot\sqrt{1-\kappa_2}\,e^{-\frac{\alpha}{2}L}\cos(k_n L) + (1-\kappa_2)e^{-\alpha L}}{1 + (1-\kappa_1)(1-\kappa_2)e^{-\alpha L} - 2\sqrt{1-\kappa_1}\cdot\sqrt{1-\kappa_2}\,e^{-\frac{\alpha}{2}L}\cos(k_n L)} \qquad (6)$$

Similarly referring to signal flow graph (SFG) of complementary circuit [7] in Fig. 7, its transfer function is represented by

$$\left|\frac{E_d}{E_i}\right|^2 = \frac{\kappa_1\kappa_2 e^{-\frac{\alpha}{2}L}}{1+(1-\kappa_1)(1-\kappa_2)e^{-\alpha L} - 2\sqrt{1-\kappa_1}\cdot\sqrt{1-\kappa_2}e^{-\frac{\alpha}{2}L}\cos(k_n L)} \tag{7}$$

In the following, new parameters will be used for simplification: $\phi = \beta L$ is the phase constant. The chaotic noise cancellation can be managed by using the specific parameters of the add/drop device, which the required signals can be retrieved by the specific users.

The schematic diagram of the chaotic noise control and cancellation is as shown in Fig. 7. The first part, the chaotic noise and control is already described in the previous section. When the chaotic signals including chaotic noise and the required signals are propagated into the connected fiber optic, then there are two parts of splitting optical powers, which one part propagates into an add/drop device, the other propagates into a throughput port with a certain coupling constant (κ_1). The other users can also retrieve the rest of the power (signals) after this coupling port. However, there is not enough coupling power or the chaotic noise is occurred throughout in the transmission line. At the required user, the specific coupling constants (κ_1 and κ_2) can be designed to retrieve the required signals. The add port is not involved in this proposed system, where more applications can be configured when the add port is involved, which will be investigated in our future work.

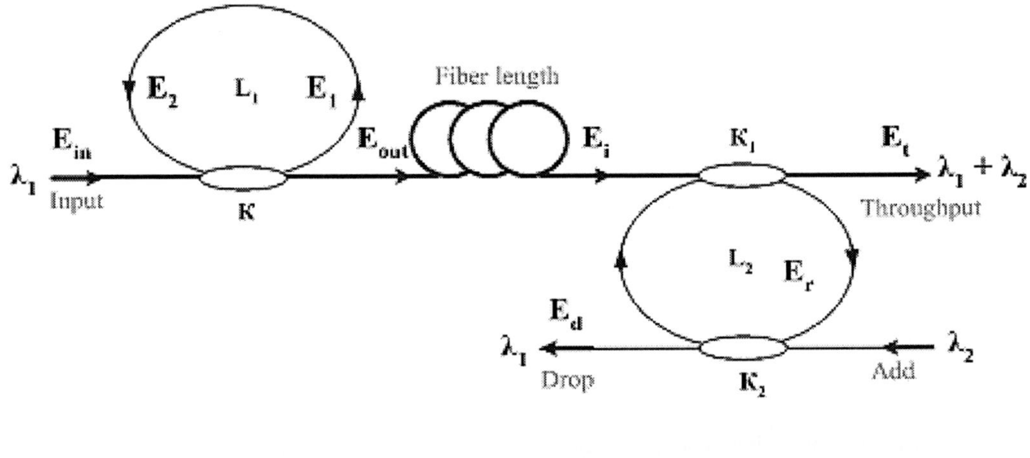

Figure 7. A schematic diagram of a chaotic noise cancellation system consists a single and double fiber couplers.

For instance, in this case the coupling constant κ_1 is 0.01, and the user coupling is κ_2=0.7. This means, there are 100 users can be linked to subscribe within the available power budget at κ_1. Notice that at κ_1, the coupling power that can keep the chaotic signal in the throughput port still valid for the other users, which is equal to 0.99. This mean the link between the chaotic generation part (service provider) and the first coupling point can be performed a subscription link. Then the providers can give the link to individual user (subscriber) at the second coupler (κ_2). At each drop port (subscriber port), the coupling ratio is roughly 1/ κ_1=N, where N is the number of subscribers. To avoid the noisy signals in the add/drop device entering into the network (throughput port), then the value of κ_1 is given to be 0.01, this means most of the power propagates into the add/drop device.

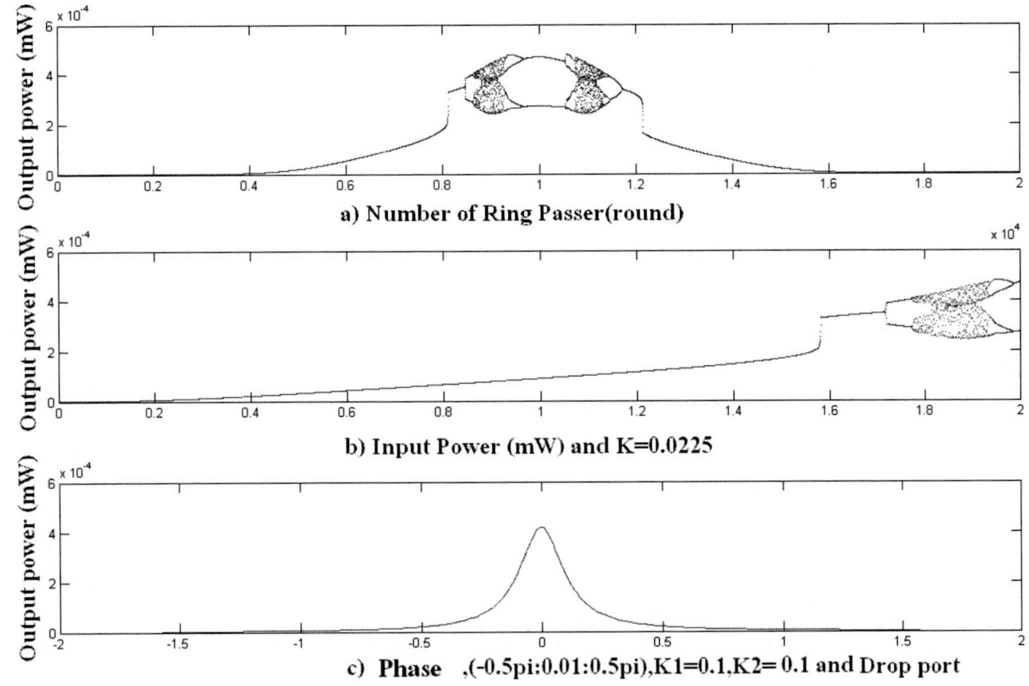

Figure 8. Shows the nonlinear behaviors of light after passing through a chaotic generation part, with the drop port output.

In Fig. 8, the bandwidth of the drop port signal is 10 GHz. Similarly, the chaotic signals and control powers are plotted in Fig. 8(a) and (b) respectively, while the output at the drop port as shown in Fig 8(c), the parameters used are $\kappa_1 = \kappa_2 = 0.1$ and $R_1 = R_2 = 10\mu m$, and the input power is 2 mW. Which means the original message is now retrieved by the required drop port i.e. specific user. In general, in some cases the drop port signal bandwidths could be narrower than the original input signals bandwidths caused by the add/drop device. This has been clearly described by Yupapin et al [6], when the bandwidth of chaotic input signal is 40 GHz. Which the drop port and throughput port signal obtained is 60%-90% of the coupling coefficients of the add/drop devices.

3. Conclusion

We have proposed the communication security via optical wireless link for mobile telephone use. The technique of message scramble with chaotic signals using a micro ring device has been described. The simulation results based on the practical device scale have shown the possibility of using such a technique and system for message security. Where the micro ring device can be fabricated and used within the mobile telephone hand set, then the scramble technique can be employed. Which means the message can be kept in secret and realized in the realistic application. However, the synchronous encryption and decryption parts may be caused by the problem of the electronic device bandwidth, which will be the subject of future research and development.

References

[1] S. Suchat, W. Khunnam and P.P. Yupapin, Quantum Key Distribution via an Optical Wireless Communication Link for Telephone Networks, *Optical Engineering*, **46**, 100502-1-100502-3 (2007).

[2] P.P. Yupapin, P. Saeung and W. Suwancharoen, Coupler-loss and Coupling-coefficient Dependent of Bistability and Stability in a Fiber Ring Resonator : Nonlinear Behaviors, *J. Nonlinear Optical Physics & Materials*, **16**, 111-118(2007).

[3] S. Mikroulis, H. Simos, E. Roditi, A. Chipouras and D. Syvridius, 40-Gb/s NRZ and RZ Operation of an all-Optical and Logic Gate based on Passive InGaAsP/InP Microring Resonator, *IEEE J. Lightwave Technol.*, **24**, 1159-1164(2006).

[4] S. Tak Chu, B. E. Little, W. Pan, T. Kaneko and Y. Kokubun, Cascaded Micro Ring Resonators for Crosstalk Reduction and Spectrum Cleanup in add-drop Filters, *J. IEEE Photonics Technology Lett.*, **11**, 1423-1425(1999).

[5] P.P. Yupapin, W. Suwanchareon and S. Suchat, Nonlinearity Penalties and Benefits of Light Traveling in a Fiber Optic Ring Resonator, *Int. J. of Light and Electron Optics*, DOI: 10.1016/j.ijleo.2007.07.009.

[6] P.P. Yupapin and W. Suwamcharoen, Chaotic Signal Generation and Cancellation using a Micro Ring Resonator Incorporating an Optical add/drop Multiplexer, *Opt. Commun.*, **280/2,** 343-350(2007).

[7] P.P. Yupapin, P. Saeung and C. Li, 2007, Characteristics of Complementary Ring-Resonator add/drop Filters Modeling by using Graphical Approach, *Opt. Commun.*, **272**, 81-86 (2007).

INDEX

#

A

B

C

D

E

F

G

H

N

O

P

S

T

U